Thermomechanics of Continua

Springer-Verlag Berlin Heidelberg GmbH

Krzysztof Wilmański

Thermomechanics of Continua

With 39 Figures and 8 Tables

Springer

Professor Dr. Ing. Krzysztof Wilmański
Forschungsgruppe Kontinuumsmechanik
Weierstrass-Institut für Angewandte Analysis und Stochastik
Mohrenstrasse 39
D-10117 Berlin, Germany

Library of Congress in Cataloging-in-Publication Data

Wilmański, Krzysztof.
Thermomechanics of continua/Krzysztof Wilmański.
p. cm.
Inlcudes bibliographical references and index.
ISBN 978-3-642-63797-1 ISBN 978-3-642-58934-8 (eBook)
DOI 10.1007/978-3-642-58934-8
1. Thermodynamics. 2. Continuum mechanics. I. Title.
QC311.W495 1998
536'.7-dc21 98-16076
CIP

ISBN 978-3-642-63797-1

Originally published by Springer-Verlag Berlin Heidelberg New York in 1998
Softcover reprint of the hardcover 1st edition 1998

Typesetting: Camera-ready by author
Cover Design: Künkel+Lopka, Heidelberg
SPIN 10652451 55/3142 - 5 4 3 2 1 0 - Printed on acid-free paper

This book is dedicated to
my friend and teacher,
Prof. Dr. Ingo Müller

Preface

The notion of continuum thermodynamics, adopted in this book, is primarily understood as a strategy for development of continuous models of various physical systems. The examples of such a strategy presented in the book have both the classical character (e.g. thermoelastic materials, viscous fluids, mixtures) and the extended one (ideal gases, Maxwellian fluids, thermoviscoelastic solids etc.). The latter has been limited intentionally to non-relativistic models; many important relativistic applications of the true extended thermodynamics will not be considered but can be found in the other sources. The notion of extended thermodynamics is also adopted in a less strict sense than suggested by the founders. For instance, in some cases we allow the constitutive dependence not only on the fields themselves but also on some derivatives. In this way, the new thermodynamical models may have some features of the usual nonequilibrium models and some of those of the extended models. This deviation from the strategy of extended thermodynamics is motivated by practical aspects; frequently the technical considerations of extended thermodynamics are so involved that one can no longer see important physical properties of the systems.

This book has a different form from that usually found in books on continuum mechanics and continuum thermodynamics. The presentation of the formal structure of continuum thermodynamics is not always as rigorous as a mathematician might anticipate and the choice of physical subjects is too disperse to make a physicist happy. However, the book contains most of the material that one needs in his/her own research of theoretical continuous models. I also made an attempt to motivate various definitions and steps in the construction of thermodynamical models. The references include the necessary basic, easily available works in this field and supplement points where some extension may be necessary. Many important parts of the book are contained in the remarks and examples. The remarks mostly refer to the topics whose various aspects in applications of continuum models have led to difficulties and frequently faulty interpretation. The examples, in many cases very simple, serve primarily the purpose of motivation.

The exercises complete the didactic image of the book. Their solutions are frequently used to clarify further the reasoning and many of them are designed to improve the skills of the reader in technical manipulations within continuum theories.

Krzysztof Wilmanski
Berlin, April 1998

Contents

1 Introduction

„...hic sunt leones...“

1.1 Preliminary Remarks

Looking at the history of thermodynamics, one might gain the impression that it not only started on the wrong foot, but that it continues to progress in a zig-zag way. For instance, thermodynamics is unusual in as much as its main measuring instrument, the thermometer, was constructed at least 100 years before we understandood what it was really measuring.

Two of the most important papers on the foundations of thermodynamics: S. CARNOT´s *„Réflexions sur la puissance, motrice du feu et sur les machines propres a développer cette puissance“* (1824) and J. R. MAYER´s *„Bemerkungen über die Kräfte der unbelebten Natur“* (1842), were rejected by J. Poggendorff, the publisher of one of the most prominent scientific journals of XIXth century – „Annalen der Physik“ and even worse, after being published elsewhere they were ignored for a long time by the scientific community. One of the most significant scientific contributions of XIXth century science to the foundations of thermodynamics, the BOLTZMANN´s H-theorem, was so vehemently criticized that it drove Boltzmann to suicide.

The history of XXth century thermodynamics does not look any better. This is due to the tremendous success of relativistic and quantum mechanics, with their important philosophical implications and spectacular practical applications. As a result, thermodynamics was declared by many physicists to be complete, if not trivial and exhausted, before it had a chance to start developing. This situation has had disastrous consequences. Without a proper understanding of its foundations and methods, thermodynamics has frequently been applied in a preposterous manner, twisted to suit the anticipated results or ignored when the results were inconvenient.

This is demonstrated well using two representative and important examples. At present we still know very little about the thermodynamical description of the so-called metastable states that appear in almost all solid state phase transformations and even less about metastable patterns, which develop in thermodynamically non-equilibrium processes. The most appealing example of the latter is life on earth. All the same, many outstanding physicists refrain from recognizing this problem referring to the classical work of W. GIBBS (1878) on equilibrium thermodynamics! The second example concerns the description of the non-Newtonian fluids, which we shall also discuss in detail in this book. Such fluids include blood, some important polymer solutions and many other substances. These fluids are often described by models that yield unstable thermodynamical equilibrium and consequently cannot exist in nature at all.

Some improvement of this rather frightening image of thermodynamics began, some 40 years ago, with the development of systematic methods of application of the second law of non-equilibrium thermodynamics to constitutive equations of continuous media. Even though the formulation of these methods limits the models to small deviations from the thermodynamical equilibrium, many continuum field theories of practical importance have achieved considerable progress, for example the thermomechanics of non-linear materials, the theory of mixtures of fluids, the relativistic thermodynamics of rarified gases, etc.

All of these models are based on three fundamental assumptions:

continuity,
local action,
thermodynamical admissibility.

The assumption of continuity means that the medium that we describe is a *three-dimensional differentiable manifold* $\mathcal{B}_0$, which is called the body, and its *current configurations* $\mathcal{B}_t$ parametrized by the time t from a certain interval $\mathcal{T}\subset\Re^1$, are measurable subsets of the three-dimensional Euclidean space of configurations. These current configurations are defined by the one-parametric family of the diffeomorphic mappings (*deformations*)

$$\mathbf{f}(\cdot,t): \quad \mathcal{B}_0 \rightarrow \mathcal{B}_t, \tag{1.1}$$

which in turn defines the so-called *function of motion.*

We shall return to the detailed discussion of these mappings in Chaps. 2 and 3. However, it should be made clear from the outset that this assumption imposes extremely strong limitations on the models. First of all, an image of an arbitrary open subset of $\mathcal{B}_0$ must also be open (the topological continuity!). This means that we cannot describe such processes as the creation of new material surfaces, i.e. we cannot describe the cutting, tearing, or the propagation of cracks, etc. Continuum theories dealing with such processes must weaken the continuity assumption in the neighbourhood of those parts of the body, where the map (1.1) becomes discontinuous. Secondly, the diffeomorphism (1.1) describing the motion of points of the body and their neighbourhoods does not permit strong mixing with changes of the neighbouring points. Such changes can be observed easily in the motion of cigarette smoke moving through the air. This limitation eliminates such important processes as the turbulent flow of fluids.

In the next section, we shall point out some other restrictions imposed by the continuity assumption and shall also demonstrate the meaning of the second assumption. The latter will not be discussed any further in this work.

The assumption of thermodynamical admissibility will be discussed in Chap. 6 and then illustrated in Chaps. 7 – 10.

1.2 Chains of Mass Points

As an illustration of the first and second assumptions, the continuity and the local action, we consider the simplest regular discrete system of mass points and the transition from this system to a one-dimensional continuum. The chain of structureless mass points as a model of a crystal lattice was first considered by P. DEBYE (1912) and by M. BORN and Th. VON KÁRMÁN (1912). Since then various modifications have been constructed. We refer the reader to the literature on its physical implementations [e.g. A. MÜNSTER (1974)] and the formal analysis [e.g. I. A. KUNIN (1975)].

We consider a chain of N mass points, which interact with each other in the elastic manner. The last condition means that the chain possesses a potential energy, Φ, whose derivatives determine the forces of interaction between the mass points. In addition we assume the chain to be uniform, which means that the mass points in their equilibrium position are spaced with the equal distance a (see: Fig.1.1.). Each point has only one degree of freedom – the displacement u, in the direction of the chain and these displacements are assumed to be small in comparison with the distance a.

Under these assumptions the potential energy can be written in the form of the series

$$\Phi = \Phi_0 + \Phi_1 + \Phi_2 + \ldots + \Phi_N = \Phi_0 + \sum_{n=1}^{N} \varphi_1(n) u_n + \frac{1}{2} \sum_{n,n'=1}^{N} \varphi_2(n,n') u_n u_{n'} + \ldots \quad (1.2)$$

The coefficients of this series are derivatives of Φ with respect to u at the equilibrium, i.e. for $u_n=0$ for $1 \le n \le N$. The constant, Φ_0, is immaterial and φ_1 must be zero due to the equilibrium condition. The other terms describe subsequently the binary, ternary,... interactions, respectively. We limit consideration to the binary interactions, i.e.

$$\Phi = \frac{1}{2} \sum_{n,n'=1}^{N} \varphi_2(n,n') u_n u_{n'} . \quad (1.3)$$

It can be seen that the assumption of uniformity of the chain yields the existence of the function

$$\varphi_2(n,n') = \psi(n-n'). \quad (1.4)$$

In addition the structure of the relation (1.3) implies immediately:

$$\psi(m) = \psi(-m), \qquad -(N-1) \le m \le (N-1). \quad (1.5)$$

We shall now investigate three simple but fundamental examples.

1. Let us consider the chain of N mass points with the equal mass, m. Each point interacts only with its nearest neighbours and the constant of interaction (the coefficient $\psi(1)$ for the arbitrary number, n, of the material point) is denoted by ζ. To preserve the symmetry of the chain with respect to an arbitrary point, we close the chain using the following **Born - von Kármán periodicity condition**

$$u_{N+1} \equiv u_1. \quad (1.6)$$

The equation of motion of an arbitrary mass point, n, has the form

$$m \frac{d^2 u_n}{dt^2} = \zeta (u_{n+1} - 2u_n + u_{n-1}), \qquad 1 \le n \le N. \quad (1.7)$$

The general solution of this equation is of the form

$$u_n(t) = A \exp(2\pi i (k n a - \omega t)), \quad (1.8)$$

which describes the monochromatic wave of the frequency ω with the speed of propagation $c=\omega/k$. Bearing the periodicity condition in mind, we have

$$k = \frac{1}{a} \frac{\ell}{N}, \qquad \ell \text{ - integer}, \quad \ell \le N. \quad (1.9)$$

Hence, the shortest wave has the length $\lambda=1/k$ (ℓ =N) = a. Substitutiton of (1.8) into (1.7) yields the following **dispersion relation**

$$\Omega = \frac{1}{\pi}\sin\left(\frac{\pi}{2}(2ka)\right), \qquad \Omega \equiv \omega\sqrt{\frac{m}{\zeta}}, \tag{1.10}$$

which is shown in Fig. 1.1. as the lower curve.

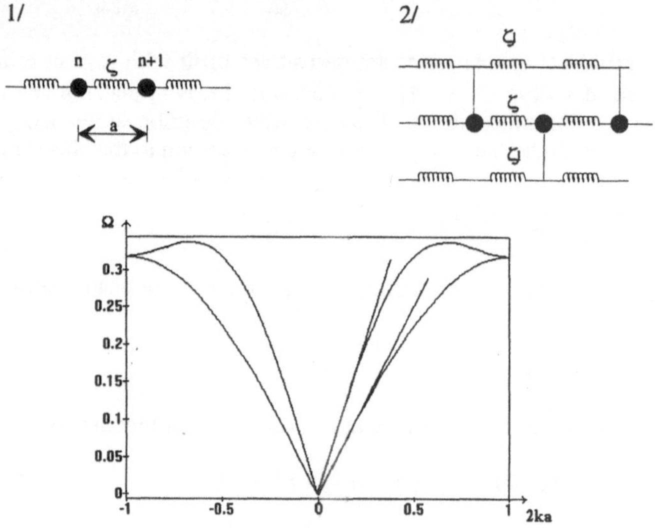

Fig. 1.1. *Dispersion curves for the uniform chains of the identical mass points.* The lower curves correspond to the chain shown in the drawing 1/ the upper curves – to the drawing 2/.

It is obvious that the chain of N mass points has N degrees of freedom, which correspond to N different values (1.9) of the wave vector (the wave number in this one dimensional case!). This, in turn, yields N different frequencies, ω, given by (1.10). Any solution of the equation of motion (1.7) can be constructed as a linear combination of at the most N different solutions (1.8). This means that the Hamiltonian of the chain is the sum of N Hamiltonians, corresponding to N different frequencies describing the so-called **normal vibrations** or **phonons**. We shall return to this point in Chap. 9.

The obvious candidate for the continuous model of the chain is the string with an elastic constant, E, and mass density, ρ. Its equation of motion has the form

$$\rho\frac{\partial^2 u}{\partial t^2} = E\frac{\partial^2 u}{\partial x^2}, \qquad u = u(x,t), \tag{1.11}$$

where u is the longitudinal displacement of the string. The monochromatic solutions of this equation are defined by the relation

$$u(x,t) = A\exp(2\pi i(kx - \omega t)). \tag{1.12}$$

Substituting this into (1.11) yields

$$\omega = ck, \qquad c \equiv \sqrt{\frac{E}{\rho}}. \tag{1.13}$$

Comparison of (1.13) with (1.10) shows that these two models coincide only for small values of ka:

$$\omega \approx a\sqrt{\frac{\zeta}{m}}k \;\Rightarrow\; c = \sqrt{\frac{E}{\rho}} = a\sqrt{\frac{\zeta}{m}}. \tag{1.14}$$

This means that the continuum is a **long-wave approximation** of the discrete structure (small values of ka).

Let us estimate the order of magnitude of this approximation. The maximum value of the frequency, ω, can be written in the form

$$\omega_{max} = \frac{1}{\pi}\sqrt{\frac{\zeta}{m}} = \frac{c}{\pi a}. \tag{1.15}$$

The typical values of the sound speed, c, in metals are about 5×10^3 m/s, the **lattice constant**, a, is of the order of 10^{-9} m, and hence $\omega_{max}\approx1.6\times10^{12}$ Hz. It can be seen in Fig. 1.1. that the tangent line in the origin corresponding to the dispersion relation (1.13) approximates the curve (1.10) well enough to 0.1 ω_{max}=100 GHz, which corresponds to the wave length app. 10^{-8} m. These estimates, in spite of the simplicity of the model, compare well with experimental observations for real materials.

The above conclusion about the long-wave approximation is very general. It can be proved, for instance, that continuous models are long-wave approximations of gases, where the role of the characteristic parameter, instead of the lattice constant, is played by the so-called **mean free path** of particles, which describes roughly the average distance of the particles and of neutral plasmas where the characteristic parameter is the **radius of the Debye ball**, i.e. of the smallest ball that can still be considered to be electrically neutral.

There are also examples of important physical systems for which any type of the continuum based on the principles of continuity and local action cannot deliver any good long-wave approximation. Plasmas with a very small number of charged particles in the Debye ball (e.g. the solar wind) and electrolytes both belong to this class. In such cases non-local continuous models are required and these are yet to be developed.

2. Let us now proceed to the second example of a chain of N mass particles in which each point interacts with the two neighbours on each side (see: Fig. 1.1., the case 2/). The constant of interaction with the second neighbour [i.e. the coefficient $\psi(2)$] will be denoted by ζ_1. As before, we introduce the Born - von Kármán periodicity condition

$$u_{N+1} = u_1, \quad u_{N+2} = u_2. \tag{1.16}$$

The equation of motion now has the form:

$$m\frac{d^2u_n}{dt^2}=\zeta\left(u_{n+1}-2u_n+u_{n-1}\right)+\zeta_1\left(u_{n+2}-2u_n+u_{n-2}\right). \tag{1.17}$$

Again we seek a monochromatic solution (normal modes of vibration) in the form (1.8). In this case easy calculations yield the following dispersion relation

$$\Omega=\frac{1}{\pi}\sin\left(\frac{\pi}{2}(2ka)\right)\sqrt{1+4\frac{\zeta_1}{\zeta}\cos^2\left(\frac{\pi}{2}(2ka)\right)}. \tag{1.18}$$

A representative dispersion curve for $\zeta_1 = 0.5\zeta$ described by (1.18) is also shown in Fig. 1.1 (upper curves). The characteristic feature of this curve is the shift of the maximum value of the frequency from the point 2ka=1 to smaller values (longer waves!).

This is considered to be an indication of long-range actions in a crystallographic lattice (e.g. in the ionic crystals). It is quite obvious that the continuum model, as described in the first example, cannot account for such effects solely as a result of the long-wave approximation – it describes the „linear" vicinity of the origin of the coordinates in Fig. 1.1.

Let us mention in passing that some attempts have been made to construct continuous models without the assumption of the local actions, i.e. accounting for at least the binary non-local interactions. The most successful and at the same time true one directly connected with the discrete structure seems to be the model of **pseudo-continua** proposed by D. ROGULA (1970), (1973), and I. A. KUNIN (1975). Details concerning its construction can be found in the book of Kunin. A sort of approximation of the non-local theories appears in the continua with higher gradients of fields as constitutive variables (e.g. Cosserat continua). These continua seem to be appropriate, for instance, in the description of liquid crystals and some non-equilibrium phase transition phenomena (hysteresis). There are still many unsolved fundamental problems, e.g. the formulation of the boundary values, which limit their practical applicability.

3. We consider another example of a chain demonstrating a feature of the discrete system which cannot be described by the usual one-component continua. Namely, we investigate a chain with equal spacing, a, of the mass points with the alternating masses m and M, M>m (see: Fig. 1.2.). For simplicity, we assume the nearest neighbour interactions. Let us also assume that we have N points of each sort, and the points with the mass, m, are numbered by even integers, and those with the mass, M, by odd integers. Also, in this case, we assume the appropriate Born - von Kármán periodicity condition.

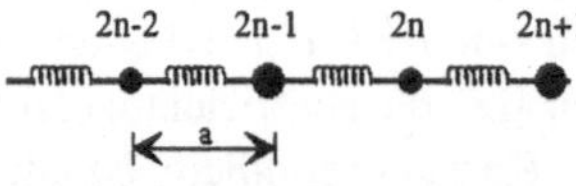

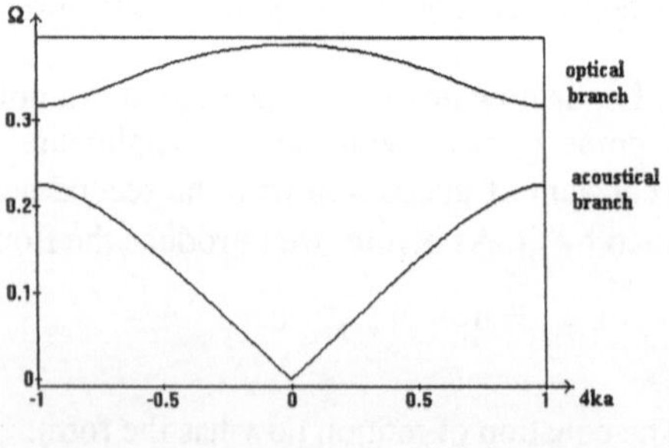

Fig. 1.2. *Dispersion curve for the uniform chain of the different mass points*

The corresponding equations of motion have the form

$$m\frac{d^2u_{2n}}{dt^2} = \zeta\left(u_{2n+1} - 2u_{2n} + u_{2n-1}\right), \qquad 1 \le n \le N,$$
$$M\frac{d^2u_{2n+1}}{dt^2} = \zeta\left(u_{2n+2} - 2u_{2n+1} + u_{2n}\right). \tag{1.20}$$

We seek the general solution of the form

$$u_{2n} = A\exp\left[2\pi i(2nka - \omega t)\right],$$
$$u_{2n+1} = B\exp\left[2\pi i((2n+1)ka - \omega t)\right] \tag{1.21}$$

The periodicity condition yields

$$k = \frac{\ell}{2Na}, \qquad \ell\text{-integer}, \quad \ell \le 2N. \tag{1.22}$$

Again, the waves cannot be shorter than a, and we can construct 2N independent solutions (normal modes of vibration, phonons) of the form (1.21).

Substitution of (1.21) into (1.20) yields the following condition for the existence of the non-trivial solutions

$$(2\pi\omega)^4 - 2\left(\frac{\zeta}{m} + \frac{\zeta}{M}\right)(2\pi\omega)^2 + 4\frac{\zeta^2}{mM}\sin^2(2\pi ka) = 0. \tag{1.23}$$

Hence,

$$\omega_1 = \frac{1}{2\pi}\sqrt{\frac{\zeta}{mM}\sqrt{(m+M) - \sqrt{(m+M)^2 - 4mM\sin^2(2\pi ka)}}},$$
$$\omega_2 = \frac{1}{2\pi}\sqrt{\frac{\zeta}{mM}\sqrt{(m+M) + \sqrt{(m+M)^2 - 4mM\sin^2(2\pi ka)}}}. \tag{1.24}$$

The corresponding dispersion curves are shown in Fig. 1.2. It can easily be checked that the first frequency corresponds to the normal vibrations with the nearest neighbours being in the same phase and, the second frequency (upper branch of the Figure 1.2.) – to the vibrations of the nearest neighbours in the opposite phase. This type of vibration appears, for instance, in ionic crystals where the neighbouring atoms have the opposite sign of the electric charge. The corresponding sharp changes of the electric moment are observed optically by reflection or by absorption. Therefore, the upper branch is called **optical** and the lower one - **acoustical**. Again, according to the considerations of the first example, the classical continuous model cannot describe this optical branch of the dispersion curve.

For comparison of the results of the above simple examples with those observed experimentally, we show in Fig. 1.3. a few dispersion curves of the real crystal lattice.

1.3 Contents of the Book

Chapters 2 and 3 of this book contain the basic geometrical and kinematical notions of the continuum model. We concentrate on their interpretation and application in thermomechanics rather than on the formal mathematical background. The latter can be found in books about modern differential geometry and mathematical foundations of mechanics such as the work of J. E. MARSDEN, T. J. R. HUGHES (1983) or J. E. MARSDEN, T. S. RATIU (1994). The motivation for not covering the more formal mathematics is that usually mathematicians have no difficulties with the proper formalization. On the other hand, engineers and physicists can experience problems with the proper interpretation of the formal notions of the geometry. Take, for example, notions such as: the rotation, which is described by the orthogonal part of the polar decomposition of the deformation gradient, the integrability conditions for the deformation gradients and also the objective time derivatives. They are often presented in the

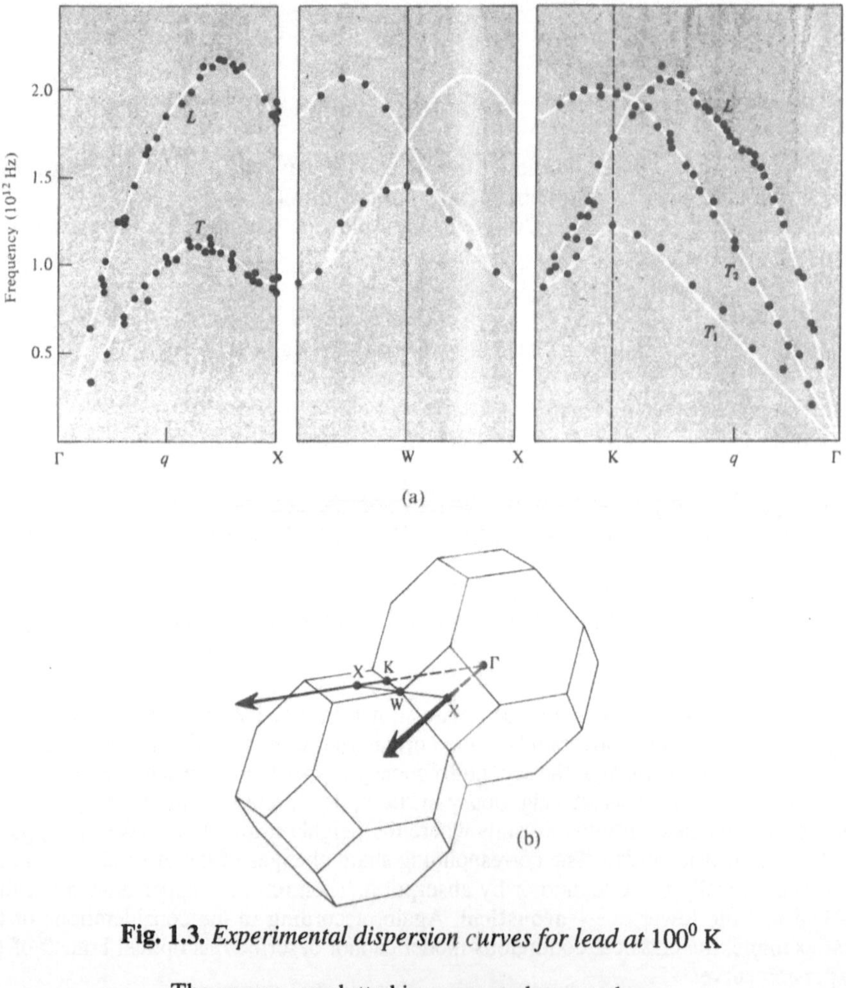

Fig. 1.3. *Experimental dispersion curves for lead at* 100^0 K

The curves are plotted in a repeated-zone scheme along the edges of the shaded triangle shown in (b) [N. W. ASHCROFT, N. D. MERMIN (1976)]

literature in a confusing manner. For this reason these topics are discussed in detail in these two chapters.

We also present, at length, the symmetry properties following from the orthogonal transformation in order to stress the difference between the objectivity and the material frame indifference (material objectivity).

Chapter 4 is devoted to the discussion of the balance laws of classical thermomechanics. We derive the local forms of these equations in the Lagrangian and Eulerian descriptions and show their structures after a transition from the inertial frame of reference to the non-inertial one.

In Chap. 5 we show the structure of the field equations for the extended thermodynamics. This chapter also includes the extensive discussion of the isotropy conditions for a few most common objects of thermomechanics. For completeness, we include the tables with isotropic representations (invariant elements and generator elements) for more complicated isotropic functions, which otherwise would have to be sought in the numerous original references. These tables were prepared by I-SHIH LIU (1988). We also introduce the theorem of T. RUGGERI on the separation of objects appearing in the hierarchical field equations of extended thermodynamics into their convective and materially objective parts. Many additional details of the proof can be found in the book by I. MÜLLER and T. RUGGERI (1993).

Chapter 6 contains the discussion of the second law of thermodynamics in the form of the entropy inequality. We show, rather extensively, the motivation of this law following from some microscopic considerations. For the exploitation of the second law within the frame of continuum theories, the method of Lagrange multipliers is presented. This method is based on the theorem of I-SHIH LIU, which we also describe in this chapter. Owing to the similarity of this method to the method of the main fields in the theory of hyperbolic systems of partial differential equations, we also present the consequences of the second law for the field equations in the case of the hyperbolicity assumption. The chapter closes with the brief presentation of the second law for the singular surface – in particular for the interfaces.

In Chap. 7 we present the full thermodynamical theory of rarified gases as it arises within the frame of extended thermodynamics. Apart from some relativistic results which will not be explored in this text, this is the only model where the full thermodynamical theory could be developed from the first principles. It demonstrates the main features of the continuum thermodynamics not influenced by the simplifying assumptions of more complicated models which we discuss further in this work.

Chapter 8 is devoted to the thermodynamical models of viscoelastic (non-Newtonian) fluids and to the viscoelastic solids. The former model has been considered for some decades and it has led to some substantial controversies concerning the stability of the thermodynamical equilibrium and the signs of the so-called normal stress coefficients (Weissenberg effect). We show that extended thermodynamics yields the solution of all those controversies in agreement with the experimental data for the most common non-Newtonian fluids. The latter model demonstrates the way in which the extended thermodynamics can be applied to solids described within the Lagrangian method.

Chapter 9 differs somewhat from the rest of the book because it concerns a medium which is, strictly speaking, not material in the usual sense of this notion within the continuum theories. We discuss the extended thermodynamics of phonons and its connection with the propagation of the so-called second sound. The purpose of these considerations is twofold. Firstly, the simplest thermodynamical explanation for the finite speed of propagation of the thermal disturbance is shown. This problem has been discussed for many years (parabolicity of the classical heat conduction equation!) and no solution which corresponds to standards of the modern continuum thermodynamics has been found. Secondly, the close connection of extended thermodynamics with the kinetic theories, which were always considered to be more fundamental than the macroscopic theories, is shown.

Chapter 10 is concerned with the extended thermodynamics for some multicomponent systems. We only present a new model for the porous materials because the mixture of fluids has been extensively discussed by I. MÜLLER AND T. RUGGERI (1993). The results of the extended thermodynamics for such a multicomponent model do not deviate substantially from those obtained by the classical thermodynamical methods. On the other hand, the Lagrangian description of the multicomponent system, as well as a new equation for porosity, make the porous material particularly attractive for the methods of extended thermodynamics.

The book ends with three Appendices. The first appendix contains a brief introduction to the equilibrium thermodynamics of homogeneous systems (Gibbs). The derivation of the classical Gibbs equation by means of the Caratheodory Theorem is shown together with the proof of existence of the absolute temperature. We briefly present the notions of the thermodynamic potentials, which can be obtained from the Gibbs equation by the Legendre transformation. We describe some properties of these potentials which render the stability of the thermodynamical equilibrium.

The second appendix contains the comprehensive collection of the fundamental relations for the curvilinear coordinates applied in some examples presented in the book.

The third appendix is devoted to the hyperbolic systems of partial differential equations. By means of a few examples, we present the basic notions of such systems.

As far as possible the references which are quoted in this book are limited to standard and easily available textbooks and monographs. In cases where the particular issue is not sufficiently covered in such reference books, we quote the original papers. For this reason, the quotations can, by no means, be considered to give credit to all researchers who have made important contributions to continuum mechanics and thermodynamics.

A remark on the notation used in the book seems to be appropriate. Some of it is standard for continuum mechanics and thermodynamics and concerns primarily vector and tensor objects. We use the absolute notation as well as the index notation with Einstein summation convention. The latter refers in most cases to Cartesian coordinates. However, there are also some deviations. For instance, models in extended thermodynamics are written in the notation used by those scientists who mainly contributed to this field. This is done on purpose as the results in this field of research are not available yet in the form of standard textbooks and must be studied from original papers. Finally, we use some basic mathematical symbols which stem from the theory of sets. They are not explained any further in the book, and if necessary, should be consulted in mathematical dictionaries.

2 Geometry

2.1 Reference and Current Configurations

For the purpose of this book we do not need to go into the whole mathematical finesse of the geometrical properties of the continuum. This can be found in numerous textbooks [e.g. J. E. MARSDEN, T. J. R. HUGHES (1983)].

In most cases considered further in these notes, the geometrical notion of the **body** can be identified with a compact measurable subset of the three-dimensional Euclidean space $\Re^3$. In spite of this simple structure, we shall often speak of the body as a differentiable manifold $\mathcal{B}_o$ without referring to its Euclidean character.

According to the remarks we have made in the Introduction, the main geometrical property of the continuum is the existence of the diffeomorphisms $\mathbf{f}(\cdot,t)$ for all values of the parameter $t \in \mathcal{T} \equiv <t_i,t_f> \subset \Re^1$, which define the **current configurations** $\mathcal{B}_t \subset \Re^3$ of the body $\mathcal{B}_o$

$$\forall\ t \in \mathcal{T}:\quad \mathbf{f}(\cdot,t): \mathcal{B}_o \to \mathcal{B}_t \equiv \mathbf{f}(\mathcal{B}_o,t) \subset \Re^3. \tag{2.1}$$

For each t from the interval $\mathcal{T}$ the mapping $\mathbf{f}(\cdot,t)$ is called the **deformation** of the body from the configuration $\mathcal{B}_o$ to the configuration $\mathcal{B}_t$.

The points of the manifold $\mathcal{B}_0$ are denoted by $\mathbf{X}$ and they are called **the material points**, whereas the points of the space $\Re^3$ containing the images $\mathcal{B}_t$ of the body $\mathcal{B}_0$ are denoted by $\mathbf{x}$. The space $\Re^3$ itself is frequently called **the configuration space or the space of motion.** With this notation in mind we can write out the relation (2.1) point by point:

$$\forall\ \mathbf{X} \in \mathcal{B}_0,\ t \in \mathcal{T}:\quad \mathbf{x} = \mathbf{f}(\mathbf{X},t) \in \mathcal{B}_t \subset \Re^3. \tag{2.2}$$

The point $\mathbf{x} \in \Re^3$, given by the relation (2.2), is called the position of the material point $\mathbf{X}$ in the current configuration, whereas the function $\mathbf{f}(\mathbf{X},\cdot)$ is often called **the motion** of the material point $\mathbf{X} \in \mathcal{B}_0$. We shall discuss some properties of this map in the next chapter.

Most properties of a specific substance, which is described by means of a continuous model, are specified for a chosen material point $\mathbf{X}$. They are defined on a neighbourhood of $\mathbf{X}$ generated by a tangent space of the manifold of the body at the material point $\mathbf{X}$. To specify these notions let us consider a curve $\mathcal{C}_0 \subset \mathcal{B}_0$ containing a chosen material point $\mathbf{X}^0$ and parametrized by S:

$$\mathcal{C}_0: \mathbf{X} = \mathbf{X}(S). \tag{2.3}$$

The tangent vector to this curve at the point $\mathbf{X}^0$

$$d\mathbf{X}^0 \equiv \partial_S \mathbf{X}(\mathbf{X}^0) dS \in \mathcal{T}_{\mathbf{X}^0}, \qquad \partial_S \mathbf{X} \equiv \frac{\partial \mathbf{X}}{\partial S}, \tag{2.4}$$

belongs, certainly, to the tangent space $\mathcal{T}_{\mathbf{X}^0}$ of the manifold $\mathcal{B}_0$ at this point. The whole space $\mathcal{T}_{\mathbf{X}^0}$ can be generated by the tangent vectors of all possible curves going through the point $\mathbf{X}^0$.

Now, by means of the map (2.2), we can construct the curve $\mathcal{C}_t$ in the current configuration $\mathcal{B}_t$

$$\mathcal{C}_t \equiv \{\mathbf{x} \,|\, \mathbf{x} = \mathbf{f}(\mathbf{X}(S), t), \mathbf{X}(S) \in \mathcal{C}_0\}. \tag{2.5}$$

Its tangent vector corresponding to the point $\mathbf{X}^0$ is then given by the formula

$$d\mathbf{x}^0 = \mathbf{F}(\mathbf{X}^0, t) d\mathbf{X}^0 = [\mathbf{F}(\mathbf{X}^0, t)\partial_S \mathbf{X}] dS, \tag{2.6}$$

where

$$\mathbf{F}(\mathbf{X}^0, t) := \frac{\partial \mathbf{f}}{\partial \mathbf{X}}(\mathbf{X}^0, t) \equiv \operatorname{Grad} \mathbf{f}(\mathbf{X}^0, t) \tag{2.7}$$

is called **the deformation gradient** at the point $\mathbf{X}^0$. Obviously, if constructed for all curves $\mathcal{C}_0$ going through the point $\mathbf{X}^0$, formula (2.6) defines the linear mapping

$$\mathbf{F}(\mathbf{X}^0, t)(\cdot): \mathcal{T}_{\mathbf{X}^0} \to \mathcal{T}_{\mathbf{x}^0}, \tag{2.8}$$

of the tangent space $\mathcal{T}_{\mathbf{X}^0}$ onto the tangent space $\mathcal{T}_{\mathbf{x}^0}$ at the point $\mathbf{x}^0$ in the current configuration $\mathcal{B}_t$ (see: Figure 2.1.).

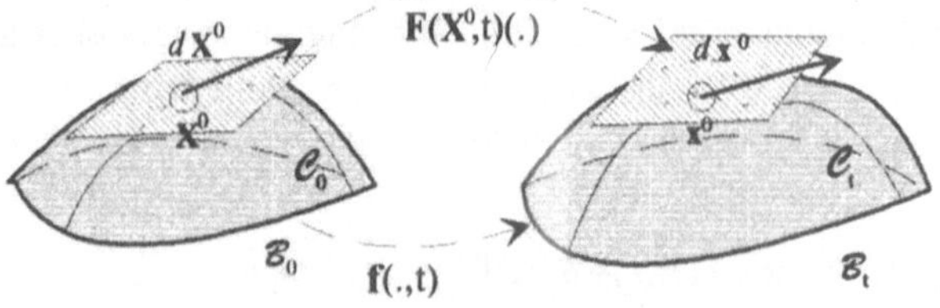

Fig. 2.1. *Transformation of the local configurations*

Any vector $\mathbf{K}(\mathbf{X}^0) \in \mathcal{T}_{\mathbf{X}^0}$ whose transformation to the tangent space of the current configuration is described by the **linear** rule (2.8), i.e.

$$\mathbf{k}(\mathbf{X}^0, t) = \mathbf{F}(\mathbf{X}^0, t)\mathbf{K}(\mathbf{X}^0), \tag{2.9}$$

is called **a material vector**. Notice that this notion is defined by the transformation rule to the current configuration and not by the choice of the vector from the tangent space $\mathcal{T}_{X^0}$, i.e. the same vector from this space can be a material vector as well as a non-material vector according to its changes in the motion of the material point.

Bearing this in mind, we can consider the deformation gradient as a **linear mapping** defining material vectors. This interpretation is particularly useful in the cases of the non-integrable deformation gradients which are, in reality, not the gradients of anything (see: Remark 2.2. below).

It should also be mentioned that many vector fields which appear in the theory of continuous media are not material. The simplest example is the field of unit normal vectors of a chosen immobile surface in the configuration space $\Re^3$ (e.g. to a membrane permeable for the material). A more sophisticated example shall be constructed in Remark 2.1.

Before we proceed with the further presentation of geometrical properties, let us describe the above objects in a chosen system of coordinates. Assuming the body is identified with its configuration $\mathcal{B}_0$ in the configuration space $\Re^3$, we can introduce a system of Cartesian coordinates prescribing three coordinates $\left\{X^\alpha\right\}_{\alpha=1,2,3}$ to each point $\mathbf{X} \in \mathcal{B}_0$. The unit orthogonal basis vectors of this system are denoted by $\mathbf{e}_\alpha$, $\alpha=1,2,3$, and certainly the covariant and contravariant basis vectors are in this case identical. We could use the same system of coordinates to describe the position $\mathbf{x}$ of the material point. Such identical coordinate systems were used frequently in early papers concerning particularly the continuum mechanics of solids. This is very often misleading. For this reason, we introduce another coordinate system $\left\{x^k\right\}_{k=1,2,3}$ to parametrize the configuration space without any reference to the body and its motion. The unit orthogonal basis vectors shall be denoted by $\mathbf{e}_k$, k=1,2,3. It is customary to call X^α the **Lagrangian coordinates** and x^k the **Eulerian coordinates**. Using these coordinates the function (2.2) has the form:

$$x^k = f^k\left(X^1, X^2, X^3, t\right), \quad k = 1,2,3, \qquad \mathbf{x} \equiv x^k \mathbf{e}_k, \quad \mathbf{X} \equiv X^\alpha \mathbf{e}_\alpha, \tag{2.10}$$

and the deformation gradient (2.7) is given by:

$$\mathbf{F} = F^k{}_\alpha \mathbf{e}_k \otimes \mathbf{e}_\alpha, \qquad F^k{}_\alpha = \frac{\partial f^k}{\partial X^\alpha}, \quad k = 1,2,3, \ \alpha = 1,2,3. \tag{2.11}$$

Certainly, the level of indices is immaterial in these Cartesian reference frames. It has been preserved solely to indicate the changes which will be necessary in the case of curvilinear coordinates.

Owing to the global character of tangent spaces to the Euclidean space (the trivial connection; all spaces are isomorphic), we could use the basis vectors $\mathbf{e}_\alpha$ and $\mathbf{e}_k$ of the Lagrangian and Eulerian coordinate systems as the bases for vectors in the tangent spaces. The relation (2.9) can now be written in the following form:

$$k^k = F^k{}_\alpha K^\alpha, \qquad \mathbf{k} = k^k \mathbf{e}_k, \quad \mathbf{K} = K^\alpha \mathbf{e}_\alpha. \tag{2.12}$$

In some examples presented later in this book we also use various curvilinear coordinate systems. The appropriate form of the equations in these systems can be obtained by means of the well-known transformation rules. For instance, the partial derivatives of the description in the Cartesian coordinates must then be replaced by the covariant derivatives. However, in the general considerations we shall always rely on the above described Cartesian systems.

Remark 2.1. *On Non-material Vectors*
Now we construct an example of a vector field which changes as a result of the motion of the body and is simultaneously non-material, i.e. it transforms in a way different from that given by (2.9).

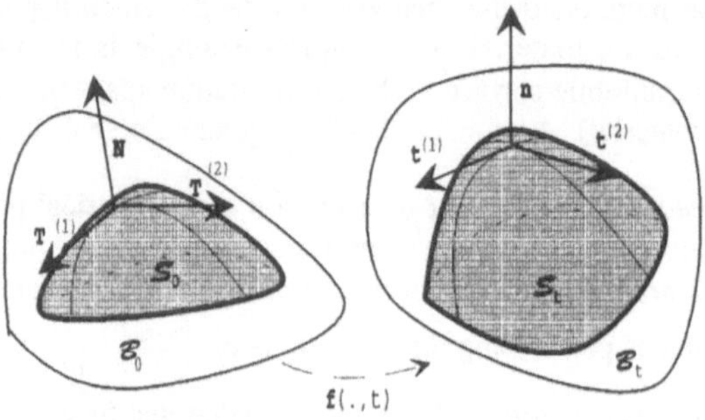

Fig. 2.2. *Transformation of the vector orthogonal to the material surface*

Let us consider a smooth surface $\mathcal{S}_0$ frozen in the body $\mathcal{B}_0$ (i.e. always containing the same material points). At a given point of this **material surface** we construct the normal vector $\mathbf{N}$ to this surface as the vector product of the two non-colinear vectors $\mathbf{T}^{(1)}$, $\mathbf{T}^{(2)}$, tangent to the surface $\mathcal{S}_0$ (Figure 2.2.):

$$\mathbf{N} = \mathbf{T}^{(1)} \times \mathbf{T}^{(2)}. \tag{2.13}$$

The vectors $\mathbf{T}^{(1)}$, $\mathbf{T}^{(2)}$ can be, certainly, chosen as the unit tangent vectors of two curves, lying on the surface $\mathcal{S}_0$. Consequently, these curves are also material. Without loss of generality we can require as well that the curves are orthogonal at the point of intersection, i.e the vectors are also orthogonal. In such a case the vector $\mathbf{N}$ is the unit vector. The vectors $\mathbf{T}^{(1)}$, $\mathbf{T}^{(2)}$ have the images in the current configuration given by the relation (2.9). They transform together with the curves lying on the material surface which means that those images – let us denote them by $\mathbf{t}^{(1)}$, $\mathbf{t}^{(2)}$, respectively – are tangent to the image $\mathcal{S}_t$ of the surface $\mathcal{S}_0$.

The unit normal vector $\mathbf{n}$ to the surface $\mathcal{S}_t$ can now be identified with the normalized vector product

$$\mathbf{n} = \frac{\mathbf{t}^{(1)} \times \mathbf{t}^{(2)}}{\upsilon}, \qquad \upsilon \equiv \left|\mathbf{t}^{(1)} \times \mathbf{t}^{(2)}\right|, \quad \mathbf{t}^{(1)} = \mathbf{F}\mathbf{T}^{(1)}, \quad \mathbf{t}^{(2)} = \mathbf{F}\mathbf{T}^{(2)}. \tag{2.14}$$

We derive the explicit form of this relation for the current normal vector of the material surface using the coordinate systems described before in this section. We have

$$\begin{aligned} n^k &= \tfrac{1}{\upsilon}\varepsilon^k{}_{lm} F^l{}_\alpha F^m{}_\beta T^{(1)\alpha} T^{(2)\beta} = \\ &= \tfrac{1}{\upsilon}\varepsilon_{nlm} F^n{}_\mu F^l{}_\alpha F^m{}_\beta \overset{-1}{F}{}^{\mu}{}_{p} T^{(1)\alpha} T^{(2)\beta} \delta^{pk}, \quad \mathbf{n} = n^k \mathbf{e}_k, \end{aligned} \tag{2.15}$$

and ε_{nlm} is the permutation symbol. We used the relation for the inverse of the deformation gradient

$$F^n{}_\mu \overset{-1}{F}{}^\mu{}_p = \delta^n{}_p, \quad \text{i.e.} \quad \mathbf{F}\mathbf{F}^{-1} = \mathbf{1}. \tag{2.16}$$

The existence of this inverse follows from the assumption that the map $\mathbf{f}(\cdot,t)$ is a diffeomorphism.

We also have

$$\varepsilon_{nlm} F^n{}_1 F^l{}_2 F^m{}_3 = \det \mathbf{F} \quad \Rightarrow \quad \varepsilon_{nlm} F^n{}_\mu F^l{}_\alpha F^m{}_\beta = \varepsilon_{\mu\alpha\beta}\, J, \tag{2.17}$$

where

$$J \equiv \det \mathbf{F} \neq 0, \tag{2.18}$$

is the Jacobi determinant. It is different from zero owing to the invertibility of $\mathbf{f}(\cdot,t)$.

Substitution of (2.17) in (2.15) yields

$$n^k = \tfrac{1}{\upsilon} J \varepsilon_{\mu\alpha\beta} \overset{-1}{F}{}^\mu{}_p T^{(1)\alpha} T^{(2)\beta} \delta^{pk} = \tfrac{1}{\upsilon} J \delta^{pk} \overset{-1}{F}{}^\mu{}_p N^\nu \delta_{\nu\mu}, \quad \mathbf{N} = N^\nu \mathbf{e}_\nu, \tag{2.19}$$

where the definition (2.13) was used.

Finally,

$$\mathbf{n} = \frac{\mathbf{F}^{-T}\mathbf{N}}{\left|\mathbf{F}^{-T}\mathbf{N}\right|}, \qquad \left|\mathbf{F}^{-T}\mathbf{N}\right| = \sqrt{\left(\mathbf{F}^{-T}\mathbf{N}\right)\cdot\left(\mathbf{F}^{-T}\mathbf{N}\right)}. \tag{2.20}$$

This formula describing the transformation of vectors orthonormal to the material surface proves that these vectors are not material – they do not transform according to the rule (2.9)•

Let us mention in passing that the distinction between material and non-material vectors would become quite natural if we formulated systematically the geometry of deformable bodies in terms of the analysis on differentiable manifolds. In the terminology of such an analysis the material vectors are simply called **vectors**, whereas the non-material vectors are called **one-forms**. A similar distinction could be made if we were using arbitrary curvilinear coordinate systems with metric tensors changing with the deformation. Then the material vectors correspond to the contravariant vectors, and the non-material vectors to the covariant vectors.

Even though the Cartesian description applied in this work may lead to certain confusions connected with the above-mentioned distinction, for the sake of simplicity of the mathematical formalism, we shall not go into the analysis on manifolds. In some places in the further presentation we shall point out the passages where caution is required.

The reader interested in the rigorous presentation of this problem is referred to the book of J. E. MARSDEN and T. J. R. HUGHES [1983] quoted already at the beginning of this chapter.

We shall return to the application of the formula (2.20) in the following.

Remark 2.2. *On Non-integrable Deformation Gradients*
In many cases of practical interest the mappings $\mathbf{f}(\cdot,t)$ cannot be assumed to be a diffeomorphism everywhere. For instance, the connection of two beams by a hinge yields this mapping to be non-differentiable at the hinge, even though the deformation gradient $\mathbf{F}$ has finite limits on both sides of this connection. Similar situations arise in the case of shock waves which yield the discontinuity of $\mathbf{F}$ on the surface which is called the wave front.

It is easy to see that these cases are covered by a continuous model if we assume that the mappings $\mathbf{f}(\cdot,t)$ are diffeomorphic only **almost everywhere** on $\mathcal{B}_0$. This means that the set of points where $\mathbf{f}(\cdot,t)$ is not diffeomorphic has a volume measure equal to zero. Further in this book we use only this weaker assumption on $\mathbf{f}(\cdot,t)$.

However, it should be mentioned that such important models as an elastic-plastic material lead to certain „deformation gradients" which are not integrable everywhere. This means that they cannot be represented by any diffeomorphic mapping of $\mathcal{B}_0$, or any other differentiable manifold, whose derivative with respect to the local coordinates of this manifold would coincide with the deformation gradient. This is, for example, the case when the integrable deformation gradient $\mathbf{F}=\partial_X\mathbf{f}$ is supposed to satisfy the following law of **multiplicative decomposition**:

$$\forall \mathbf{X} \in \mathcal{B}_0: \quad \mathbf{F}(\mathbf{X},t)(\cdot) = \mathbf{F}^e(\mathbf{X},t)(\cdot) \circ \mathbf{F}^p(\mathbf{X},t)(\cdot), \tag{2.21}$$

where $\mathbf{F}^e$ and $\mathbf{F}^p$ are called the **elastic** and **plastic local configurations**, respectively, and they are given by certain constitutive (material) relations. The mappings of formula (2.21) are supposed to act on vector spaces: $\mathbf{F}^p$ on the tangent space $\mathcal{T}_X$ and $\mathbf{F}^e$ on the vector space which is the image of $\mathcal{T}_X$ with respect to the mapping $\mathbf{F}^p$. The latter vector space is in the case of elastic-plastic material not a tangent space of any differentiable manifold diffeomorphic, even locally, with $\mathcal{B}_0$. Hence, rather common claims in the literature on this subject that this vector space is induced by a sufficiently small neighbourhood of a material point in its intermediate (so-called relaxed) configuration, which is considered as a differentiable manifold, and that $\mathbf{F}^p$ is the gradient with respect to the local coordinates in this intermediate configuration, do not make any sense. Such notions as the coordinates, and consequently, the differentiation with respect to such coordinates cannot be introduced in these intermediate configurations at all. It is easy to see, even in the simplest examples of homogeneous deformation, that the collection of material points $\mathbf{X} \in \mathcal{B}_0$ with the differential structure, defined for each $\mathbf{X}$ by the vector spaces $\{\mathbf{F}^p(\mathbf{X},t)(\mathcal{T}_X)\}$, does not form any differentiable manifold any more.

For this reason, the mappings $\mathbf{F}(\mathbf{X},t)(\cdot)$, $\mathbf{F}^e(\mathbf{X},t)(\cdot)$, $\mathbf{F}^p(\mathbf{X},t)(\cdot)$, and the like are called by R. A. TOUPIN [see: W. NOLL, R. A. TOUPIN, C.-C. WANG (1968)] the **local configurations**. In the case of differentiable manifolds, when they are identical with deformation gradients, Toupin calls them **faithful**.

As an example illustrating these remarks, let us consider a one-dimensional tensile deformation of the material defined by the following relations (see: Fig. 2.3.):

a/ for $\dot{\varepsilon} > 0$ (loading)

$$\sigma = \begin{cases} 2E\varepsilon & \text{for } \varepsilon \le \frac{\sigma_Y}{E}, \\ E\varepsilon + \sigma_Y & \text{for } \frac{\sigma_Y}{E} < \varepsilon, \end{cases} \tag{2.22}$$

b/ for $\dot{\varepsilon} < 0$ (unloading starting at $\varepsilon = \varepsilon_{max}$)

$$\sigma = \begin{cases} 2E\varepsilon & \text{for } \varepsilon_{max} \leq \frac{\sigma_Y}{E}, \\ 2E\left(\varepsilon - \frac{\sigma_Y}{E}\right) & \text{for } \frac{\sigma_Y}{E} < \varepsilon_{max}, \end{cases} \tag{2.23}$$

the deformation gradient being

$$\mathbf{F} = F^k{}_\alpha \mathbf{e}_k \otimes \mathbf{e}_\alpha, \quad \left(F^k{}_\alpha\right) = \begin{pmatrix} 1 & 0 & 0 \\ 0 & 1 & 0 \\ 0 & 0 & 1+\varepsilon \end{pmatrix}, \quad 0 < \varepsilon << 1. \tag{2.24}$$

In the above relations σ is the force acting in the x^3-direction per unit area perpendicular to x^3 (normal stresses). σ_Y is a constant which we relate to a certain limit stress (the yield stress), E is an elastic constant (Young modulus). ε_{max} describes the maximum value of ε in the loading process ($\dot{\varepsilon} > 0$) before the unloading ($\dot{\varepsilon} < 0$) starts, $\dot{\varepsilon}$ being the rate of changes of ε.

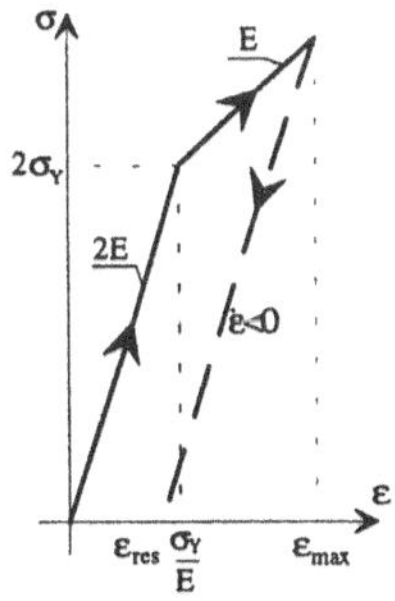

Fig. 2.3. *Stress-strain relation for a non-homogeneous elastic plastic rod*

It is obvious that the deformation gradient (2.24) can be prescribed to the homogeneous motion of the infinite strip with the distributed constant loading σ and $-\sigma$ on the upper and the lower surface, respectively. It corresponds also to the approximation of the description of motion of the x^3-homogeneous thin rod. In both these cases **F** is faithful.

In contrast to a rather common belief, the above definition of the material does not specify uniquely the plastic deformations. We illustrate this in Fig. 2.4. with the example of a system of two rods of the same length l and of the same area of cross-section A. If this system was replaced by a single rod („homogenization“!), then its resultant properties would be identical with those of Fig. 2.3., but the residual deformation ε_{res} after the full unloading (σ=0) would not be identical with that ε_p, which one would be inclined to call the plastic deformation. This is due to the fact that the state σ=0 corresponds in our example to the self-equilibrated state $\sigma^{(1)} = -\sigma^{(2)} = 0.5\varepsilon_p E$ ($\sigma^{(1)} + \sigma^{(2)} = 0$) which produces the elastic deformation of both rods. This deformation is needed to construct the compatible structure again from the non-compatible system of fully relaxed rods shown in the lower part of Fig. 2.4. Then $\varepsilon_{res} = \varepsilon_p + {}^1/_E \sigma^{(2)} = 0.5\varepsilon_p$.

In the above considerations the plastic deformation was defined as that appearing after removing the clamp (see: the relaxed rods in Fig. 2.4.):

$$\left(\overset{p}{F}{}^{\xi}{}_{\alpha}\right)=\begin{pmatrix}1&0&0\\0&1&0\\0&0&1+\varepsilon_p\end{pmatrix},\quad \left(\overset{e}{F}{}^{k}{}_{\xi}\right)=\begin{pmatrix}1&0&0\\0&1&0\\0&0&1+\varepsilon-\varepsilon_p\end{pmatrix},\quad |\varepsilon|,|\varepsilon_p|<<1, \tag{2.25}$$

$$\mathbf{F}=\mathbf{F}^{e}\mathbf{F}^{p},\quad \mathbf{F}^{e}=\overset{e}{F}{}^{k}{}_{\xi}\mathbf{e}_k\otimes\mathbf{e}_{\xi},\quad \mathbf{F}^{p}=\overset{p}{F}{}^{\xi}{}_{\alpha}\mathbf{e}_{\xi}\otimes\mathbf{e}_{\alpha},\qquad \mathbf{e}_{\xi}\cdot\mathbf{e}_{\zeta}=\delta_{\xi\zeta}.$$

Neither of these local configurations can be identified with any deformation gradient. The material after „removing the clamps" splits into isolated points, and there exists no differentiable manifold made of these points whose neighbourhoods, however small, would produce deformation gradients coinciding with (2.25)•

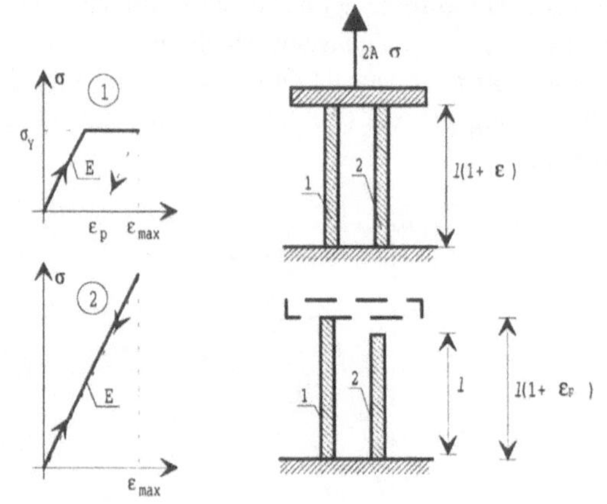

Fig. 2.4. *The model of a non-homogeneous rod whose properties follow from the combination of two homogeneous rods of the same cross-section area* A

If the local configuration **F**(·,t) is given as a field on the manifold $\mathcal{B}_0$ then it can be checked easily whether it is faithful or not. In the former case, it must be connected with the deformation of the body **f**(·,t) which means that it satisfies the relation (2.7) almost everywhere on $\mathcal{B}_0$. If we assume in such a case that the local configuration **F** is differentiable, then it must fulfil the following condition:

$$\operatorname{Grad}\mathbf{F}=\left(\operatorname{Grad}\mathbf{F}\right)^{T^{23}},\quad \text{i.e.}\quad \frac{\partial F^{k}{}_{\alpha}}{\partial X^{\beta}}=\frac{\partial F^{k}{}_{\beta}}{\partial X^{\alpha}}, \tag{2.26}$$

which follows immediately from (2.7). This is one of the two **integrability conditions** for the local configuration **F**. The second one has a kinematical character and it shall be considered later in this book.

Exercise 2.1. Show that the first of the following local configurations

$$\overset{(1)}{\mathbf{F}} = \overset{(1)}{F}{}^{k}{}_{\alpha}\mathbf{e}_k \otimes \mathbf{e}_\alpha, \qquad \overset{(2)}{\mathbf{F}} = \overset{(2)}{F}{}^{k}{}_{\alpha}\mathbf{e}_k \otimes \mathbf{e}_\alpha,$$

$$\left(\overset{(1)}{F}{}^{k}{}_{\alpha}\right) \equiv \begin{pmatrix} X^1X^2 & \frac{1}{2}\left(X^1\right)^2 & 0 \\ \frac{1}{3}X^1\left(X^2\right)^3 & X^1\left(X^2\right)^2 & 0 \\ 0 & 0 & 1 \end{pmatrix}, \quad \left(\overset{(2)}{F}{}^{k}{}_{\alpha}\right) \equiv \begin{pmatrix} X^1X^2 & \left(X^1\right)^2 & 0 \\ X^1\left(X^2\right)^3 & X^1\left(X^2\right)^2 & 0 \\ 0 & 0 & 1 \end{pmatrix},$$

is faithful and that the second one is not•

Exercise 2.2. Prove the following identities

$$\mathrm{Div}\left(J\,\mathbf{F}^{-1}\right) = 0, \qquad \mathrm{div}\left(J^{-1}\,\mathbf{F}\right) = 0, \qquad \mathrm{grad} \equiv \mathbf{F}^{-T}\mathrm{Grad}\,\bullet$$

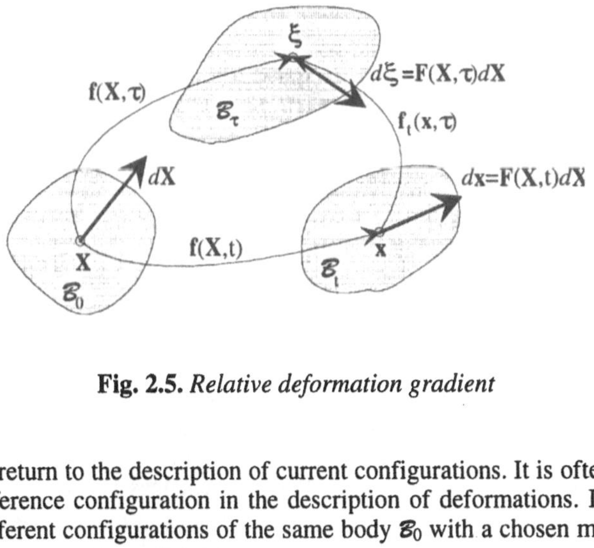

Fig. 2.5. *Relative deformation gradient*

Now let us return to the description of current configurations. It is often necessary to change the reference configuration in the description of deformations. In Fig. 2.5. we show three different configurations of the same body $\mathcal{B}_0$ with a chosen material point **X** and a chosen material vector $d\mathbf{X}$. We have

$$\xi = \mathbf{f}(\mathbf{X},\tau), \quad \tau \in 7, \qquad \mathbf{x} = \mathbf{f}(\mathbf{X},t), \quad t \in 7. \tag{2.27}$$

Owing to the invertibility of $\mathbf{f}(\cdot,t)$ for all $t \in 7$, we can also write

$$\mathbf{x} = \mathbf{f}\left[\mathbf{f}^{-1}(\xi,\tau),t\right] \equiv \mathbf{f}_\tau(\xi,t). \tag{2.28}$$

The map $\mathbf{f}_\tau(\cdot,t)$ of $\mathcal{B}_\tau \equiv \mathbf{f}(\mathcal{B}_0,\tau)$ onto $\mathcal{B}_t$ certainly defines the deformation of the new reference configuration $\mathcal{B}_\tau$ of the body $\mathcal{B}_0$ onto the configuration $\mathcal{B}_t$. Similarly,

$$\xi = \mathbf{f}_t(\mathbf{x},\tau) \equiv \mathbf{f}\left[\mathbf{f}^{-1}(\mathbf{x},t),\tau\right], \qquad \mathbf{f}_t(\cdot,\tau) = \mathbf{f}_\tau^{-1}(\cdot,t), \tag{2.29}$$

defines the deformation to the configuration $\mathcal{B}_\tau$ relative to the reference configuration $\mathcal{B}_t$.

The deformation gradients related to these mappings have then the form

$$\begin{aligned}
&d\xi = \mathbf{F}(\mathbf{X},\tau)d\mathbf{X}, && \mathbf{F}(\mathbf{X},\tau) \equiv \frac{\partial \mathbf{f}}{\partial \mathbf{X}}(\mathbf{X},\tau),\\
&d\mathbf{x} = \mathbf{F}(\mathbf{X},t)d\mathbf{X}, && \\
&d\mathbf{x} = \mathbf{F}_\tau(\mathbf{X},t)d\xi, && \mathbf{F}_\tau(\mathbf{X},t) \equiv \frac{\partial \mathbf{f}_\tau}{\partial \xi}(\xi,t),\\
&d\xi = \mathbf{F}_t(\mathbf{x},\tau)d\mathbf{x}, && \mathbf{F}_t(\mathbf{x},\tau) \equiv \frac{\partial \mathbf{f}_t}{\partial \mathbf{x}}(\mathbf{x},\tau).
\end{aligned} \tag{2.30}$$

Hence, we have the following relations for the relative deformation gradients:

$$\begin{aligned}
&\mathbf{F}(\mathbf{X},t) = \mathbf{F}_\tau(\mathbf{f}(\mathbf{X},\tau),t)\mathbf{F}(\mathbf{X},\tau), && \text{i.e.} \quad \mathbf{F}(t) = \mathbf{F}_\tau(t)\mathbf{F}(\tau),\\
&\mathbf{F}(\mathbf{X},\tau) = \mathbf{F}_t(\mathbf{f}(\mathbf{X},t),\tau)\mathbf{F}(\mathbf{X},t), && \text{i.e.} \quad \mathbf{F}(\tau) = \mathbf{F}_t(\tau)\mathbf{F}(t),\\
&\mathbf{F}_\tau(\xi,t) = \mathbf{F}_t^{-1}(\mathbf{f}_\tau(\xi,t),\tau), && \\
&\mathbf{F}_t(\mathbf{x},\tau) = \mathbf{F}(\mathbf{f}^{-1}(\mathbf{x},t),\tau)\mathbf{F}^{-1}(\mathbf{f}^{-1}(\mathbf{x},t),t) && \text{i.e.} \quad \mathbf{F}_t(\tau) = \mathbf{F}(\tau)\mathbf{F}^{-1}(t),\\
&\mathbf{F}_t(\mathbf{x},t) = \mathbf{1}, \qquad \text{etc.}
\end{aligned} \tag{2.31}$$

In the above relations, we also indicate the short-hand notation for the deformation gradients frequently used in the literature in which the appropriate spatial variables are dropped. In the sequel, we will use this notation if it does not lead to any confusion.

2.2 Polar Decomposition of the Deformation Gradient

Local changes in the geometry of continuous bodies can be described, as is usually done in differential geometry, by the changes in the metric tensor. In the Euclidean space it is particularly simple. Let us consider a chosen material point **X** and an infinitesimal material vector $d\mathbf{X}$ from the tangent space $\mathcal{T}_\mathbf{X}$. The changes in the length of three such linearly independent vectors describe the local changes in the geometry. In the current configuration $\mathcal{B}_t$ they follow from the relation describing the current length of the infinitesimal material vector

$$ds^2 \equiv d\mathbf{x}\cdot d\mathbf{x} = (\mathbf{F}d\mathbf{X})\cdot(\mathbf{F}d\mathbf{X}) = \mathbf{C}\cdot(d\mathbf{X}\otimes d\mathbf{X}), \tag{2.32}$$

where

$$\mathbf{C} \equiv \mathbf{F}^T\mathbf{F}, \qquad \mathbf{C}^T = \mathbf{C}, \tag{2.33}$$

is **the right Cauchy-Green deformation tensor**. We see, that in contrast to the deformation gradient **F** with generally nine independent components in its matrix representation, the changes in the metric properties, following from the change of configuration, are described by six independent components of the deformation tensor **C**.

Consequently, the question of what is the meaning of the missing three components arises. The answer to this question is given by the **polar decomposition theorem**. This yields a decomposition, which is unique in the case of a deformation of continuous bodies, of the deformation gradient **F** into the orthogonal part **R** and the symmetric tensor **U** according to the formula

$$\mathbf{F} = \mathbf{R}\,\mathbf{U}, \qquad \mathbf{R}^{-1} = \mathbf{R}^{T}, \qquad \mathbf{U}^{T} = \mathbf{U}. \tag{2.34}$$

Substitution of this relation in (2.33) yields

$$\mathbf{C} = (\mathbf{RU})^{T}(\mathbf{RU}) = \mathbf{U}^{2}, \tag{2.35}$$

i.e. the orthogonal part **R** of **F** is immaterial for the changes in the metric tensor. However, the tensor **R** can be calculated, if needed, after the solution of the initial-boundary value problem for the field equations has been found.

In continuum mechanics, it is customary to call the symmetric tensor **U** the **right stretch tensor**, and **R** the **tensor of rotation.**

Let us note that **U** maps tangent spaces $\mathcal{T}_X$ into itself (otherwise it could not be a symmetric tensor!), whereas **R** maps tangent spaces $\mathcal{T}_X$ into tangent spaces $\mathcal{T}_x$ of the current configuration $\mathcal{B}_t$. Therefore, the latter is sometimes called a two-point tensor and its rules of transformation are different from the left from those from the right-hand side.

We proceed to prove the theorem of polar decomposition. First of all, we need certain properties of the deformation gradient **F** which distinguish this gradient from the usual square matrix. Let us consider the material infinitesimal parallelepiped in the reference configuration $\mathcal{B}_0$ spanned by the vectors $dX^1\,\mathbf{e}_1$, $dX^2\mathbf{e}_2$, and $dX^3\mathbf{e}_3$. Its volume is given by the relation

$$dV = \left[\left(dX^1\mathbf{e}_1\right)\times\left(dX^2\mathbf{e}_2\right)\right]\cdot\left(dX^3\mathbf{e}_3\right) \equiv dX^1 dX^2 dX^3. \tag{2.36}$$

Owing to the assumption that the parallelepiped is material, its current image has the volume given by the following relation:

$$dv = \left[\left(dX^1\mathbf{F}\,\mathbf{e}_1\right)\times\left(dX^2\mathbf{F}\,\mathbf{e}_2\right)\right]\cdot\left(dX^3\mathbf{F}\,\mathbf{e}_3\right) = J\,dV\left(\mathbf{F}^{-1}\,\mathbf{e}_3\right)\cdot\left(\mathbf{F}\,\mathbf{e}_3\right) = J\,dV. \tag{2.37}$$

The derivation of this relation is similar to the derivation of the relation (2.19). It suffices to substitute $\mathbf{e}_1$ for $\mathbf{T}^{(1)}$, $\mathbf{e}_2$ for $\mathbf{T}^{(2)}$ and $\mathbf{e}_3$ for **N**. Consequently, the determinant J of the deformation gradient measures the volume changes of material infinitesimal elements. For this reason it must be **positive** for material media.

On the other hand, if the theorem of the polar decomposition holds true, we have, according to (2.34),

$$0 < J = \det\mathbf{F} = \det\mathbf{R}\,\det\mathbf{U} \quad\Rightarrow\quad \begin{cases}\det\mathbf{U} > 0 & \text{for} \quad \det\mathbf{R} = 1,\\ \det\mathbf{U} < 0 & \text{for} \quad \det\mathbf{R} = -1.\end{cases} \tag{2.38}$$

It is easy to see that solely the first condition can hold for continuous media without violating the neighbourhood relations of the continuum. For instance, in the second case, the local inversion of material would be admissible. Hence, the orthogonal part of the deformation tensor, the tensor of rotation **R**, must be **proper.**

Now we consider the eigenvalue problems for the tensors **U** and **C**. Let us denote by $\lambda_U^{(A)}$, A=1,2,3,the eigenvalues, and by $\mathbf{K}^{(A)}$, A=1,2,3, the normalized (unit) eigenvectors of the right stretch tensor **U**, i.e.

$$\left(\mathbf{U}-\lambda_U^{(A)}\mathbf{1}\right)\mathbf{K}^{(A)}=0, \qquad A=1,2,3, \qquad \mathbf{K}^{(A)}\cdot\mathbf{K}^{(B)}=\delta^{AB}. \tag{2.39}$$

All three eigenvalues are, certainly, real due to the symmetry of the tensor **U**. They are called the **principal stretches**. If we choose the eigenvectors as the local basis vectors, then we can write the right stretch tensor **U** in the form of the following *spectral representation*:

$$\mathbf{U}=\sum_{A=1}^{3}\lambda_U^{(A)}\mathbf{K}^{(A)}\otimes\mathbf{K}^{(A)} \quad \Rightarrow \quad \det\mathbf{U}=\lambda_U^{(1)}\lambda_U^{(2)}\lambda_U^{(3)}. \tag{2.40}$$

Consequently, the determinant of **U** is positive either for all principal stretches being positive or for two of them being negative. Physically, as we will see later, the positive stretches have a simple physical interpretation. On the other hand, the negative stretches with preserved directions of the eigenvectors would yield in some heterogeneous deformations the transitions through the zero values of the determinant in (2.40). This would yield a locally vanishing material volume in the process of deformation, i.e. infinite values of mass densities. Such processes are forbidden within continuum mechanics. Hence, we can formulate the following **axiom of continuity:** *all principal stretches are positive.*

Let us turn our attention to the eigenvalue problem for the right Cauchy-Green deformation tensor **C**.

$$\mathbf{U}\left(\mathbf{U}-\lambda_U^{(A)}\mathbf{1}\right)\mathbf{K}^{(A)}=\left(\mathbf{C}-\left(\lambda_U^{(A)}\right)^2\mathbf{1}\right)\mathbf{K}^{(A)}=0 \quad \Rightarrow \quad \lambda_C^{(A)}=\left(\lambda_U^{(A)}\right)^2, \tag{2.41}$$

i.e. the eigenvalues $\lambda_C^{(A)}$ of **C** are squares of the eigenvalues $\lambda_U^{(A)}$ of **U** and both tensors have the same eigenvectors. Moreover, the eigenvalues of the tensor **C** are all real and positive because this tensor is symmetric, and according to the relation (2.32), positive definite. Consequently, the eigenvalues of the tensor **U** can be found uniquely if the eigenvalues of the tensor **C** are known.

The above properties yield the following procedure for determining the polar decomposition of the given deformation gradient **F**.

1. Construct the right Cauchy-Green deformation tensor **C** according to the relation (2.33),
2. Find the solution of the eigenvalue problem for the tensor **C**,
3. Construct the right stretch tensor **U** according to the relation

$$\mathbf{U}=\sum_{A=1}^{3}\sqrt{\lambda_C^{(A)}}\,\mathbf{K}^{(A)}\otimes\mathbf{K}^{(A)}, \tag{2.42}$$

4. Find the tensor of rotation **R** from the relation

$$\mathbf{R}=\mathbf{F}\mathbf{U}^{-1}. \tag{2.43}$$

Certainly, this construction is unique.

Let us notice that a similar construction in the general theory of square matrices is not unique because we can choose different combinations of signs for the square root in (2.42). The uniqueness in our case follows from the continuity assumption.

The eigenvalue problem for the right Cauchy-Green tensor **C** yields the important notion of **principal invariants** which we use frequently further on in this book. Namely, the secular equation for the eigenvalues can be written in the following explicit form:

$$\det(\mathbf{C}-\lambda_C \mathbf{1}) \equiv (\lambda_C)^3 - I^C(\lambda_C)^2 + II^C(\lambda_C) - III^C = 0, \tag{2.44}$$

where

$$\begin{aligned} I^C &= \operatorname{tr}\mathbf{C} = \lambda_C^{(1)} + \lambda_C^{(2)} + \lambda_C^{(3)}, \\ II^C &= \tfrac{1}{2}\left(I^{C2} - \operatorname{tr}\mathbf{C}^2\right) = \lambda_C^{(1)}\lambda_C^{(2)} + \lambda_C^{(1)}\lambda_C^{(3)} + \lambda_C^{(2)}\lambda_C^{(3)}, \\ III^C &\equiv J^2 = \det\mathbf{C} = \lambda_C^{(1)}\lambda_C^{(2)}\lambda_C^{(3)}. \end{aligned} \tag{2.45}$$

The roots of the equation (2.44) are, certainly, identical to the eigenvalues $\lambda_C^{(A)}$, A=1,2,3. The scalars I^C, II^C, and III^C are called invariants because a change in the basis changes the representation of the eigenvectors in the spectral representation of the tensor **C**, but not the eigenvalues of this tensor, and in turn, these are given uniquely by means of the invariants. Conversely, according to (2.45), the invariants are uniquely given in terms of the eigenvalues as well.

Example 2.1. *Deformation of a Prism in the Homogeneous Extension*
In order to illustrate the above considerations, we investigate the deformation function **f** given by the following relations (see: Fig. 2.6.):

$$x = \frac{1}{\sqrt{\lambda}}X, \quad y = \frac{1}{\sqrt{\lambda}}Y, \quad z = \lambda Z, \qquad \lambda = \text{const.} > 0, \tag{2.46}$$

where (X,Y,Z) denote the Lagrangian Cartesian coordinates and (x,y,z) the Eulerian Cartesian coordinates. We have

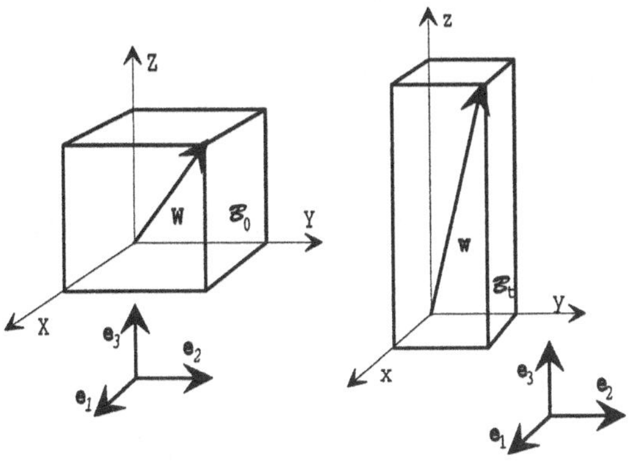

Fig. 2.6. *Extension of a prism*

$$\mathbf{F} = F^k{}_\alpha \mathbf{e}_k \otimes \mathbf{e}_\alpha, \qquad \mathbf{C} = C^{\alpha\beta} \mathbf{e}_\alpha \otimes \mathbf{e}_\beta,$$

$$\left(F^k{}_\alpha\right) = \begin{pmatrix} \frac{1}{\sqrt{\lambda}} & 0 & 0 \\ 0 & \frac{1}{\sqrt{\lambda}} & 0 \\ 0 & 0 & \lambda \end{pmatrix} \Rightarrow \left(C^{\alpha\beta}\right) = \begin{pmatrix} \frac{1}{\lambda} & 0 & 0 \\ 0 & \frac{1}{\lambda} & 0 \\ 0 & 0 & \lambda^2 \end{pmatrix}. \tag{2.47}$$

The solution of the eigenvalue problem for **C** yields immediately

$$\mathbf{U} = U^{\alpha\beta} \mathbf{e}_\alpha \otimes \mathbf{e}_\beta, \quad \left(U^{\alpha\beta}\right) = \begin{pmatrix} \frac{1}{\sqrt{\lambda}} & 0 & 0 \\ 0 & \frac{1}{\sqrt{\lambda}} & 0 \\ 0 & 0 & \lambda \end{pmatrix}, \qquad \mathbf{R} = \mathbf{1},$$
$$\mathbf{K}^{(1)} = \mathbf{e}_1, \quad \mathbf{K}^{(2)} = \mathbf{e}_2, \quad \mathbf{K}^{(3)} = \mathbf{e}_3. \tag{2.48}$$

Certainly, the lack of rotations in this example (**R=1**) does not mean that all material vectors preserve the same directions during such a deformation. We see easily from the relation (2.39) that only the eigenvectors of the stretch tensor **U**, and consequently the eigenvectors of the tensor **C**, rotate around the eigenvector of the orthogonal tensor **R**, corresponding to the single real eigenvalue $\lambda_R=1$ by the angle determined by the complex eigenvalues of **R**. Namely, for the eigenvectors $\mathbf{K}^{(A)}$ of **U** we have

$$\mathbf{F}\mathbf{K}^{(A)} = \mathbf{R}\mathbf{U}\mathbf{K}^{(A)} = \lambda_U^{(A)} \mathbf{R}\mathbf{K}^{(A)}, \tag{2.49}$$

with

$$\left(\mathbf{F}\mathbf{K}^{(A)}\right) \cdot \left(\mathbf{F}\mathbf{K}^{(A)}\right) = \left(\lambda_U^{(A)}\right)^2 \left(\mathbf{R}\mathbf{K}^{(A)}\right) \cdot \left(\mathbf{R}\mathbf{K}^{(A)}\right) = \left(\lambda_U^{(A)}\right)^2 \mathbf{K}^{(A)} \cdot \mathbf{K}^{(A)}. \tag{2.50}$$

Hence, for **R=1** the vectors $\mathbf{K}^{(A)}$ change the length but not the direction.

In our example, the eigenvectors $\mathbf{K}^{(A)}$ coincide with the edges of the prism [compare the relations (2.48)] and they do not rotate. However, any other material vector changes its direction in spite of the „lack of rotations". For instance, the vector

$$\mathbf{W} = \mathbf{e}_1 + \mathbf{e}_2 + \mathbf{e}_3, \tag{2.51}$$

transforms according to the formula (see: Fig. 2.6.)

$$\mathbf{w} = \mathbf{F}\mathbf{W} = \mathbf{U}\mathbf{W} = \frac{1}{\sqrt{\lambda}}\mathbf{e}_1 + \frac{1}{\sqrt{\lambda}}\mathbf{e}_2 + \lambda \mathbf{e}_3, \tag{2.52}$$

and its cosines of angles with the axes of coordinates change in the following way:

$$\left(\frac{1}{\sqrt{3}}, \frac{1}{\sqrt{3}}, \frac{1}{\sqrt{3}}\right) \to \left(\frac{1}{\sqrt{\lambda + \frac{2}{\sqrt{\lambda^2}}}}, \frac{1}{\sqrt{\lambda + \frac{2}{\sqrt{\lambda^2}}}}, \frac{\sqrt{\lambda}}{\sqrt{\lambda + \frac{2}{\sqrt{\lambda^2}}}}\right). \tag{2.53}$$

In the relation (2.52) the second equality holds solely in the sense of the matrix relation because the vector **w** belongs to the space $\mathcal{T}_x$, and the vector **UW** belongs to $\mathcal{T}_X$ ■

Example 2.2. *Deformation of a Prism in Simple Shearing*
Now we consider the geometry of a somewhat more sophisticated deformation of the prism called **simple shearing**. The deformation in this case is supposed to be given by the relations (see: Fig. 2.7.)

$$x = X, \quad y = Y + \kappa Z, \quad z = Z, \qquad \kappa \equiv \tan\varphi = \text{const.}, \quad 0 \le \varphi \le \tfrac{\pi}{2}, \tag{2.54}$$

where (X,Y,Z) denote again the Lagrangian coordinates, and (x,y,z) the Eulerian coordinates. It follows that

$$\left(F^k{}_\alpha\right) = \begin{pmatrix} 1 & 0 & 0 \\ 0 & 1 & \kappa \\ 0 & 0 & 1 \end{pmatrix} \Rightarrow \left(C^{\alpha\beta}\right) = \begin{pmatrix} 1 & 0 & 0 \\ 0 & 1 & \kappa \\ 0 & \kappa & 1+\kappa^2 \end{pmatrix}. \tag{2.55}$$

The solution of the eigenvalue problem for **C** yields, after easy calculations,

$$\begin{aligned} &\lambda_C^{(1)} = 1, \quad \lambda_C^{(2)} = \left(\frac{1-\sin\alpha}{\cos\alpha}\right)^2, \quad \lambda_C^{(3)} = \left(\frac{1+\sin\alpha}{\cos\alpha}\right)^2, \\ &\mathbf{K}^{(1)} = \mathbf{e}_1, \quad \mathbf{K}^{(2)} = -\frac{1}{\sqrt{2}}\frac{\cos\alpha}{\sqrt{1-\sin\alpha}}\mathbf{e}_2 + \frac{1}{\sqrt{2}}\sqrt{1-\sin\alpha}\,\mathbf{e}_3, \\ &\mathbf{K}^{(3)} = \frac{1}{\sqrt{2}}\frac{\cos\alpha}{\sqrt{1+\sin\alpha}}\mathbf{e}_2 + \frac{1}{\sqrt{2}}\sqrt{1+\sin\alpha}\,\mathbf{e}_3, \end{aligned} \tag{2.56}$$

where

$$\tan\alpha \equiv \tfrac{1}{2}\kappa \equiv \tfrac{1}{2}\tan\varphi. \tag{2.57}$$

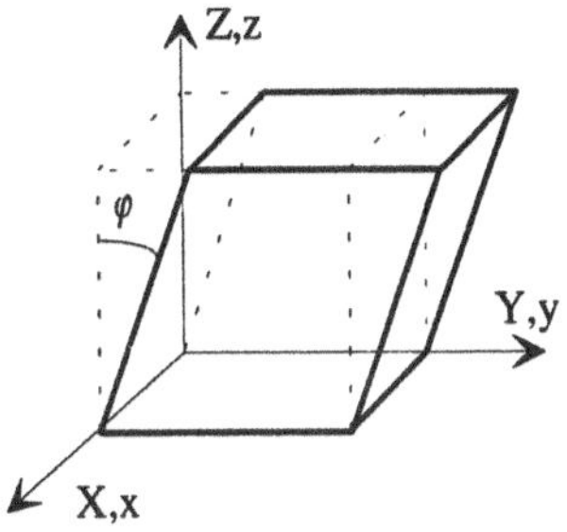

Fig. 2.7. *The geometry of a simple shearing*

Consequently, we obtain the following right stretch tensor:

$$\left(U^{\alpha\beta}\right)=\begin{pmatrix}1 & 0 & 0\\ 0 & \cos\alpha & \sin\alpha\\ 0 & \sin\alpha & \dfrac{2}{\cos\alpha}-\cos\alpha\end{pmatrix}. \tag{2.58}$$

It remains to calculate the tensor of rotation **R**. Bearing the relation (2.43) in mind, we arrive at

$$\left(R^{k}{}_{\alpha}\right)=\begin{pmatrix}1 & 0 & 0\\ 0 & \cos\alpha & \sin\alpha\\ 0 & -\sin\alpha & \cos\alpha\end{pmatrix}. \tag{2.59}$$

The real eigenvector of **R** coincides, certainly, with the direction $\mathbf{e}_1$ (x-axis) and the angle α, defining the complex eigenvalues $\exp(\pm i\alpha)$ of **R**, describes the angle of rotation of the eigenvectors of **U** around this axis. However, these eigenvectors no longer have such a simple geometrical interpretation as was the case in the example of extension. It is easy to see from the formula (2.56) for the eigenvectors that they do not coincide with the directions of the edges of the prism. In Fig. 2.8. we illustrate these properties by the following numerical example:

$$\begin{aligned}&\kappa\equiv\tan\varphi=\tfrac{3}{8}\ \Rightarrow\ \varphi=20.56^0\ \Rightarrow\ \alpha=10.62^0,\\ &\lambda_U^{(1)}=1,\ \ \lambda_U^{(2)}=0.8299,\ \ \lambda_U^{(3)}=1.2049.\end{aligned} \tag{2.60}$$

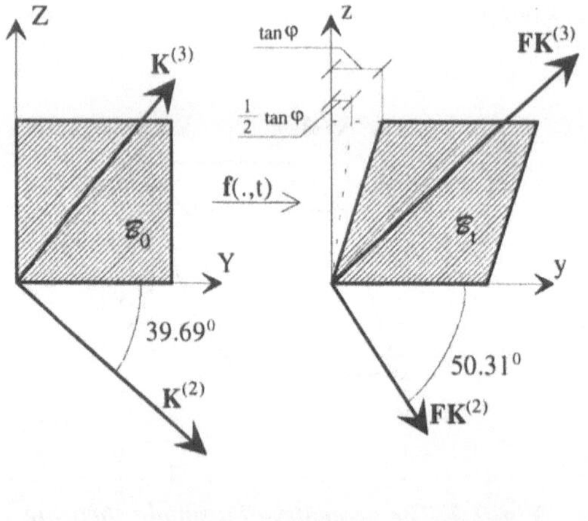

Fig. 2.8. *A numerical example of the simple shearing*

It is seen that the eigenvectors of the stretch tensor do not coincide in this case with any characteristic material vector. In contrast to the linear theory of shearing ($\tan\varphi\cong\varphi$, $\alpha\cong{}^1/_2\varphi$), the inverse rotation $\mathbf{R}^{-1}$ does not produce a current configuration symmetrical with respect to the diagonal of the y- and z-axes ■

Example 2.3. *A Polar Decomposition and Elastic-plastic Deformation*
An extensive character of the simple examples above is caused by frequent confusions arising in practical applications of the polar decomposition theorem. Now let us consider a more general example of the deformation gradient in the case of elastic-plastic materials which we have mentioned in Remark 2.2.

According to the relation (2.21), we have for non-singular local configurations

$$\mathbf{F} = \mathbf{R}\,\mathbf{U} = \left(\mathbf{R}^e\mathbf{U}^e\right)\left(\mathbf{R}^p\mathbf{U}^p\right), \qquad \mathbf{F}^e = \mathbf{R}^e\mathbf{U}^e, \quad \mathbf{F}^p = \mathbf{R}^p\mathbf{U}^p. \tag{2.61}$$

It is easy to see that $\mathbf{R}^{eT}\mathbf{F}$ is usually not symmetric as

$$\left(\mathbf{U}^e\mathbf{R}^p\mathbf{U}^p\right)^T = \mathbf{U}^p\mathbf{R}^{pT}\mathbf{U}^e \neq \mathbf{U}^e\mathbf{R}^p\mathbf{U}^p. \tag{2.62}$$

Hence, owing to the uniqueness of the polar decomposition, neither the rotation tensor $\mathbf{R}$ and the tensor of elastic rotations $\mathbf{R}^e$ nor the total stretch tensor $\mathbf{U}$ and $\mathbf{U}^e\mathbf{R}^p\mathbf{U}^p$ coincide. Even in the case of a lack of plastic rotations, i.e. $\mathbf{R}^p=\mathbf{1}$, it would usually not follow that $\mathbf{U}=\mathbf{U}^e\mathbf{U}^p$. The tensor on the right-hand side, in contrast with the tensor on the left-hand side, is not even symmetric.

Let us check the properties of the right Cauchy-Green tensor. We have

$$\begin{aligned} \mathbf{C} = \mathbf{F}^T\mathbf{F} = \mathbf{U}^2 &= \mathbf{U}^p\mathbf{R}^{pT}\mathbf{U}^e\mathbf{R}^{eT}\mathbf{R}^e\mathbf{U}^e\mathbf{R}^p\mathbf{U}^p = \\ &= \mathbf{U}^p\mathbf{R}^{pT}\mathbf{U}^{e2}\mathbf{R}^p\mathbf{U}^p = \mathbf{U}^p\mathbf{R}^{pT}\mathbf{C}^e\mathbf{R}^p\mathbf{U}^p, \qquad \mathbf{C}^e \equiv \mathbf{U}^{e2}. \end{aligned} \tag{2.63}$$

Hence,

$$\mathbf{U} = \left(\mathbf{U}^p\mathbf{R}^{pT}\mathbf{C}^e\mathbf{R}^p\mathbf{U}^p\right)^{\frac{1}{2}}. \tag{2.64}$$

This relation shows that the total stretch tensor $\mathbf{U}$ depends not only on the elastic and plastic stretch tensors $\mathbf{U}^e$ and $\mathbf{U}^p$, but also on the plastic rotation $\mathbf{R}^p$. In spite of the experimental evidence (e.g. the so-called deformation induced *textures* of polycrystals) and numerous theoretical results of the structural (semimicroscopical) plasticity of polycrystals, confirmed by experiments on metals, most available macroscopical theories of large plastic deformations neglect the influence of $\mathbf{R}^p$ on the total deformation. Moreover, in many cases a justification of this flaw in the theory has been attempted by dubious invariance arguments (e.g. „objectivity principle in the intermediate configuration“)

Exercise 2.3. Show that the product $\mathbf{T}=\mathbf{R}_1\mathbf{U}_1\mathbf{U}_2$ of the following tensors:

$$\begin{aligned} \mathbf{R}_1 &= \tfrac{1}{\sqrt{2}}\left(\mathbf{e}_1\otimes\mathbf{e}_1 - \mathbf{e}_1\otimes\mathbf{e}_2 + \mathbf{e}_2\otimes\mathbf{e}_1 + \mathbf{e}_2\otimes\mathbf{e}_2\right) + \mathbf{e}_3\otimes\mathbf{e}_3, \\ \mathbf{U}_1 &= \mathbf{e}_1\otimes\mathbf{e}_1 + \tfrac{1}{2}\left(\mathbf{e}_1\otimes\mathbf{e}_2 + \mathbf{e}_2\otimes\mathbf{e}_1\right) + \mathbf{e}_2\otimes\mathbf{e}_2 + \mathbf{e}_3\otimes\mathbf{e}_3, \\ \mathbf{U}_2 &= 2\,\mathbf{e}_1\otimes\mathbf{e}_1 + \mathbf{e}_1\otimes\mathbf{e}_2 + \mathbf{e}_2\otimes\mathbf{e}_1 + \mathbf{e}_2\otimes\mathbf{e}_2 + \mathbf{e}_3\otimes\mathbf{e}_3, \end{aligned}$$

has the following orthogonal and symmetric parts:

$$\begin{aligned} \mathbf{R} &= 0.614\left(\mathbf{e}_1\otimes\mathbf{e}_1 + \mathbf{e}_2\otimes\mathbf{e}_2\right) - 0.789\left(\mathbf{e}_1\otimes\mathbf{e}_2 - \mathbf{e}_2\otimes\mathbf{e}_1\right) + \mathbf{e}_3\otimes\mathbf{e}_3, \\ \mathbf{U} &= 2.729\,\mathbf{e}_1\otimes\mathbf{e}_1 + 1.674\left(\mathbf{e}_1\otimes\mathbf{e}_2 + \mathbf{e}_2\otimes\mathbf{e}_1\right) + 1.302\,\mathbf{e}_2\otimes\mathbf{e}_2 + \mathbf{e}_3\otimes\mathbf{e}_3 \bullet \end{aligned}$$

2.3 Some Other Measures of Deformation

Apart from the right Cauchy-Green tensor **C** and the right stretch tensor **U** used to describe the changes in the geometry of the body induced by the deformation, it is frequently convenient to employ other equivalent deformation measures.

Directly from the formula (2.32), we see that it can be convenient to introduce a relative measure which vanishes in the reference configuration. Namely,

$$d\mathbf{x}\cdot d\mathbf{x}-d\mathbf{X}\cdot d\mathbf{X}=d\mathbf{X}\cdot\left(\mathbf{F}^{\mathrm{T}}\mathbf{F}-\mathbf{1}\right)d\mathbf{X}\equiv 2\mathbf{E}\cdot\left(d\mathbf{X}\otimes d\mathbf{X}\right), \tag{2.66}$$

where we have defined the following symmetric tensor:

$$\mathbf{E}\equiv\tfrac{1}{2}\left(\mathbf{F}^{\mathrm{T}}\mathbf{F}-\mathbf{1}\right)=\tfrac{1}{2}\left(\mathbf{C}-\mathbf{1}\right), \tag{2.67}$$

which is called the **Green-St.Venant deformation tensor**. In contrast to **C**, which is equal to **1** in $\mathcal{B}_0$-configuration, the tensor above is equal to zero in this configuration.

We can also use the inverse tensors expressing the changes of the geometry relative to the current configuration $\mathcal{B}_t$. We have

$$d\mathbf{X}\cdot d\mathbf{X}=\left(\mathbf{F}^{-1}d\mathbf{x}\right)\cdot\left(\mathbf{F}^{-1}d\mathbf{x}\right)=d\mathbf{x}\cdot\left(\mathbf{F}^{-\mathrm{T}}\mathbf{F}^{-1}\right)d\mathbf{x}=d\mathbf{x}\cdot\left(\mathbf{F}\mathbf{F}^{\mathrm{T}}\right)^{-1}d\mathbf{x}. \tag{2.68}$$

The tensor

$$\mathbf{B}\equiv\mathbf{F}\mathbf{F}^{\mathrm{T}} \tag{2.69}$$

is called the **left Cauchy-Green deformation tensor**. It is easy to check that this tensor has the principal invariants identical with the invariants (2.40) of the right Cauchy-Green tensor **C**. Consequently, their eigenvalues are also identical. Certainly, the eigenvectors cannot be identical because the eigenvectors of the tensor **C** are defined in the tangent space $\mathcal{T}_X$ at the point **X** of the reference configuration, whereas the eigenvectors of the tensor **B** are defined in the tangent space $\mathcal{T}_x$ at the point **x** of the current configuration.

The inverse of the tensor **B** is denoted by **c** and it describes the metric tensor of the configuration $\mathcal{B}_0$ relative to $\mathcal{B}_t$

$$d\mathbf{X}\cdot d\mathbf{X}=\mathbf{c}\cdot\left(d\mathbf{x}\otimes d\mathbf{x}\right),\qquad \mathbf{c}\equiv\mathbf{B}^{-1}. \tag{2.70}$$

We can, of course, introduce also in this case the relative measure which vanishes in the reference configuration $\mathcal{B}_0$. Namely,

$$d\mathbf{x}\cdot d\mathbf{x}-d\mathbf{X}\cdot d\mathbf{X}=\left(\mathbf{1}-\mathbf{c}\right)\cdot\left(d\mathbf{x}\otimes d\mathbf{x}\right)\equiv 2\mathbf{e}\cdot\left(d\mathbf{x}\otimes d\mathbf{x}\right), \tag{2.71}$$

and the tensor

$$\mathbf{e} \equiv \tfrac{1}{2}(\mathbf{1}-\mathbf{c}) \tag{2.72}$$

is called the **Almansi-Hamel deformation tensor**. Both the Green-St.Venant and Almansi-Hamel deformation tensors coincide in the case of small deformations, and they are commonly used in the classical linear theory of elasticity.

Let us finally mention that the left Cauchy-Green tensor, and consequently, all other measures of deformation related to the current configuration $\mathcal{B}_t$ follow from the polar decomposition of the deformation gradient **F** with the inverse order of the symmetrical and orthogonal parts. Namely, we can show in the similar manner as before, that under the continuity assumption, there exists a unique decomposition of the deformation gradient

$$\mathbf{F} = \mathbf{VR}, \qquad \mathbf{R}^{-1} = \mathbf{R}^{T}, \quad \mathbf{V} = \mathbf{V}^{T}, \tag{2.73}$$

where **R** is the **same** orthogonal tensor which we had in the formula (2.34), and the symmetric tensor **V** is called the **left stretch tensor**. It is instructive to write the relation (2.73) in coordinates. We have

$$\begin{aligned} &\mathbf{F} = F^{k}{}_{\alpha}\mathbf{e}_{k} \otimes \mathbf{e}_{\alpha}, \quad \mathbf{R} = R_{k\alpha}\mathbf{e}_{k} \otimes \mathbf{e}_{\alpha}, \quad \mathbf{V} = V^{kl}\mathbf{e}_{k} \otimes \mathbf{e}_{l}, \\ &F^{k}{}_{\alpha} = V^{kl}R_{l\alpha}, \qquad \overset{-1}{R}_{\alpha k} = R_{k\alpha}, \quad V^{kl} = V^{lk}. \end{aligned} \tag{2.74}$$

The eigenvalue problem for **V** is, certainly, conjugated to that of **U**. Namely, if we denote by λ_V and $\mathbf{k}_V$ the eigenvalues and the eigenvectors of **V**, respectively, then

$$\mathbf{R}(\mathbf{U}-\lambda_U\mathbf{1})\mathbf{K} = \left(\mathbf{RUR}^{T}\mathbf{R}-\lambda_U\mathbf{R}\right)\mathbf{K} = (\mathbf{V}-\lambda_U\mathbf{1})\mathbf{RK} = 0, \tag{2.75}$$

i.e.

$$\lambda_V = \lambda_U, \quad \mathbf{k} = \mathbf{RK}, \tag{2.76}$$

where the relation

$$\mathbf{V} = \mathbf{R\,U\,R}^{T}, \tag{2.77}$$

was used. Hence, the eigenvalues of **U** and **V** coincide and the eigenvectors differ by the rotation **R** mapping $\mathcal{T}_X$ into $\mathcal{T}_x$.

Bearing (2.73) in mind, we immediately obtain

$$\mathbf{B} = \mathbf{V}^{2} \qquad \text{i.e.} \qquad B_{kl} = V_{k}{}^{m}V_{ml}. \tag{2.78}$$

Tab. 2.1. *Some measures of deformation*
[references are quoted after C. TRUESDELL, R. A. TOUPIN (1960)]

Name	Definition		Eigen-values	Eigen-vectors	Author
Right Cauchy - - Green	**C**	$\mathbf{F}^T\mathbf{F}$	λ^2	**K**	G.GREEN; On the propagation of light in crystallized media (1841)
Left Cauchy - - Green (Finger)	**B**	$\mathbf{F}\mathbf{F}^T$	λ^2	**k=FK**	J.FINGER; Über die allgemeinsten Beziehungen zwischen Deformationen und den zugehörigen Spannungen in aeolotropen und isotropen Substanzen (1894)
Right stretch	**U**	$\mathbf{C}^{1/2}$	λ	**K**	EULER? CAUCHY?
Left stretch	**V**	$\mathbf{B}^{1/2}$	λ	**k**	EULER? CAUCHY?
Cauchy	**c**	$\mathbf{B}^{-1}$	$1/\lambda^2$	**k**	L.A.CAUCHY; Sur la condensation et la dilatation des corps solides (1827)
Green-St. Venant (Lagrange)	**E**	$0.5(\mathbf{C}-\mathbf{1})$	$0.5(\lambda^2-1)$	**K**	G.GREEN; On the propagation of light in crystallized media (1841) A.J.C.B DE ST.VENANT; Sur les pressions qui se développent à l'intérieur des corps colides lorsque les déplacements de leurs points... (1844)
Almansi - Hamel (Euler)	**e**	$0.5(\mathbf{1}-\mathbf{c})$	$0.5(1-1/\lambda^2)$	**k**	E.ALMANSI; Sulle deformazioni finite dei solidi elastici isotropi (1911) G.HAMEL; Elementare Mechanik (1912)
Piola	$\mathbf{C}^{-1}$		$1/\lambda^2$	**K**	G.PIOLA; La meccanica de' corpi naturalmente estesi trattata col calcol delle variazioni (1833)

For the purpose of this book we shall use primarily the two fundamental measures of deformation - the left and right Cauchy-Green deformation tensors.

Some other deformation tensors appearing in various applications, for instance the logarithmic Hencky measures **H** and **h**, can be found in textbooks on continuum mechanics [e.g. C. TRUESDELL, R. A. TOUPIN [1960)]. In Table 2.1. we have collected the most common tensors and their eigenvalues and eigenvectors.

Exercise 2.4. Find the deformation measures **E**, **V**, **B**, and **e** for the examples of a homogeneous extension of a prism and a simple shearing of a prism•

Exercise 2.5. Using the Eulerian cylindrical coordinates (r,ϑ,z), show that the deformation

$$r=\sqrt{2AX},\quad \vartheta=BY,\quad z=\frac{Z}{AB}-BCY,\quad AB\neq 0,\qquad A,B,C-\text{const.}, \tag{2.79}$$

yields the following left Cauchy-Green tensor:

$$\left(B^{kl}\right)=\begin{pmatrix} \dfrac{A^2}{r^2} & 0 & 0 \\ 0 & B^2 & -B^2C \\ 0 & -B^2C & B^2C^2+\dfrac{1}{A^2B^2} \end{pmatrix}, \qquad \mathbf{B}=B^{kl}\mathbf{g}_k\otimes\mathbf{g}_l, \tag{2.80}$$

where $\mathbf{g}_k$, k=1,2,3, are the covariant basis vectors of the Eulerian cylindrical coordinates and (X,Y,Z) are the Lagrangian Cartesian coordinates. Prove that the principal invariants of this tensor have the form

$$\begin{aligned} I^B &= \frac{A^2}{r^2}+B^2r^2+\frac{1}{A^2B^2}, \\ II^B &= \frac{r^2}{A^2}+\frac{1}{r^2}\left(\frac{1}{B^2}+A^2B^2C^2\right)+A^2B^2, \\ III^B &= 1, \end{aligned} \tag{2.81}$$

Show that in the general case the principal invariants of the tensors **B** and **C** are identical•

Exercise 2.6. Using the Lagrangian cylindrical coordinates (R,Θ,Z) and the Eulerian cylindrical coordinates (r,ϑ,z), show that the deformation

$$\begin{aligned} &r=\sqrt{AR^2+B}, \quad \vartheta=C\Theta+DZ, \quad z=E\Theta+FZ, \\ &A(CF-DE)=1, \quad A,B,C,D,E,F-\text{const.}, \end{aligned} \tag{2.82}$$

yields the following left Cauchy-Green tensor:

$$\left(B^{kl}\right)=\begin{pmatrix} \dfrac{A^2R^2}{r^2} & 0 & 0 \\ 0 & \dfrac{C}{R^2}+D^2 & \dfrac{CE}{R^2}+DF \\ 0 & \dfrac{CE}{R^2}+DF & \dfrac{E^2}{R^2}+F^2 \end{pmatrix}, \tag{2.83}$$

with the principal invariants

$$I^B = \frac{A^2R^2}{r^2} + r^2\left(\frac{C^2}{R^2} + D^2\right) + F^2 + \frac{E^2}{R^2},$$

$$II^B = \frac{r^2}{A^2R^2} + \frac{A^2}{r^2}\left(E^2 + F^2R^2\right) + A^2\left(C^2 + D^2R^2\right), \tag{2.84}$$

$$III^B = 1.$$

Find the form of the tensor of rotation in the particular case (torsion of cylinder)

$$A = 1, \quad B = 0, \quad C = 1, \quad E = 0, \quad F = 1, \tag{2.85}$$

i.e. for the deformation

$$r = R, \quad \vartheta = \Theta + DZ, \quad z = Z \bullet \tag{2.86}$$

Remark 2.3. *On Universal Solutions - Part I*
The examples of deformations above belong to a very important special class of so-called **universal solutions**. These solutions constitute the foundation of the experimental verification of constitutive relations for various materials in statical conditions. We shall return to this point further on in this book. Now let us solely list all deformations of this type. The most important of them were found by R. S. RIVLIN (1948) and J. L. ERICKSEN (1954). An extensive discussion can be found in the book of C. TRUESDELL (1972).

There are five families of universal solutions. We describe them using the following notation. The capital letters denote the Lagrangian coordinates: (X,Y,Z) are rectangular Cartesian; (R,Θ,Z) cylindrical polar; (R,Θ,Φ) spherical polar. The small letters denote Eulerian coordinates: (x,y,z); (r,ϑ,z); (r,ϑ,φ) with a similar meaning as before. (see: Appendix B).

Family 1: Bending, stretching, and shearing of a rectangular block

$$r = \sqrt{2AX}, \quad \vartheta = BY, \quad z = \frac{Z}{AB} - BCY, \qquad AB \neq 0. \tag{2.87}$$

Family 2: Straightening, stretching, and shearing of a sector of a circular-cylindrical tube

$$x = \tfrac{1}{2}AB^2R^2, \quad y = \frac{\Theta}{AB}, \quad z = \frac{Z}{B} + \frac{C\Theta}{AB}, \qquad AB \neq 0. \tag{2.88}$$

Family 3: Inflation or eversion, bending, torsion, extension, and shearing of a sector of a circular-cylindrical tube

$$r = \sqrt{AR^2 + B}, \quad \vartheta = C\Theta + DZ, \quad z = E\Theta + FZ, \qquad A(CF - DE) = 1. \tag{2.89}$$

Family 4: Inflation or eversion of a sector of a spherical shell

$$r = \left(\pm R^3 + A\right)^{\frac{1}{3}}, \quad \vartheta = \pm\Theta, \quad \varphi = \Phi. \tag{2.90}$$

Family 5: Inflation, stretching, and shearing of a circular cylinder

$$r = AR, \quad \vartheta = B\ln R + C\Theta, \quad z = DZ, \qquad A^2CD = 1. \tag{2.91}$$

The important rule of these solutions is connected, as we will see further on, with the lack of superposition principles in non-linear theories•

3 Kinematics

3.1 Description of Motion

In the previous chapter we discussed the geometrical properties of a chosen current configuration $\mathcal{B}_t$ under the assumption that the parameter t describing time changes of the body is kept constant, i.e. the diffeomorphism **f** was considered as a deformation map **f**(·,t). Now we turn our attention to problems arising in the case of motion, i.e. the diffeomorphism **f** is discussed as the map **f**(**X**,·) for a chosen point $\mathbf{X} \in \mathcal{B}_0$. We assume that **f**(**X**,·) is twice differentiable. Then

$$\forall t \in \mathcal{T}: \quad \mathbf{v}(\mathbf{X},t) = \frac{\partial \mathbf{f}}{\partial t}(\mathbf{X},t), \tag{3.1}$$

is called the **velocity** at the material point **X**, and

$$\forall t \in \mathcal{T}: \quad \mathbf{a} = \frac{\partial^2 \mathbf{f}}{\partial t^2}(\mathbf{X},t), \tag{3.2}$$

is the **acceleration** at the material point **X**.

Both these fields are defined on the reference configuration $\mathcal{B}_0$. This is frequently not very convenient as the example of the classical fluid mechanics shows. Therefore, we shall now discuss the following two questions:

1. What does the description of motion look like after a change in the reference configuration, and in particular, what is the form of this description when the current configuration $\mathcal{B}_t$ is chosen as the reference configuration?
2. What does the description of motion look like after a time-dependent transformation of the reference system in the present configuration (the change of observer)?

To answer the first question we consider again three configurations (see: Sect. 2.1.): the reference configuration $\mathcal{B}_0$, the current configuration $\mathcal{B}_t$, and another current configuration $\mathcal{B}_\tau$ with

$$\begin{aligned} \xi &= \mathbf{f}(\mathbf{X},\tau), \quad \xi \in \mathcal{B}_\tau \equiv \mathbf{f}(\mathcal{B}_0,\tau), \\ \mathbf{x} &= \mathbf{f}(\mathbf{X},t), \quad \mathbf{x} \in \mathcal{B}_t \equiv \mathbf{f}(\mathcal{B}_0,t). \end{aligned} \tag{3.3}$$

Now let us fix the instant of time t and allow the parameter τ to change. According to (2.28), we have

$$\xi = \mathbf{f}_t(\mathbf{x},\tau) \equiv \mathbf{f}\left(\mathbf{f}^{-1}(\mathbf{x},t),\tau\right), \tag{3.4}$$

and at the same time, the relation (3.1) yields for the velocity field

$$\mathbf{v} = \mathbf{v}\left(\mathbf{f}^{-1}(\mathbf{x},t),t\right) = \mathbf{v}(\mathbf{x},t). \tag{3.5}$$

For the velocity field $\mathbf{v}(\mathbf{x},t)$, given at any place $\mathbf{x}$ of the current configurations, the relations (3.4) and (3.5) yield then the following differential equation with $\mathbf{x}$ as a parameter

$$\frac{d\xi}{d\tau} = \mathbf{v}(\xi,\tau), \qquad \xi(\mathbf{x},\tau)\big|_{\tau=t} = \mathbf{x}. \tag{3.6}$$

The solution to this problem, i.e. the function (3.4) is called the **trajectory** of the material point $\mathbf{X}=\mathbf{f}^{-1}(\mathbf{x},t)$, whose current position at the instant of time t coincides with $\mathbf{x}$. This type of formulation of the problem of motion is typical for theories of fluids for which a natural past configuration of a reference $\mathcal{B}_0$ cannot be chosen. The equation (3.6) then defines the **streamline** for the velocity field $\mathbf{v}$ passing through the point $\mathbf{x}$ at the instant of time t.

The second derivative of the function (3.4) with respect to τ defines, according to the definition (3.2), the acceleration at the configuration $\mathcal{B}_\tau$. Bearing the equation (3.6) in mind, we have

$$\mathbf{a}(\xi,\tau) = \frac{\partial \mathbf{v}}{\partial \tau} + \left(\mathbf{v}\cdot \operatorname{grad}_\xi\right)\mathbf{v}, \qquad \operatorname{grad}_\xi \mathbf{v} \equiv \frac{\partial \mathbf{v}}{\partial \xi} = \frac{\partial v^k}{\partial \xi^l}\mathbf{e}_k \otimes \mathbf{e}_l, \tag{3.7}$$

and for $\tau=t$ we deduce

$$\mathbf{a}(\mathbf{x},t) = \frac{\partial \mathbf{v}}{\partial t} + (\mathbf{v}\cdot \operatorname{grad})\mathbf{v}(\mathbf{x},t) \equiv \dot{\mathbf{v}}(\mathbf{x},t), \tag{3.8}$$

where grad denotes the derivative with respect to $\mathbf{x}$ (the so-called spatial gradient). The time derivative appearing in (3.8), denoted by the dot, is called **material** because it is calculated along the trajectory $\xi=\mathbf{f}_t(\mathbf{x},\tau)$ of the material point $\mathbf{X}=\mathbf{f}^{-1}(\mathbf{x},t)$ at $\tau=t$.

If the fields appearing in a chosen description of motion are defined as functions of positions $\mathbf{x}$ and time t, as it is the case in the formula (3.8), then the description is called **Eulerian** or **spatial**.

Exercise 3.1. Given an acceleration field in the spatial description

$$\mathbf{a}(\mathbf{x},t) = k^2x^1\mathbf{e}_1 + k^2x^2\mathbf{e}_2, \qquad k = \text{const.}, \tag{3.9}$$

determine the deformation $\mathbf{x}=\mathbf{f}(\mathbf{X},t)$ of this motion with the following initial conditions

$$\begin{aligned} x^1\left(X^1,X^2,0\right) &= X^1, \quad \dot{x}^1\left(X^1,X^2,0\right) = kX^1, \\ x^2\left(X^1,X^2,0\right) &= X^2, \quad \dot{x}^2\left(X^1,X^2,0\right) = -kX^2. \end{aligned} \tag{3.10}$$

Find the velocity field in the material (Lagrangian) and the spatial (Eulerian) descriptions.

Answer:

$$x^1 = X^1 e^{kt}, \quad x^2 = X^2 e^{-kt},$$
$$v^1 = kX^1 e^{kt}, \quad v^2 = -kX^2 e^{-kt}, \tag{3.11}$$
$$v^1 = kx^1, \quad v^2 = -kx^2 \bullet$$

Exercise 3.2. Given a velocity field

$$\mathbf{v}(\mathbf{x},t) = u\left(x^2\right)\mathbf{e}_1, \tag{3.12}$$

determine the relative deformation $\xi = \mathbf{f}_t(\mathbf{x},\tau)$ of this motion. Show that

$$\mathbf{F}_t(\mathbf{x},\tau) = \mathbf{1} + (\tau - t)\kappa\mathbf{N}, \tag{3.13}$$

where

$$\kappa \equiv \frac{du}{dx^2}, \quad \mathbf{N} \equiv \mathbf{e}_1 \otimes \mathbf{e}_2. \tag{3.14}$$

Answer:

$$\xi^1 = x^1 + (\tau - t)u\left(x^2\right), \qquad \xi^2 = x^2 \bullet \tag{3.15}$$

Exercise 3.3. Let the motion be given by the relation

$$\mathbf{x} = X^1 \exp\left(t^2\right)\mathbf{e}_1 + X^2 \exp(t)\mathbf{e}_2 + X^3\mathbf{e}_3. \tag{3.16}$$

Determine the streamlines of that motion.

Answer.

$$\xi(\mathbf{x},\tau) = x^1 \exp((\tau + t)(\tau - t))\mathbf{e}_1 + x^2 \exp(\tau - t)\mathbf{e}_2 + x^3\mathbf{e}_3 \bullet \tag{3.17}$$

In Sect. 3.3. we return to the second question formulated above.

3.2 Time Changes of Some Geometric Objects

All geometric objects, which we discussed in Chap. 2, can be calculated from the deformation gradient $\mathbf{F}$. For this reason, we begin with the investigation of the time derivative of the relative deformation gradient $\mathbf{F}_t(\mathbf{x},\tau)$. For a chosen and fixed value of time we have

$$\frac{\partial}{\partial \tau}\mathbf{F}_t(\mathbf{x},\tau) \equiv \dot{\mathbf{F}}_t(\mathbf{x},\tau) = \dot{\mathbf{R}}_t(\mathbf{x},\tau)\mathbf{U}_t(\mathbf{x},\tau) + \mathbf{R}_t(\mathbf{x},\tau)\dot{\mathbf{U}}_t(\mathbf{x},\tau), \tag{3.18}$$

where the polar decomposition theorem was used for $\mathbf{F}_t(\tau)$. For $\tau = t$ we introduce the following notation:

$$\mathbf{L}(\mathbf{x},t) \equiv \dot{\mathbf{F}}_t(\mathbf{x},\tau=t), \quad \mathbf{D}(\mathbf{x},t) \equiv \dot{\mathbf{U}}_t(\mathbf{x},\tau=t), \quad \mathbf{W}(\mathbf{x},t) \equiv \dot{\mathbf{R}}_t(\mathbf{x},\tau=t), \tag{3.19}$$

with the obvious relation

$$\mathbf{F}_t(\mathbf{x},t) \equiv \mathbf{1}, \quad \mathbf{R}_t(\mathbf{x},t) \equiv \mathbf{1}, \quad \mathbf{U}_t(\mathbf{x},t) \equiv \mathbf{1}. \quad \Rightarrow \quad \mathbf{L}(\mathbf{x},t) = \mathbf{D}(\mathbf{x},t) + \mathbf{W}(\mathbf{x},t). \tag{3.20}$$

Bearing the definition $(2.30)_4$ of $\mathbf{F}_t$ in mind, we have

$$\begin{aligned} \dot{\mathbf{F}}_t(\mathbf{x},\tau=t) &= \frac{\partial}{\partial \tau}\left(\frac{\partial \mathbf{f}_t}{\partial \mathbf{x}}(\mathbf{x},\tau)\right)\Bigg|_{\tau=t} = \frac{\partial}{\partial \mathbf{x}}\left(\frac{\partial \mathbf{f}_t}{\partial t}(\mathbf{x},\tau)\right)\Bigg|_{\tau=t} = \\ &= \frac{\partial \mathbf{v}}{\partial \mathbf{x}}(\mathbf{x},\tau)\Big|_{\tau=t} = \operatorname{grad}\mathbf{v}(\mathbf{x},t). \end{aligned} \tag{3.21}$$

Hence, $\mathbf{L}$ is the **velocity gradient** in the spatial description

$$\mathbf{L} = \operatorname{grad}\mathbf{v}. \tag{3.22}$$

Simultaneously, bearing the relation $(2.31)_2$ in mind, we find

$$\begin{aligned} \frac{\partial \mathbf{F}_t}{\partial \tau}(\mathbf{x},\tau)\Big|_{\tau=t} &= \frac{\partial}{\partial \tau}\left[\mathbf{F}\left(\mathbf{f}^{-1}(\mathbf{x},t),\tau\right)\mathbf{F}^{-1}\left(\mathbf{f}^{-1}(\mathbf{x},t),t\right)\right]\Big|_{\tau=t} = \\ &= \dot{\mathbf{F}}\left(\mathbf{f}^{-1}(\mathbf{x},t),\tau\right)\mathbf{F}^{-1}\left(\mathbf{f}^{-1}(\mathbf{x},t),t\right)\Big|_{\tau=t} = \dot{\mathbf{F}}(\mathbf{X},t)\mathbf{F}^{-1}(\mathbf{X},t)\Big|_{\mathbf{X}=\mathbf{f}^{-1}(\mathbf{x},t)}. \end{aligned} \tag{3.23}$$

Hence

$$\mathbf{L}(\mathbf{x},t) = \dot{\mathbf{F}}(\mathbf{X},t)\mathbf{F}^{-1}(\mathbf{X},t)\Big|_{\mathbf{X}=\mathbf{f}^{-1}(\mathbf{x},t)}. \tag{3.24}$$

Let us return to the definitions (3.19). Owing to the symmetry of $\mathbf{U}_t(\mathbf{x},\tau)$ for all τ, we have

$$\mathbf{D} = \mathbf{D}^T. \tag{3.25}$$

Simultaneously,

$$\frac{\partial}{\partial \tau}\left[\mathbf{R}_t(\mathbf{x},\tau)\mathbf{R}_t^T(\mathbf{x},\tau)\right] = 0 = \frac{\partial \mathbf{R}_t(\mathbf{x},\tau)}{\partial \tau}\mathbf{R}_t^T(\mathbf{x},\tau) + \mathbf{R}_t(\mathbf{x},\tau)\frac{\partial \mathbf{R}_t^T(\mathbf{x},\tau)}{\partial \tau}. \tag{3.26}$$

Hence,

$$\mathbf{W} = -\mathbf{W}^T. \tag{3.27}$$

According to the relation (3.22), the conditions (3.25) and (3.27) yield immediately

$$\mathbf{D} = \tfrac{1}{2}\left(\mathbf{L} + \mathbf{L}^T\right), \qquad \mathbf{W} = \tfrac{1}{2}\left(\mathbf{L} - \mathbf{L}^T\right). \tag{3.28}$$

The second rank tensor $\mathbf{D}$ is called the **stretching tensor (rate of the deformation tensor)** and $\mathbf{W}$ the **spin tensor**.

Now we make use of the polar decomposition theorem in the (3.24) and deduce

$$\mathbf{L} = (\mathbf{RU})^{\bullet}(\mathbf{RU})^{-1} = \boldsymbol{\Omega} + \mathbf{R}\dot{\mathbf{U}}\mathbf{U}^{-1}\mathbf{R}^{T}, \tag{3.29}$$

where

$$\boldsymbol{\Omega} \equiv \dot{\mathbf{R}}\mathbf{R}^{T} = -\boldsymbol{\Omega}^{T} \tag{3.30}$$

is the **rate of the rotation tensor.**

Substitution of (3.29) in (3.28) leads to the **Euler-Cauchy-Stokes decomposition** of the velocity gradient

$$\begin{aligned} &\mathbf{D} = \tfrac{1}{2}\mathbf{R}\left(\dot{\mathbf{U}}\mathbf{U}^{-1} + \mathbf{U}^{-1}\dot{\mathbf{U}}\right)\mathbf{R}^{T}, \\ &\mathbf{W} = \boldsymbol{\Omega} + \tfrac{1}{2}\mathbf{R}\left(\dot{\mathbf{U}}\mathbf{U}^{-1} - \mathbf{U}^{-1}\dot{\mathbf{U}}\right)\mathbf{R}^{T}, \\ &\mathbf{L} = \mathbf{D} + \mathbf{W}. \end{aligned} \tag{3.31}$$

Since $\mathbf{W}$ is skew-symmetric it can be represented as an axial vector $\mathbf{w}$. Its components are usually defined as follows:

$$\mathbf{w} = w^{k}\mathbf{e}_{k}, \quad w^{k} \equiv \tfrac{1}{2}\varepsilon^{k}{}_{lm}W^{ml} = \tfrac{1}{2}\varepsilon^{k}{}_{lm}\frac{\partial v^{m}}{\partial x^{l}}, \quad \text{i.e.} \quad W_{kl} = \varepsilon_{lkm}w^{m}. \tag{3.32}$$

Hence

$$\mathbf{w} = \tfrac{1}{2}\operatorname{curl}\mathbf{v}. \tag{3.33}$$

This vector is called the **vorticity** in hydrodynamics. We shall not use this notion further in this book.

It should be mentioned that the rate of the deformation tensor $\mathbf{D}$ is not equal to the material time derivative of any of the deformation measures which were discussed in Chap. 2. For instance,

$$\begin{aligned} &\dot{\mathbf{C}} = \left(\mathbf{F}^{T}\mathbf{F}\right)^{\bullet} = \mathbf{F}^{T}\left(\mathbf{F}^{-T}\dot{\mathbf{F}}^{T} + \dot{\mathbf{F}}\mathbf{F}^{-1}\right)\mathbf{F} = 2\mathbf{F}^{T}\mathbf{D}\mathbf{F}, \\ &\dot{\mathbf{B}} = \left(\mathbf{F}\mathbf{F}^{T}\right)^{\bullet} = \dot{\mathbf{F}}\mathbf{F}^{-1}\mathbf{F}\mathbf{F}^{T} + \mathbf{F}\mathbf{F}^{T}\mathbf{F}^{-T}\dot{\mathbf{F}}^{T} = \mathbf{L}\mathbf{B} + \mathbf{B}\mathbf{L}^{T} = \\ &\quad = \mathbf{D}\mathbf{B} + \mathbf{B}\mathbf{D} + \mathbf{W}\mathbf{B} - \mathbf{B}\mathbf{W}. \end{aligned} \tag{3.34}$$

On the other hand, if we introduce the relative Cauchy-Green tensor

$$\mathbf{C}_{t}(\tau) = \mathbf{F}_{t}^{T}(\tau)\mathbf{F}_{t}(\tau), \tag{3.35}$$

then its time derivative for $\tau=t$ has the following form:

$$\left.\frac{\partial \mathbf{C}_{t}(\tau)}{\partial \tau}\right|_{\tau=t} = \left.\left(\frac{\partial \mathbf{F}_{t}^{T}(\tau)}{\partial \tau}\mathbf{F}_{t}(\tau) + \mathbf{F}_{t}^{T}(\tau)\frac{\partial \mathbf{F}_{t}(\tau)}{\partial \tau}\right)\right|_{\tau=t} = \mathbf{L}^{T} + \mathbf{L} = 2\mathbf{D}. \tag{3.36}$$

Consequently, the stretching tensor $\mathbf{D}$ is indeed a measure of the rate of deformation in the current configuration in the Eulerian description [compare $(3.19)_2$].

In a similar manner we can compute higher time derivatives of $\mathbf{C}_t(\tau)$. We define the following symmetric tensors of second rank:

$$\mathbf{A}_n(t) \equiv \mathbf{C}_t^{(n)}(t) = \frac{\partial^n}{\partial \tau^n} \mathbf{C}_t(\tau)\Big|_{\tau=t}, \qquad n = 1,2,3,\ldots . \tag{3.37}$$

The tensor $\mathbf{A}_n(t)$ is called the **Rivlin-Ericksen tensor of the order** n. Rivlin-Ericksen tensors were introduced in the construction of the non-linear viscoelasticity, and in particular, in the description of non-Newtonian fluids.

Exercise 3.4. Consider the velocity field (3.12). Show that

$$\mathbf{A}_1 = \kappa\left(\mathbf{N} + \mathbf{N}^T\right), \qquad \mathbf{A}_2 = 2\kappa^2 \mathbf{N}^T \mathbf{N}, \qquad \mathbf{A}_3 = 0 \bullet \tag{3.38}$$

Exercise 3.5. Prove the following relation

$$\mathbf{a} = \frac{\partial \mathbf{v}}{\partial t} + \frac{1}{2} \operatorname{grad} v^2 + 2\mathbf{W}\mathbf{v}, \qquad v^2 \equiv \mathbf{v} \cdot \mathbf{v} \bullet \tag{3.39}$$

Exercise 3.6. Construct the velocity gradient **L**, the rate of the deformation tensor **D**, the spin tensor **W**, the rate of the rotation tensor **Ω**, and the vorticity **w** in the case of simple shearing (Example 2.1.).

Answer:

$$\begin{aligned} &\mathbf{L} = \frac{\dot{\varphi}}{\cos^2 \varphi} \mathbf{e}_2 \otimes \mathbf{e}_3, \quad \mathbf{D} = \frac{1}{2} \frac{\dot{\varphi}}{\cos^2 \varphi} (\mathbf{e}_2 \otimes \mathbf{e}_3 + \mathbf{e}_3 \otimes \mathbf{e}_2), \\ &\mathbf{W} = \frac{1}{2} \frac{\dot{\varphi}}{\cos^2 \varphi} (\mathbf{e}_2 \otimes \mathbf{e}_3 - \mathbf{e}_3 \otimes \mathbf{e}_2), \qquad \mathbf{w} = -\frac{1}{2} \frac{\dot{\varphi}}{\cos^2 \varphi} \mathbf{e}_1, \\ &\boldsymbol{\Omega} = 2 \frac{\dot{\varphi}}{1 + 3\cos^2 \varphi} (\mathbf{e}_2 \otimes \mathbf{e}_3 + \mathbf{e}_3 \otimes \mathbf{e}_2), \qquad \dot{\varphi} \equiv \frac{d\varphi}{dt} \bullet \end{aligned} \tag{3.40}$$

3.3 Change of the Reference Frame

Now we return to the second question raised at the beginning of this chapter.

The **reference system** or the **frame of reference** can be interpreted physically as an **observer** who measures the position in the configuration space and the time with a ruler and with a clock, respectively. If two observers use different rulers and clocks, they come up with different results for the same position and the same instant of time. However, within the frame of Newtonian (non-relativistic) mechanics, the **distance** between two points of the configuration space and the **time lapse** between two events should be the same in both measurements*). We impose this requirement on changes of the reference system.

Let **x** be a generic point of the configuration space $\Re^3$ and t an instant of time. The pair (**x**,t) is sometimes called an **event**. The reference system is then defined as the map

$$\Phi'(\mathbf{x},t) = \left(\Phi_t'(\mathbf{x}), T_{\Phi'}(t)\right) \equiv (\mathbf{x}', t'), \tag{3.41}$$

*) These arguments are slightly misleading. The problem of invariance with respect to a change of observers in continuum mechanics has been extensively discussed during the last 25 years and it has been shown [see, for instance, I. MÜLLER (1985)] that this requirement is only a good approximation and not a universal principle of Newtonian Mechanics. We shall return to this problem a few times in the following [see: Remark 5.2.].

where $(\mathbf{x}',t')$ are the results of measurements of the (´)-observer.

Now let us choose another reference system

$$\Phi^*(\mathbf{x},t) = \left(\Phi^*{}_t(\mathbf{x}), T_{\Phi^*}(t)\right) \equiv \left(\mathbf{x}^*, t^*\right), \tag{3.42}$$

which corresponds to measurements of (*)-observer (see: Fig. 3.1.). The map

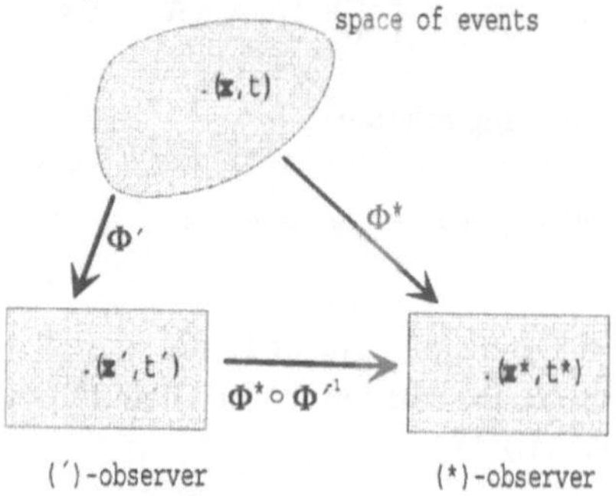

Fig. 3.1. *Change of the reference system*

$$\Phi^* \circ \Phi'^{-1}: \quad (\mathbf{x}',t') \mapsto \left(\mathbf{x}^*, t^*\right), \tag{3.43}$$

is called the **change of the reference system** or the **change of frame** from Φ' to Φ^*.

According to the previous remarks, we require that:

1. $\Phi_t^* \circ \Phi_t'^{-1}$ is an isometry, i.e. for any two positions $\mathbf{x}^{(1)}$ and $\mathbf{x}^{(2)}$

$$\left|\Phi_t^*\left(\mathbf{x}^{(1)}\right) - \Phi_t^*\left(\mathbf{x}^{(2)}\right)\right| = \left|\Phi_t'\left(\mathbf{x}^{(1)}\right) - \Phi_t'\left(\mathbf{x}^{(2)}\right)\right|, \tag{3.44}$$

2. $T_{\Phi^*} \circ T_{\Phi'}^{-1}$ preserves the time interval, i.e. for two instants of time, $t^{(1)}$ and $t^{(2)}$, we must have

$$T_{\Phi^*}\left(t^{(1)}\right) - T_{\Phi^*}\left(t^{(2)}\right) = T_{\Phi'}\left(t^{(1)}\right) - T_{\Phi'}\left(t^{(2)}\right). \tag{3.45}$$

The above requirements yield for the Euclidean space the following most general transformation rule for the change of the reference system

$$\begin{aligned} \mathbf{x}^* &= \mathbf{O}(t)\left(\mathbf{x}' - \mathbf{x}^\circ\right) + \mathbf{c}(t), \\ t^* &= t' + a, \end{aligned} \tag{3.46}$$

where $\mathbf{O} \in \mathcal{O}$, $\mathcal{O}$ is the orthogonal group, i.e. $\mathbf{O}^T = \mathbf{O}^{-1}$, $\mathbf{x}^o$, $\mathbf{c}(t)$ are arbitrary vectors, and a is an arbitrary scalar.

The change of frame (3.46) is a time-dependent rigid transformation. This means that if we change the motion of the body by this transformation imposing an additional rigid body motion, the measurements of distances and time lapses will be the same as in the original motion. The transformation (3.46) is often called a **Euclidean transformation.** The collection of all Euclidean transformations denoted by Σ is called the **Euclidean class.**

Owing to the global character of tangent spaces $\mathcal{T}_X$ to the Euclidean space $\mathfrak{R}^3$, we can identify a vector $\mathbf{r} \in \mathcal{T}_X$ with the difference of two positions, say $\mathbf{x}^{(1)}$ and $\mathbf{x}^{(2)}$,

$$\mathbf{r} = \mathbf{x}^{(2)} - \mathbf{x}^{(1)}, \qquad \mathbf{r} \in \mathcal{T}_X, \quad \mathbf{x}^{(1)}, \mathbf{x}^{(2)} \in \mathfrak{R}^3. \tag{3.47}$$

This identification induces the rule of transformation for vectors connected with the Euclidean transformation (3.46). Let us denote this transformation by $\mathbf{O}_{(1)}$. According to (3.46), we have

$$\begin{aligned} \mathbf{r}^* &= \mathbf{O}_{(1)}(\mathbf{r}') = \mathbf{x}^{*(2)} - \mathbf{x}^{*(1)} = \mathbf{O}_{(1)}\left(\Phi_t'\left(\mathbf{x}^{(2)}\right) - \Phi_t'\left(\mathbf{x}^{(1)}\right)\right) = \\ &= \mathbf{O}\left(\mathbf{x}'^{(2)} - \mathbf{x}'^{(1)}\right) = \mathbf{O}\mathbf{r}', \end{aligned} \tag{3.48}$$

i.e.

$$\mathbf{O}_{(1)}(\cdot) \equiv \mathbf{O}(t)(\cdot). \tag{3.49}$$

More generally, let $\mathcal{S}_n$ denote the space of tensors $\mathbf{T}$ of the n-th rank. Then the change of frame (3.46) induces a linear map on $\mathcal{S}_n$, denoted by $\mathbf{O}_{(n)}$, in the following way. For any $\mathbf{r}'^{(1)}, \ldots, \mathbf{r}'^{(n)}$ we define

$$\mathbf{O}_{(n)}\left(\mathbf{r}'^{(1)} \otimes \ldots \otimes \mathbf{r}'^{(n)}\right) = \mathbf{r}^{*(1)} \otimes \ldots \otimes \mathbf{r}^{*(n)}. \tag{3.50}$$

In particular, for n=2 we have

$$\mathbf{O}_{(2)}\left(\mathbf{r}'^{(1)} \otimes \mathbf{r}'^{(2)}\right) = \left(\mathbf{O}(t)\mathbf{r}'^{(1)}\right) \otimes \left(\mathbf{O}(t)\mathbf{r}'^{(2)}\right) = \mathbf{O}(t)\left(\mathbf{r}'^{(1)} \otimes \mathbf{r}'^{(2)}\right)\mathbf{O}^T(t). \tag{3.51}$$

Hence,

$$\mathbf{r}^{*(1)} \otimes \mathbf{r}^{*(2)} = \mathbf{O}(t)\left(\mathbf{r}'^{(1)} \otimes \mathbf{r}'^{(2)}\right)\mathbf{O}^T(t), \tag{3.52}$$

and for a tensor $\mathbf{T}$ of second rank we conclude

$$\mathbf{T}^* = \mathbf{O}(t)\mathbf{T}'\mathbf{O}^T(t). \tag{3.53}$$

From the physical viewpoint, we can say that **r**′and **r***, **T**′and **T***, are the same vector and the same tensor observed by the (′)-observer and by the (*)-observer, respectively. For this reason, we shall further identify the space of configurations $\mathfrak{R}^3$ and the time with the space of the observer corresponding to the identity map Φ=id. That is to say that we identify the position **x** and the instant of time t with those measured by a certain chosen observer.

Let us notice that $\mathbf{O}_{(n)}$ depends only on the rotational part **O**(t) of the change of frame. Therefore, we may also refer to $\mathbf{O}_{(n)}$ as the **linear map on $\mathcal{S}_n$ induced** by the **rotation O**(t).

It should also be noticed that a scalar independent of the observer should have the same value in any frame. Formally, we can write that for n=0, $\mathbf{O}_{(0)}$ defined on $\mathcal{S}_0$, should be the identity, i.e. $\mathbf{O}_{(0)}=1$.

For a chosen $\mathcal{S}_n$, let μ be a map of reference systems on $\mathcal{S}_n$

$$\Phi \mapsto \mu(\Phi) \in \mathcal{S}_n. \tag{3.54}$$

We call μ(Φ) the value **observed** in the reference system Φ. Such a map is called an **observable** quantity.

For a chosen subclass of changes of frame Ξ⊂Σ, an observable quantity μ is said to be **frame-indifferent with respect to** Ξ if

$$\forall\, \Phi, \Phi^* \text{ s.t. } \Phi^* \circ \Phi^{-1} \in \Xi: \quad \mu(\Phi^*) = \mathbf{O}_{(n)}\mu(\Phi). \tag{3.55}$$

In the case Ξ=Σ we say that μ is **objective** with respect to Euclidean transformations.

In mechanics we know that motions depend on the observer, and consequently, the velocity and the acceleration are not, generally, objective. Now we proceed to discuss this dependence on the choice of observer in some detail.

According to our identification, we can choose one of the observers as an identity map. Let us consider the change of frame: id →Φ*. For a given motion **f**(**X**,·), we have

$$\mathbf{x} = \mathbf{f}(\mathbf{X}, t), \quad \mathbf{x}^* = \mathbf{f}^*(\mathbf{X}, t^*). \tag{3.56}$$

Bearing (3.46) in mind, we obtain

$$\mathbf{f}^*(\mathbf{X}, t^*) = \mathbf{O}(t)\left[\mathbf{f}(\mathbf{X}, t) - \mathbf{x}^0\right] + \mathbf{c}(t), \quad t^* = t + a. \tag{3.57}$$

Hence, it follows that

$$\dot{\mathbf{x}}^* = \frac{\partial \mathbf{f}^*}{\partial t^*} = \dot{\mathbf{O}}(t)\left(\mathbf{x} - \mathbf{x}^0\right) + \mathbf{O}(t)\dot{\mathbf{x}} + \dot{\mathbf{c}}(t), \tag{3.58}$$

or

$$\mathbf{O}(t)\dot{\mathbf{x}} = \dot{\mathbf{x}}^* - \Omega\,(t)\left(\mathbf{x}^* - \mathbf{c}\right) - \dot{\mathbf{c}}, \tag{3.59}$$

where

is called the tensor of **relative angular velocity** of the Φ^*-frame with respect to the id-frame. The vector $\dot{\mathbf{c}}$ is called the **relative translational velocity** of those two frames.

Obviously, the relation (3.59) is not of the form (3.55), and hence, the velocity is not an objective quantity.

Taking again the time derivative in (3.58), we obtain

$$\mathbf{O}(t)\ddot{\mathbf{x}} = \ddot{\mathbf{x}}^* - 2\Omega\left(\dot{\mathbf{x}}^* - \dot{\mathbf{c}}\right) + \Omega^2\left(\mathbf{x}^* - \mathbf{c}\right) - \dot{\Omega}\left(\mathbf{x}^* - \mathbf{c}\right) - \ddot{\mathbf{c}}. \tag{3.61}$$

Therefore, the acceleration $\ddot{\mathbf{x}}$ is also not an objective quantity. The additional terms in (3.61), which distinguish this transformation rule from (3.55), have the following names:

$2\Omega\left(\dot{\mathbf{x}}^* - \dot{\mathbf{c}}\right)$	- **Coriolis acceleration,**
$-\Omega^2\left(\mathbf{x}^* - \mathbf{c}\right)$	- **centrifugal acceleration,**
$\dot{\Omega}\left(\mathbf{x}^* - \mathbf{c}\right)$	- **Euler acceleration,**
$\ddot{\mathbf{c}}.$	- **relative translational acceleration.**

Among all Euclidean transformations one can choose those which transform the acceleration in the objective manner. Then we have

$$\ddot{\mathbf{x}}^* = \mathbf{O}\ddot{\mathbf{x}} \quad \Leftrightarrow \quad \ddot{\mathbf{c}} = 0, \quad \Omega = 0 \quad \Leftrightarrow \quad \mathbf{c}(t) = \mathbf{V}t + \mathbf{c}', \quad \mathbf{O}(t) = \mathbf{O}, \tag{3.62}$$

V, **c**, and **O** being time-independent. The change of frame defined by such constants

$$\mathbf{x}^* = \mathbf{O}\left(\mathbf{x} - \mathbf{x}^0\right) + \mathbf{V}t + \mathbf{c}, \quad t^* = t + a, \tag{3.63}$$

is called a **Galilean transformation**. The collection of all Galilean transformations, denoted by Γ, is called the **Galilean class.**

Certainly, the acceleration is objective with respect to Γ whereas the velocity is not.

We complete this section with the objectivity properties of some other quantities which appeared in Chap. 2. Again, we consider two frames discussed for the velocity and acceleration. In the case of the deformation gradient (2.7), we have in those two frames

$$\mathbf{F}(\mathbf{X},t) = \frac{\partial \mathbf{f}}{\partial \mathbf{X}}(\mathbf{X},t), \quad \mathbf{F}^*(\mathbf{X},t) = \frac{\partial \mathbf{f}^*}{\partial \mathbf{X}}(\mathbf{X},t).$$

Bearing the relation (3.57) in mind, we obtain

$$\mathbf{F}^*(\mathbf{X},t) = \mathbf{O}(t)\mathbf{F}(\mathbf{X},t). \tag{3.64}$$

Since **F** is a second rank tensor, it is not objective [see: (3.53)].

By means of the relation (3.64) we can easily deduce the rules of transformation for the tensors of stretch **U**, **V**, and the rotation **R**. Namely, the polar decomposition in (3.64) yields

$$\mathbf{R}^*\mathbf{U}^* = \mathbf{O}\mathbf{R}\mathbf{U}, \qquad \mathbf{V}^*\mathbf{R}^* = \mathbf{O}\mathbf{V}\mathbf{R}, \tag{3.65}$$

and from the uniqueness of the polar decomposition we conclude immediately,

$$\mathbf{R}^* = \mathbf{O}\mathbf{R}, \qquad \mathbf{U}^* = \mathbf{U}, \qquad \mathbf{V}^* = \mathbf{O}\mathbf{V}\mathbf{O}^{\mathrm{T}}. \tag{3.66}$$

Hence, **R** and **U** are not objective in the sense of the definition (3.55), but **V** is objective. Further on in this book we return to this problem of invariance for the stretch tensor **U**.

For the measures of deformation we have

$$\mathbf{C}^* = \mathbf{U}^{*2} = \mathbf{U}^2 = \mathbf{C}, \qquad \mathbf{B}^* = \mathbf{V}^{*2} = \mathbf{O}\mathbf{V}\mathbf{O}^{\mathrm{T}}\mathbf{O}\mathbf{V}\mathbf{O}^{\mathrm{T}} = \mathbf{O}\mathbf{B}\mathbf{O}^{\mathrm{T}}. \tag{3.67}$$

Hence, the left Cauchy-Green tensor **B** is objective, and the right Cauchy-Green tensor **C** is not.

Differentiation of (3.64) with respect to time leads to

$$\dot{\mathbf{F}}^* = \dot{\mathbf{O}}\mathbf{F} + \mathbf{O}\dot{\mathbf{F}} = \mathbf{O}\dot{\mathbf{F}} + \boldsymbol{\Omega}\,\mathbf{O}\mathbf{F}. \tag{3.68}$$

Then, according to (3.24), we obtain the following rule of transformation for the velocity gradient:

$$\mathbf{L}^* = \dot{\mathbf{F}}^*\mathbf{F}^{*-1} = \left(\mathbf{O}\dot{\mathbf{F}} + \boldsymbol{\Omega}\,\mathbf{O}\mathbf{F}\right)\mathbf{F}^{-1}\mathbf{O}^{\mathrm{T}} = \mathbf{O}\mathbf{L}\mathbf{O}^{\mathrm{T}} + \boldsymbol{\Omega}\,, \tag{3.69}$$

and for the rate of deformation and the spin

$$\mathbf{D}^* = \mathbf{O}\mathbf{D}\mathbf{O}^{\mathrm{T}}, \qquad \mathbf{W}^* = \mathbf{O}\mathbf{W}\mathbf{O}^{\mathrm{T}} + \boldsymbol{\Omega}\,. \tag{3.70}$$

It follows that **L** and **W** are not objective, whereas **D** is objective.

Exercise 3.7. Let ψ, **q** and **T** be an objective scalar, an objective vector, and an objective second rank tensor, respectively.

1. Show that their gradients are objective quantities.
2. Show that $\dot{\psi}$ is objective, whereas $\dot{\mathbf{q}}$ and $\dot{\mathbf{T}}$ are not.
3. Show that

$$\frac{\mathrm{D_J}\mathbf{q}}{\mathrm{D}t} \equiv \dot{\mathbf{q}} - \mathbf{W}\mathbf{q}, \qquad \frac{\mathrm{D_J}\mathbf{T}}{\mathrm{D}t} \equiv \dot{\mathbf{T}} - \mathbf{W}\mathbf{T} + \mathbf{T}\mathbf{W}, \tag{3.71}$$

are objective. These derivatives are usually called **corotational (Jaumann-Zaremba)** time derivatives •

Exercise 3.8. Show that

1. the Rivlin-Ericksen tensor $\mathbf{A}_n$ is objective,
2. the relative right stretch tensor $\mathbf{U}_t$ is objective, whereas the relative rotation tensor $\mathbf{R}_t$ is not objective •

Remark 3.1. *On Objective Time Derivatives*
An extensive amount of literature has been devoted to the problem of objective time derivatives indicated in Exercise 3.6. This problem has featured in numerous papers on the construction of models of viscous fluids. A similar problem had to be faced within the framework of models of large plastic deformations later on, and different propositions have appeared which increase the confusion within the subject even to the present day.

The solution to this problem was found about 100 years ago by S. LIE [e.g. see: W. SLEBODZINSKI (1970)]. Here we present a few particularly simple cases which have bearing on further considerations. We limit our attention to **Lie derivatives with respect to the velocity field v**. In general, these derivatives can be introduced with respect to arbitrary vector fields.

Now let us consider the arbitrary material vector field **Q**(**X**) on $\mathcal{B}_0$. We denote its current image as follows:

$$\mathbf{q}(\mathbf{x},t) \equiv \mathbf{F}(\mathbf{X},t)\mathbf{Q}(\mathbf{X})\Big|_{\mathbf{X}=\mathbf{f}^{-1}(\mathbf{x},t)}. \tag{3.72}$$

For any other configuration $\mathcal{B}_\tau$ we have

$$\begin{aligned}\mathbf{q}(\xi,\tau) &= \mathbf{F}(\mathbf{X},\tau)\mathbf{Q}(\mathbf{X})\Big|_{\mathbf{X}=\mathbf{f}^{-1}(\xi,\tau)} = \\ &= \mathbf{F}(\mathbf{X},\tau)\mathbf{F}^{-1}(\mathbf{X},t)\mathbf{q}(\mathbf{x},t)\Big|_{\mathbf{X}=\mathbf{f}^{-1}(\mathbf{x},t),\,\mathbf{x}=\mathbf{f}_t^{-1}(\xi,\tau)},\end{aligned} \tag{3.73}$$

with

$$\xi = \mathbf{f}\left(\mathbf{f}^{-1}(\mathbf{x},t),\tau\right) \equiv \mathbf{f}_t(\mathbf{x},\tau), \tag{3.74}$$

i.e.

$$\mathbf{q}(\mathbf{x},t) = \mathbf{F}_t^{-1}(\tau)\mathbf{q}(\xi,\tau). \tag{3.75}$$

The definition $(2.31)_2$ was used in this relation.

The Lie derivative of the material vector field **q**(**x**,t), with respect to the velocity field **v**, is defined by the relation

$$\mathcal{L}_v\mathbf{q}(\mathbf{x},t) = \left(\frac{d}{d\tau}\mathbf{F}_t^{-1}(\tau)\mathbf{q}(\xi,\tau)\right)\Bigg|_{\tau=t}. \tag{3.76}$$

Bearing the identity

$$\frac{d}{d\tau}\left(\mathbf{F}_t(\tau)\mathbf{F}_t^{-1}(\tau)\right) = 0 = \dot{\mathbf{F}}_t(\tau)\mathbf{F}_t^{-1}(\tau) + \mathbf{F}_t(\tau)\dot{\mathbf{F}}_t^{-1}(\tau), \tag{3.77}$$

in mind, we obtain

$$\mathcal{L}_v\,\mathbf{q}(\mathbf{x},t) = \frac{\partial\mathbf{q}}{\partial t} + (\mathbf{v}\cdot\text{grad})\mathbf{q} - \mathbf{L}\,\mathbf{q} \equiv \dot{\mathbf{q}} - \mathbf{L}\,\mathbf{q}. \tag{3.78}$$

In a particular case where **q**=**v** we have, of course,

$$\mathcal{L}_v \mathbf{v} = \frac{\partial \mathbf{v}}{\partial t}. \tag{3.79}$$

Comparison of (3.78) with the formula (3.71) now yields

$$\mathcal{L}_v \mathbf{q} = \frac{D_J \mathbf{q}}{Dt} - \mathbf{D}\mathbf{q}, \tag{3.80}$$

and, owing to the objectivity of **D**, we see that the Lie derivative is objective when the field **q** is objective.

For the material tensors of the second rank **T** whose transformation rule, corresponding to the formula (3.75) for vectors, has the following form:

$$\mathbf{T}(\mathbf{x},t) = \mathbf{F}_t^{-1}(\tau)\mathbf{T}(\xi,\tau)\mathbf{F}_t^{-T}(\tau), \qquad \xi = \mathbf{f}_t(\mathbf{x},\tau), \tag{3.81}$$

we have the following definition of the Lie derivative with respect to the velocity field **v**:

$$\mathcal{L}_v \mathbf{T}(\mathbf{x},t) = \left. \left(\frac{d}{d\tau} \mathbf{F}_t^{-1}(\tau)\mathbf{T}(\xi,\tau)\mathbf{F}_t^{-T}(\tau) \right) \right|_{\tau=t}. \tag{3.82}$$

Hence, after easy calculations, we obtain

$$\mathcal{L}_v \mathbf{T} = \dot{\mathbf{T}} - \mathbf{L}\mathbf{T} - \mathbf{T}\mathbf{L}^T. \tag{3.83}$$

This is the famous **Oldroyd derivative** appearing in the theories of Maxwellian fluids.

Bearing $(3.71)_2$ in mind, we arrive at

$$\mathcal{L}_v \mathbf{T} = \frac{D_J \mathbf{T}}{Dt} - \mathbf{D}\mathbf{T} - \mathbf{T}\mathbf{D}. \tag{3.84}$$

Therefore, the Lie derivative is again objective when **T** itself is objective.

Let us complete this remark with the formula for the Lie derivative in the case of a non-material vector. For simplicity let us consider the non-material vector **N(X)**, $\mathbf{X} \in \mathcal{B}_0$, whose deformation-induced transformation is given by the relation [compare (2.20) without the normalization of the vector **n**]

$$\mathbf{n}(\mathbf{x},t) = \mathbf{F}^{-T}(\mathbf{X},t)\mathbf{N}(\mathbf{X})\big|_{\mathbf{X}=\mathbf{f}^{-1}(\mathbf{x},t)}. \tag{3.85}$$

For the intermediate configuration $\mathcal{B}_\tau$ we then obtain

$$\mathbf{n}(\xi,\tau) = \mathbf{F}^{-T}(\mathbf{X},\tau)\mathbf{N}(\mathbf{X})\big|_{\mathbf{X}=\mathbf{f}^{-1}(\xi,\tau)} \quad \Rightarrow \quad \mathbf{N} = \mathbf{F}^T(\tau)\mathbf{n}(\xi,\tau)\big|_{\xi=\mathbf{f}(\mathbf{X},\tau)}. \tag{3.86}$$

Hence

$$\mathbf{n}(\mathbf{x},t) = \mathbf{F}^{-T}(t)\mathbf{F}^T(\tau)\mathbf{n}(\xi,\tau)\big|_{\substack{\mathbf{X}=\mathbf{f}^{-1}(\mathbf{x},t)\\ \xi=\mathbf{f}(\mathbf{X},\tau)}} \equiv \mathbf{F}_t^T(\tau)\mathbf{n}(\xi,\tau)\big|_{\xi=\mathbf{f}_t^{-1}(\mathbf{x},\tau)}, \tag{3.87}$$

and the Lie derivative with respect to the velocity field **v**,

$$\mathcal{L}_v \mathbf{n} = \left(\frac{d}{d\tau} \mathbf{F}_t^T(\tau)\, \mathbf{n}(\xi, \tau) \right) \Bigg|_{\tau=t}, \tag{3.88}$$

finally has the form

$$\mathcal{L}_v \mathbf{n} = \dot{\mathbf{n}} + \mathbf{L}^T \mathbf{n}. \tag{3.89}$$

It is easy to check that the Lie derivatives commute with contractions. For instance, in the case of the material tensor field **T** and the non-material vector field **n**, we have

$$\begin{aligned} \mathcal{L}_v (\mathbf{Tn}) &= \dot{\mathbf{T}}\mathbf{n} + \mathbf{T}\dot{\mathbf{n}} - \mathbf{LTn} - \mathbf{TL}^T\mathbf{n} + \mathbf{TL}^T\mathbf{n} = \\ &= \dot{\mathbf{T}}\mathbf{n} + \mathbf{T}\dot{\mathbf{n}} - \mathbf{LTn} = (\mathbf{Tn})^\bullet - \mathbf{L}(\mathbf{Tn}), \end{aligned} \tag{3.90}$$

the last part of this relation being the Lie derivative of the material vector (**Tn**).

The physical meaning of the Lie derivative with respect to the velocity field **v** is quite simple. All three definitions (3.76), (3.82), and (3.88) have the common feature that we perform the differentiation along trajectories of the material points [see: (3.6)]. Hence, these derivatives measure time changes of fields as seen by the observer moving together with the material point, i.e. moving with velocity **v** and rotating with material vectors attached to the chosen trajectory. This property explains their objective character – the dot derivative is due to the translation of material point along the trajectory, and the terms with the velocity gradient **L** are due to the material rotation•

4 Balance Equations

4.1 Preliminary Remarks

The main purpose of various theories of continuous media is to find fields of interest either as functions of time and material points $\mathbf{X}$ belonging to the body $\mathcal{B}_0$ (Lagrangian description) or as functions of time and positions $\mathbf{x}$ belonging to the current configuration $\mathcal{B}_t$ (Eulerian description). These fields are supposed to be solutions of field equations, which must be specified for a chosen class of materials.

Within the frame of continuum thermodynamics, it is customary to construct the field equations on the basis of certain balance equations. The balance equations have, in contrast to the field equations, a material independent character and require the so-called constitutive relations, specifying the material (*closure problem*) such that a closed set of differential equations for chosen fields is obtained.

Let us stress right away that the strategy described above does not exhaust all possibilities considered in the contemporary theories of continua. On the one hand, continuum models are considered in which some field equations do not follow from the balance equations. Such is the case, for instance, in the theories based on certain **evolution equations** (e.g. for the extent of chemical reactions in mixtures or for the volume fractions in some multicomponent models of porous materials). On the other hand, some materials seem to require **non-linear functional constitutive relations**. If this is the case, the fields are not described by field equations, which would form a set of partial differential equations.

Further on in this book we shall not consider such cases. We limit our attention solely to such thermodynamical problems of continua which are described by partial differential equations.

In this chapter we present the main features of the balance equations of continua. As indicated by various kinetic (submacroscopic) theories, these equations consist of three main parts: time changes of a quantity under consideration, its flux owing to interactions with the external parts of the system, and the internal production. The production term vanishes in the particular cases of conservation laws. Moreover, if it does not vanish identically, according to the second law of thermodynamics which we present further in this book, this term must possess certain relaxation properties in order to lead to a stable equilibrium state. We shall specify these features in the following.

Even though we expose in this chapter some elements of the axiomatic foundations of the continuum balance equations, the presentation has a rather lax character. Its main purpose is to show the limitations of the continuum models. Mathematical details, particularly those connected with the measure theory, can be found in the literature quoted further on in this work.

4.2 Global Balance Equations

Now we proceed to present the general balance equation in the Lagrangian description [e.g. K. WILMANSKI (1980)]. To this aim we need the notion of a subbody. Among all subsets of $\mathcal{B}_0$ we distinguish a family $\mathcal{A}$ of sets, closed in the natural topology of $\mathcal{B}_0$, which is the Boolean algebra by the following binary operations:

$$\begin{aligned} &\forall \mathcal{P}_1, \mathcal{P}_2 \in \mathcal{A}: \quad \mathcal{P}_1 \vee \mathcal{P}_2 \equiv \mathcal{P}_1 \cup \mathcal{P}_2 \in \mathcal{A}, \\ &\forall \mathcal{P}_1, \mathcal{P}_2 \in \mathcal{A}: \quad \mathcal{P}_1 \wedge \mathcal{P}_2 \equiv \mathrm{cl}(\mathrm{int}\mathcal{P}_1 \cap \mathrm{int}\,\mathcal{P}_2) \in \mathcal{A}, \end{aligned} \tag{4.1}$$

where cl and int denote the closure and the interior in the topology of $\mathcal{B}_0$, respectively. Each of these sets is called a **subbody** of $\mathcal{B}_0$. All members of $\mathcal{A}$ are assumed to be volume measurable, and it is assumed that their **boundaries**

$$\forall \mathcal{P} \in \mathcal{A}: \quad \partial\mathcal{P} \equiv \mathcal{P} \cap \mathcal{P}^e, \quad \mathrm{int}\mathcal{P} \cap \mathrm{int}\mathcal{P}^e = \varnothing, \quad \mathcal{P} \vee \mathcal{P}^e \equiv \mathcal{B}_0, \tag{4.2}$$

are orientable and surface measurable within an appropriate Boolean family of surface elements. Details of this structure of subbodies can be found in numerous papers and textbooks*) and we do not need to look into any further details in this book.

Now let us consider a scalar **additive** function $\Phi(\cdot,t)$, defined on $\mathcal{A}$ for each instant of time $t \in \mathcal{T}$, satisfying the following **axiom of continuity** with respect to the volume measure V:

$$\Phi(\cdot,t): \quad \mathcal{A} \to \Re^1 \quad \text{s.t.} \quad \exists \alpha > 0 \forall \mathcal{P} \in \mathcal{A}: |\Phi(\mathcal{P},t)| \le \alpha\, V(\mathcal{P}). \tag{4.3}$$

This assumption means, that the function $\Phi(\cdot,t)$ possesses a Radon-Nikodym derivative which we denote by $\varphi(\cdot,t)$ and call the scalar **field**. We have

$$\forall \mathcal{P} \in \mathcal{A}: \quad \Phi(\mathcal{P},t) = \int_{\mathcal{P}} \varphi(\mathbf{X},t)\,dV, \tag{4.4}$$

where the integral here and in further relations is the Lebesgue integral with respect to the indicated measure.

Similar axioms are supposed to be satisfied by other (vector or tensor) fields appearing further on in our considerations.

Any scalar field $\varphi(\cdot,t)$ is supposed to satisfy the following **balance equation**

$$\forall \mathcal{P} \in \mathcal{A}: \quad \frac{d}{dt}\int_{\mathcal{P}} \varphi(\mathbf{X},t)\,dV = \Psi(\mathcal{P},\mathcal{P}^e,t) + \Phi^*(\mathcal{P},t), \tag{4.5}$$

where

$$\begin{aligned} &\exists \beta,\gamma > 0 \forall \mathcal{P} \in \mathcal{A}: \quad |\Psi(\mathcal{P},\mathcal{P}^e,t)| \le \beta S(\partial\mathcal{P}) + \gamma V(\mathcal{P}), \\ &\exists \delta > 0 \forall \mathcal{P} \in \mathcal{A}: \quad |\Phi^*(\mathcal{P},t)| \le \delta V(\mathcal{P}), \end{aligned} \tag{4.6}$$

$\Psi(\cdot,\cdot,t)$ being biadditive and called the **flux** of $\Phi(\cdot,t)$, and $\Phi^*(\cdot,t)$ being additive and called the **production** of $\Phi(\cdot,t)$.

---

*) e.g. C. TRUESDELL (1972), K. WILMANSKI (1972), (1974).

The continuity assumption (4.6) concerning the area of the surface $S(\partial\mathcal{P})$ must be strengthened to describe surface interaction between two separate subbodies $\mathcal{P}_1$, $\mathcal{P}_2$ (i.e. for $\mathcal{P}_1 \wedge \mathcal{P}_2 = \varnothing$). Namely, one requires the continuity of $\Psi(\mathcal{P}_1, \mathcal{P}_2, t)$ with respect to the area of $\partial\mathcal{P}_1 \cap \partial\mathcal{P}_2$. Certain delicate points of measure theory concerning this assumption are still being discussed in the literature. Simultaneously, the classical form of the continuity assumption (4.6) is too restrictive to account for entropy production on singular surfaces. We will discuss this problem separately in the chapter on the entropy inequality.

Bearing the above assumptions in mind, one can write the general global balance equation for φ in the following form:

$$\forall \mathcal{P} \in \mathcal{A}: \quad \frac{d}{dt}\int_{\mathcal{P}} \varphi(\mathbf{X},t)\,dV = \oint_{\partial\mathcal{P}} \psi_S(\mathbf{X},t;\partial\mathcal{P})\,dS + \int_{\mathcal{P}} \left(\psi_V(\mathbf{X},t) + \varphi^*(\mathbf{X},t)\right) dV, \tag{4.7}$$

where $\psi_S(\cdot,t)$, $\psi_V(\cdot,t)$, and $\varphi^*(\cdot,t)$ are called the **flux density** of φ, the **supply** of φ and the **production** of φ, respectively. The distinction between ψ_V and φ^* can only be made clear by their physical interpretation which we present in the following.

As we mentioned, the balance equation of the form (4.7) is assumed to hold for fields $\varphi(\mathbf{X},t) \in \mathcal{S}_n$, n=0,1,2,... and not only for scalar quantities. Here we shall skip the general axioms yielding (4.7) for vector and tensor fields.

Using the extension mentioned above of the continuity axiom (4.6), W. NOLL (1959) has proved that the flux density depends on the choice of the surface at almost every point $\mathbf{X} \in \mathcal{B}_0$ only through the **orientation** of this surface. This is sometimes [e.g. for the interaction forces] referred to as **Cauchy's postulate**. This means that the function $\psi_S(\mathbf{X},t;\partial\mathcal{P}^*)$ has the same value for all surfaces $\partial\mathcal{P}^*$ containing $\mathbf{X}$, which have the same tangent plane at $\mathbf{X}$, i.e.

$$\psi_S(\mathbf{X},t;\partial\mathcal{P}) = \psi_S(\mathbf{X},t;\partial\mathcal{P}^*), \tag{4.8}$$

if the normal vector $\mathbf{N}(\mathbf{X})$ is identical for $\partial\mathcal{P}$ and $\partial\mathcal{P}^*$ at $\mathbf{X}$. This statement means that the surface integral can be written in the form:

$$\oint_{\partial\mathcal{P}} \psi_S(\mathbf{X},t;\partial\mathcal{P})\,dS = \oint_{\partial\mathcal{P}} \psi_S(\mathbf{X},t;\mathbf{N})\,dS, \tag{4.9}$$

$\mathbf{N}$ being chosen to be **exterior** relative to $\mathcal{P}$.

Now we can immediately prove

Cauchy's Fundamental Lemma. Suppose that $\psi_S(\cdot,t;\mathbf{N})$ is a continuous function in $\mathcal{P} \in \mathcal{A}$, and $\frac{\partial\varphi}{\partial t}(\cdot,t)$, $\psi_V(\cdot,t)$, $\varphi^*(\cdot,t)$ are bounded in $\mathcal{P} \in \mathcal{A}$. Then the balance equation (4.8) implies that

$$\psi_S(\mathbf{X},t;-\mathbf{N}) = -\psi_S(\mathbf{X},t;\mathbf{N}), \tag{4.10}$$

for any $\mathbf{X} \in \mathcal{P}$ and any unit vector $\mathbf{N}$.

Proof: For any $\mathbf{X}\in\mathcal{P}$ and any unit vector $\mathbf{N}$, we consider a small pillbox $\mathcal{P}_\varepsilon\subset\mathcal{P}$ of thickness ε and centred at $\mathbf{X}$ such that its flat surfaces $\mathcal{S}^+$, $\mathcal{S}^-$ have the exterior unit normal vectors $\mathbf{N}$ and $-\mathbf{N}$, respectively (see: Figure 4.1.). Then (4.8) implies

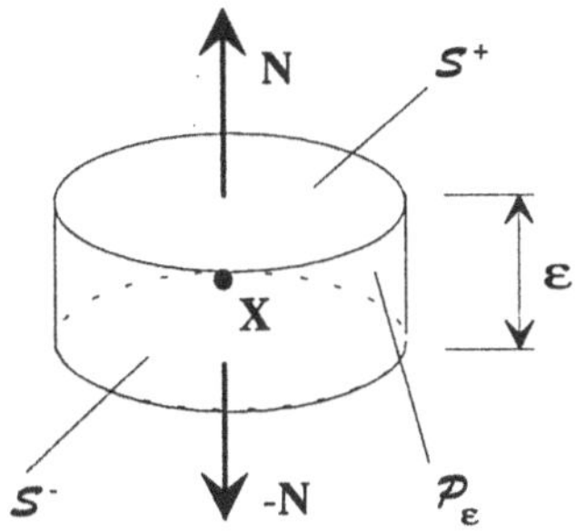

Fig. 4.1. *Pillbox used in the proof of Cauchy's Lemma*

$$\lim_{\varepsilon\to 0}\int_{\mathcal{P}_\varepsilon}\left(\frac{\partial\varphi}{\partial t}-\psi_V-\varphi^*\right)dV=\lim_{\varepsilon\to 0}\oint_{\partial\mathcal{P}_\varepsilon}\psi_S(\mathbf{X},t;\mathbf{N})\,dS.$$

According to our construction, the second part of this relation becomes

$$\oint_{\mathcal{S}}\left[\psi_S(\mathbf{X},t;\mathbf{N})+\psi_S(\mathbf{X},t;-\mathbf{N})\right]dS=0.$$

Now the continuity of the integrand and arbitrariness of $\mathcal{S}$ yield (4.10) ■

Different balance laws, appearing in the continuum thermodynamics, concern a mapping φ with values in different spaces $\mathcal{S}_n$ of tensors of the n-th rank, n=0,1,2,... . Consequently, for each φ, the mappings ψ_S, ψ_V, φ^* have values in the same $\mathcal{S}_n$ spaces. Now we prove the existence of the linear mapping $\psi(\mathbf{X},t)(\cdot)$ on the vector space $\mathcal{V}^3$, whose values belong to $\mathcal{S}_{n+1}$, and which determine uniquely the flux ψ_S.

Cauchy's Fundamental Theorem*). Under the assumptions of Cauchy's Lemma for $\psi_S(\mathbf{X},t;\mathbf{N})\in\mathcal{S}_n$, there exists a field $\psi(\mathbf{X},t)\in\mathcal{S}_{n+1}$ such that

$$\psi_S(\mathbf{X},t;\mathbf{N})=\psi(\mathbf{X},t)\mathbf{N}. \tag{4.11}$$

Proof: According to (4.10), $\psi_S(\mathbf{X},t;\cdot)$ is a function defined on the set of unit vectors $\mathbf{N}$. We can extend this function to the whole vector space $\mathcal{V}^3$ in the following manner:

*) The following proof of Cauchy's Theorem is assigned to WALTER NOLL by M. E. GURTIN (1972).

$$\psi_S^{ext}(\mathbf{X},t;\mathbf{w}) \equiv \begin{cases} |\mathbf{w}|\,\psi_S\left(\mathbf{X},t;\dfrac{\mathbf{w}}{|\mathbf{w}|}\right) & \text{if} \quad \mathbf{w} \neq 0, \\ 0 & \text{if} \quad \mathbf{w} = 0, \end{cases} \tag{4.12}$$

for any vector $\mathbf{w} \in \mathcal{V}^3$. Now we show that $\psi_S^{ext}(\mathbf{X},t;\cdot)$ is a linear mapping, i.e. we prove the following properties:

$$\begin{aligned} &1) \quad \forall \alpha \in \Re^1, \mathbf{w} \in \mathcal{V}^3: \quad \psi_S^{ext}(\mathbf{X},t;\alpha\,\mathbf{w}) = \alpha\,\psi_S^{ext}(\mathbf{X},t;\mathbf{w}), \\ &2) \quad \forall \mathbf{w}_1, \mathbf{w}_2 \in \mathcal{V}^3: \quad \psi_S^{ext}(\mathbf{X},t;\mathbf{w}_1 + \mathbf{w}_2) = \psi_S^{ext}(\mathbf{X},t;\mathbf{w}_1) + \psi_S^{ext}(\mathbf{X},t;\mathbf{w}_2). \end{aligned} \tag{4.13}$$

Let us notice that for either $\alpha=0$, or $\mathbf{w}=0$, or $\alpha>0$, $\mathbf{w}\neq 0$ the relation (4.13) follows directly from the definition (4.12). Therefore, we confine our interest to the case $\alpha<0$ and $\mathbf{w}\neq 0$. We have

$$\begin{aligned} \psi_S^{ext}(\mathbf{X},t;\alpha\,\mathbf{w}) &= \psi_S^{ext}(\mathbf{X},t;-|\alpha|\,\mathbf{w}) = |\alpha|\,\psi_S^{ext}(\mathbf{X},t;-\mathbf{w}) = \\ &= -|\alpha|\,\psi_S^{ext}(\mathbf{X},t;\mathbf{w}) = \alpha\,\psi_S^{ext}(\mathbf{X},t;\mathbf{w}), \end{aligned}$$

which proves 1).

In the case of linearly dependent vectors $\mathbf{w}_1$, $\mathbf{w}_2$, the property 2) reduces to 1). Therefore, now we assume $\mathbf{w}_1$ and $\mathbf{w}_2$ to be linearly independent. Let

$$\mathbf{w}_3 = -(\mathbf{w}_1 + \mathbf{w}_2). \tag{4.14}$$

Let us consider a triangular block $\mathcal{P}_\delta \in \mathcal{A}$, $\mathcal{P}_\delta \subset \mathcal{P}$, containing $\mathbf{X}$, with the faces $\mathcal{S}_1$, $\mathcal{S}_2$, $\mathcal{S}_3$ normal to $\mathbf{w}_1$, $\mathbf{w}_2$, $\mathbf{w}_3$, respectively, and the two parallel end triangles $\mathcal{S}_4$ and $\mathcal{S}_5$ apart by the distance δ (Fig. 4.2.). Let ε be the height of the triangles $\mathcal{S}_4$ and $\mathcal{S}_5$, and A_i, $i=1,2,3$ be the areas of $\mathcal{S}_i$. From the construction of the block we have

$$\frac{A_1}{|\mathbf{w}_1|} = \frac{A_2}{|\mathbf{w}_2|} = \frac{A_3}{|\mathbf{w}_3|}. \tag{4.15}$$

The balance equation, written for $\mathcal{P}_\delta$, yields

$$\begin{aligned} \frac{1}{\varepsilon}\int_{\mathcal{P}_\delta}\left(\frac{\partial\varphi}{\partial t} - \psi_V - \varphi^*\right) dV - \frac{1}{\varepsilon}\int_{\mathcal{S}_4\cup\mathcal{S}_5} \psi_S(\mathbf{X},t;\mathbf{N})\,dS = \\ = \frac{1}{\varepsilon}\sum_{i=1}^{3}\int_{\mathcal{S}_i} \psi_S\left(\mathbf{X},t;\frac{\mathbf{w}_i}{|\mathbf{w}_i|}\right) dS. \end{aligned}$$

It is easy to see that $V(\mathcal{P}_\delta)$ and A_4, A_5 are of the order ε^2 (Fig. 4.2.), whereas A_i, $i=1,2,3$ is of order ε [compare (4.15)]. Hence, we obtain

$$\lim_{\varepsilon\to 0}\sum_{i=1}^{3}\frac{1}{\varepsilon}\int_{\mathcal{S}_i}\psi_S\left(\mathbf{X},t;\frac{\mathbf{w}_i}{|\mathbf{w}_i|}\right) dS = 0. \tag{4.16}$$

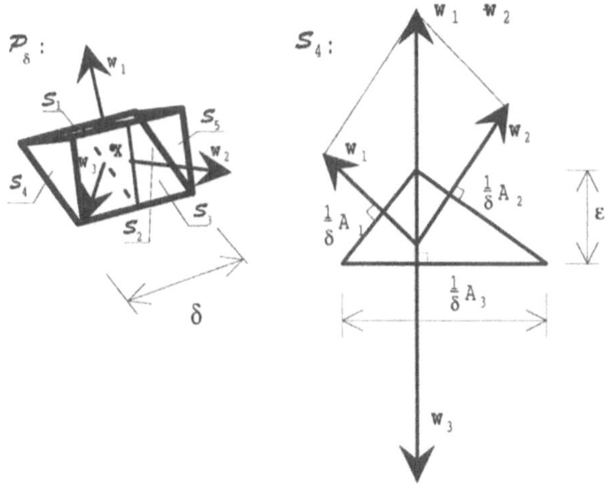

Fig. 4.2. *Triangular block used in the proof of Cauchy's Theorem*

Now let us apply the mean value theorem to the above relation. We have

$$\lim_{\varepsilon\to 0}\sum_{i=1}^{3}\frac{1}{\varepsilon}A_i\psi_S\left(\mathbf{X}^{(i)},t;\frac{\mathbf{w}_i}{|\mathbf{w}_i|}\right)=0,$$

where $\mathbf{X}^{(i)}\in \mathcal{S}_i$. Bearing (4.15) in mind, we finally arrive at

$$\lim_{\varepsilon\to 0}\sum_{i=1}^{3}|\mathbf{w}_i|\psi_S\left(\mathbf{X}^{(i)},t;\frac{\mathbf{w}_i}{|\mathbf{w}_i|}\right)=0.$$

According to Cauchy's Lemma and the relation (4.14), we obtain the assertion 2) and, consequently, the theorem is proved ■

Bearing the above theorem in mind, we write the global balance equation in the following final form:

$$\forall \mathcal{P}\in\mathcal{A}:\quad \frac{d}{dt}\int_{\mathcal{P}}\varphi(\mathbf{X},t)d\mathrm{V}=\oint_{\partial\mathcal{P}}\psi(\mathbf{X},t)\mathbf{N}(\mathbf{X})d\mathrm{S}+ \\ +\int_{\mathcal{P}}\psi_V(\mathbf{X},t)d\mathrm{V}+\int_{\mathcal{P}}\varphi^*(\mathbf{X},t)d\mathrm{V}. \tag{4.17}$$

4.3 Local Balance Equations

The formula (4.17) does not expose local properties of the field φ smearing out its discontinuities appearing on sets of zero volume measure. For this reason, we discuss separately subbodies not containing such discontinuities, and those which do contain orientable surfaces across which φ suffers a finite discontinuity.

Let us begin with the first case. We consider a descending family of subbodies $\{\mathcal{P}_i\}_{i=1}^{\infty} \subset \mathcal{A}$ on which φ, ψ, ψ_V, φ* are continuous such that

$$\bigcap_{i=1}^{\infty} \mathcal{P}_i = \{\mathbf{X}\}, \quad i > j \Rightarrow \mathcal{P}_i \wedge \mathcal{P}_j = \mathcal{P}_i, \tag{4.18}$$

{**X**} being a one-point set consisting of the chosen material point. We can write (4.17) in the following form:

$$\forall \mathcal{P}_i: \quad \frac{1}{V(\mathcal{P}_i)} \int_{\mathcal{P}_i} \left[\frac{\partial \varphi}{\partial t} - \operatorname{Div} \psi - \psi_V - \varphi^* \right] dV = 0, \tag{4.19}$$

where the Stokes (divergence) Theorem was applied to the surface integral. Div denotes the divergence operator with respect to **X**.

Taking the limit in (4.19) over the descending family of bodies, we obtain

$$\frac{\partial \varphi}{\partial t} = \operatorname{Div} \psi + \psi_V + \varphi^*. \tag{4.20}$$

This relation is called the **local balance equation** for the field φ at its **regular point** $\mathbf{X} \in \mathcal{B}_0$.

Now let us turn our attention to subbodies containing points of discontinuity of the field φ. We limit consideration to the cases in which these points form an orientable two-dimensional differentiable manifold $\mathcal{S}_0(t)$, given by the equation

$$G(\mathbf{X}, t) = 0, \quad \mathbf{X} \in \mathcal{B}_0. \tag{4.21}$$

The unit normal vector **N** of this surface is then given by the relation

$$\mathbf{N} = \frac{\operatorname{Grad} G}{|\operatorname{Grad} G|} = \mathbf{N}(\mathbf{X}, t), \quad \operatorname{Grad} G \equiv \frac{\partial G}{\partial \mathbf{X}}, \tag{4.22}$$

and the velocity in the direction of this normal vector, the so-called **speed of propagation**, has the form

$$U = -\frac{\dfrac{\partial G}{\partial t}}{|\operatorname{Grad} G|}. \tag{4.23}$$

Furthermore, we do not consider any effects connected with the gliding of the surface $\mathcal{S}_0(t)$ in the tangential direction. Hence, the formula (4.23) gives the only non-trivial component of the surface velocity. Let us notice that this velocity is measured in relation to the material points, as the surface $\mathcal{S}_0(t)$ represents the image of a real surface from the configuration space, obtained by the projection using the inverse function of the deformation: $\mathbf{f}^{-1}(\mathbf{x},t)$. Later we will return to the spatial description of this surface and its motion in $\mathbf{f}(\mathcal{B}_0,t) \subset \Re^3$.

Again let us introduce a descending family of subbodies $\{\mathcal{P}_i\}_{i=1}^{\infty} \subset \mathcal{A}$, such that (see: Figure 4.3.)

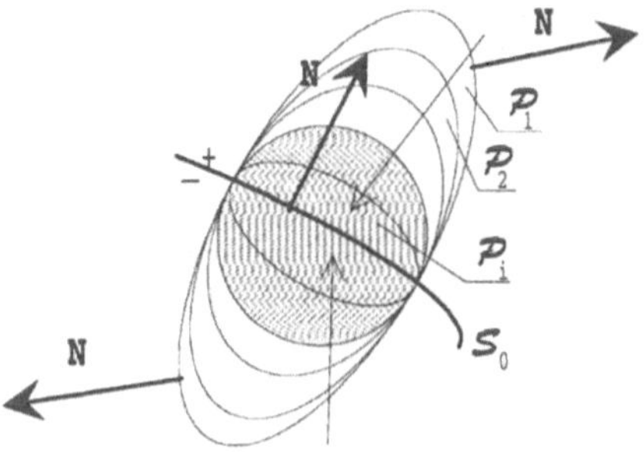

Fig. 4.3. *Family of subbodies descending to the singular surface* $\mathcal{S}_0(t)$

$$\forall\, i,j:\ \mathcal{P}_i \cap \mathcal{S}_0(t) = \mathcal{P}_j \cap \mathcal{S}_0(t),\quad \bigcap_{i=1}^{\infty} \mathcal{P}_i = \mathcal{P}_1 \cap \mathcal{S}_0(t). \tag{4.24}$$

The balance equation (4.17) can be written in the form

$$\begin{aligned}\forall \mathcal{P}_i:\ \frac{d}{dt}\left\{\int_{\mathcal{P}_i^+} \varphi(\mathbf{X},t)\,dV + \int_{\mathcal{P}_i^-} \varphi(\mathbf{X},t)\,dV\right\} = \\ = \int_{\partial\mathcal{P}_i^+ \setminus \mathcal{S}_0(t)} \psi(\mathbf{X},t)\,\mathbf{N}(\mathbf{X},t)\,dS + \int_{\partial\mathcal{P}_i^- \setminus \mathcal{S}_0(t)} \psi(\mathbf{X},t)\,\mathbf{N}(\mathbf{X},t)\,dS + \\ + \int_{\mathcal{P}_i} \left[\psi_V(\mathbf{X},t) + \varphi^*(\mathbf{X},t)\right] dV,\end{aligned} \tag{4.25}$$

where $\mathcal{P}_i^+$ and $\mathcal{P}_i^-$ denote the subbodies lying instantaneously at the positive and negative sides of $\mathcal{S}_0(t)$, respectively, and satisfying the condition

$$\mathcal{P}_i^+ \vee \mathcal{P}_i^- = \mathcal{P}_i \quad \mathcal{P}_i^+ \wedge \mathcal{P}_i^- = \varnothing. \tag{4.26}$$

The left-hand side of the relation (4.25) yields

$$\begin{aligned}\frac{d}{dt}\left\{\int_{\mathcal{P}_i^+} \varphi\,dV + \int_{\mathcal{P}_i^-} \varphi\,dV\right\} = \int_{\mathcal{P}_i} \frac{\partial \varphi}{\partial t}\,dV + \int_{\mathcal{P}_i \cap \mathcal{S}_0(t)} \left(\varphi^+ U \mathbf{N}\right)\cdot(-\mathbf{N})\,dS + \\ + \int_{\mathcal{P}_i \cap \mathcal{S}_0(t)} \left(\varphi^- U \mathbf{N}\right)\cdot \mathbf{N}\,dS.\end{aligned} \tag{4.27}$$

We assume that the limits φ^+ and φ^- of the field φ, calculated at the positive and negative side of the surface $\mathcal{S}_0(t)$, are finite.

Exercise 4.1. Prove the relation (4.27)•

After substituting (4.27) in (4.25), and taking the limit over the whole descending family of subbodies, we obtain

$$\int_{\mathcal{P}_1 \cap \mathcal{S}_0(t)} \left(-[[\varphi]]U - [[\psi]]\mathbf{N}\right) dS = 0, \tag{4.28}$$

where the boundedness of $\frac{\partial \varphi}{\partial t}$, ψ_V, and φ^* was assumed. Furthermore, we introduce the following notation:

$$[[\varphi]] = \varphi^+ - \varphi^-, \quad [[\psi]] = \psi^+ - \psi^-. \tag{4.29}$$

If $[[\varphi]]$, $[[\psi]]$, and U are continuous on $\mathcal{S}_0(t)$, then (4.28) yields the **Kotchine's condition**:

$$[[\varphi]]U + [[\psi]]N = 0, \tag{4.30}$$

which is also called the **local balance equation on a singular surface (the jump condition, the dynamic compatibility condition)**.

4.4 Local Conservation Laws

Four quantities play a particular role for thermomechanical processes owing to the fact that they satisfy the balance equations without the production terms φ^*. These quantities are mass, momentum, angular momentum, and energy. The balance equations of these quantities are called **conservation laws**. They are so called owing to the conservation property of the body, say $\mathcal{B}_{is}$, which does not interact in any way with the external world. We say that the body $\mathcal{B}_{is}$ is **isolated**, and this means that neither a flux through the boundary $\partial\mathcal{B}_{is}$, nor a supply is possible. Then we have

$\frac{d}{dt}\int_{\mathcal{B}_{is}} \rho_0 \, dV = 0$	**- Mass conservation,**
$\frac{d}{dt}\int_{\mathcal{B}_{is}} \rho_0 \mathbf{v} \, dV = 0$	**- Momentum conservation,**
$\frac{d}{dt}\int_{\mathcal{B}_{is}} \rho_0 (\mathbf{x} - \mathbf{x}_0) \times \mathbf{v} \, dV = 0$	**- Angular momentum (moment of momentum) conservation,**
$\frac{d}{dt}\int_{\mathcal{B}_{is}} \rho_0 \left(\frac{1}{2}v^2 + \varepsilon\right) dV = 0$	**- Energy conservation,**

(4.31)

where ρ_0 is the **mass density** in the reference configuration $\mathcal{B}_{is}$, ε denotes the **specific internal energy**, and $\mathbf{x}_0$ is an arbitrary reference point of the configuration space. The angular momentum is assumed to be identical to the moment of momentum, which restricts consideration to the so-called **non-polar** continua. Such continua do not possess any additional mechanical degrees of freedom apart from a translation of material points. The extension of the moment of momentum to more general models is not difficult, but we shall not consider it in this book.

The reference system Φ, in which the conservation laws for an isolated system have the form (4.31), is called **inertial**. Later we will show the structure of conservation laws in arbitrary reference systems $\Phi \in \Sigma$. One can, for instance, expect that in a rotating system the change of momentum should not vanish, e.g. centrifugal forces can be expected on the right-hand side of the relation $(4.31)_2$. Indeed, this is the case, as we will see in the following.

If the subbody $\mathcal{P}$ is not isolated, then the global conservation laws following from (4.17) contain surface and volume interactions with the external world. These laws have the following form for the four quantities mentioned above:

$$\begin{aligned}
&\frac{d}{dt}\int_{\mathcal{P}} \rho_0 \, dV = 0,\\
&\frac{d}{dt}\int_{\mathcal{P}} \rho_0 \mathbf{v} \, dV = \oint_{\partial\mathcal{P}} \mathbf{PN} \, dS + \int_{\mathcal{P}} \rho_0 \mathbf{b} \, dV,\\
&\frac{d}{dt}\int_{\mathcal{P}} \rho_0 (\mathbf{x}-\mathbf{x}_0)\times \mathbf{v} \, dV = \oint_{\partial\mathcal{P}} (\mathbf{x}-\mathbf{x}_0)\times(\mathbf{PN}) \, dS + \int_{\mathcal{P}} \rho_0 (\mathbf{x}-\mathbf{x}_0)\times \mathbf{b} \, dV,\\
&\frac{d}{dt}\int_{\mathcal{P}} \rho_0 \left(\tfrac{1}{2}\mathrm{v}^2 + \varepsilon\right) dV = \oint_{\partial\mathcal{P}} (\mathbf{v}\cdot\mathbf{PN} - \mathbf{Q}\cdot\mathbf{N}) \, dS + + \int_{\mathcal{P}} \rho_0 (\mathbf{v}\cdot\mathbf{b} + \mathrm{r}) \, dV,
\end{aligned} \tag{4.32}$$

where the flux of momentum $\mathbf{t}_N \equiv \mathbf{PN}$, the tensor of the second rank $\mathbf{P}$, $\mathbf{b}$, $\mathbf{Q}$, and r are called **traction, the first Piola-Kirchhoff stress tensor, body forces, heat flux vector,** and **supply of energy**, respectively.

Bearing the local relations (4.20) and (4.30) in mind, we easily arrive at the following **local conservation laws**:

(4.33) - at any regular point $\mathbf{X} \in \mathcal{B}_0$	(4.34) -at any point $\mathbf{X}$ of a singular surface (the so-called **Rankine-Hugoniot conditions**)
$\frac{\partial \rho_0}{\partial t} = 0,$	$U[[\rho_0]] = 0,$
$\rho_0 \mathbf{a} = \mathrm{Div}\,\mathbf{P} + \rho_0 \mathbf{b},$	$U[[\rho_0 \mathbf{v}]] + [[\mathbf{P}]]\mathbf{N} = 0,$
$\mathbf{FP}^T = \mathbf{PF}^T,$	identity,
$\rho_0 \frac{\partial}{\partial t}\left(\frac{1}{2}\mathrm{v}^2 + \varepsilon\right) = \mathrm{Div}\left(\mathbf{P}^T\mathbf{v} - \mathbf{Q}\right) +$ $+\rho_0 \mathbf{b}\cdot\mathbf{v} + \rho_0 \mathrm{r},$	$U\left[\left[\rho_0\left(\frac{1}{2}\mathrm{v}^2 + \varepsilon\right)\right]\right] + \left[\left[\mathbf{P}^T\mathbf{v} - \mathbf{Q}\right]\right]\cdot\mathbf{N} = 0,$
(4.39) $\frac{\partial \mathbf{F}}{\partial t}(\mathbf{X},t) = \mathrm{Grad}\,\mathbf{v}(\mathbf{X},t),$	(4.42) $U[[\mathbf{F}]] + [[\mathbf{v}]]\otimes\mathbf{N} = 0,$

provided the motion **f** is continuous everywhere.

Exercise 4.2. Prove the relation $(4.33)_3$•

Let us add a few remarks on the physical interpretation of the equations above. The equation $(4.33)_1$ means, certainly, that mass cannot be produced at any material point $\mathbf{X} \in \mathcal{B}_0$. However, the mass may be an arbitrary differentiable function of $\mathbf{X}$ (the heterogeneity of the material). The equation $(4.33)_2$ is often referred to as **Cauchy's first law**, whereas the equation $(4.33)_3$ is called **Cauchy's second law**. As we see further on, the latter describes the symmetry of the so-called Cauchy stress tensor which means that the mechanical reactions on arbitrary cross-sections (contact surfaces) are described by six components of the second rank tensor in the case of non-polar materials.

The equation $(4.33)_4$ can also be written in the form

$$\rho_0 \frac{\partial \varepsilon}{\partial t} + \mathbf{v} \cdot \left(\rho_0 \frac{\partial \mathbf{v}}{\partial t} \right) = \operatorname{tr}\left(\mathbf{P}\dot{\mathbf{F}}^T\right) - \operatorname{Div}\mathbf{Q} + \rho_0 r + \mathbf{v} \cdot \left(\operatorname{Div}\mathbf{P} + \rho_0 \mathbf{b}\right), \tag{4.35}$$

and bearing $(4.33)_2$ in mind, we obtain

$$\rho_0 \frac{\partial \varepsilon}{\partial t} + \operatorname{Div}\mathbf{Q} = \operatorname{tr}\left(\mathbf{P}\dot{\mathbf{F}}^T\right) + \rho_0 r. \tag{4.36}$$

This relation is sometimes called the **balance equation of internal energy**. In the classical thermodynamics, it is identical to the first law of thermodynamics for thermomechanical processes. The term $\operatorname{tr}\left(\mathbf{P}\dot{\mathbf{F}}^T\right)$ describes the **power of stresses**, and as a result of its presence, the internal energy, in contrast to the total energy, is not a conserved quantity.

In the case $U \equiv 0$, the surface $\mathcal{S}_0$ is obviously immobile with respect to the material. Consequently, it is a material surface. The most important example of such a surface is the boundary of a body. For such surfaces the equation $(4.34)_2$ reduces to

$$[[\mathbf{P}]]\mathbf{N} = 0, \tag{4.37}$$

i.e. the traction must be continuous. This relation is called **Poisson's condition**, and it forms the basis for the so-called **stress boundary conditions.**

On the other hand, the equation $(3.34)_3$ together with (4.37) implies

$$(\mathbf{PN}) \cdot [[\mathbf{v}]] - [[\mathbf{Q}]] \cdot \mathbf{N} = 0. \tag{4.38}$$

To exploit this relation, we must return to the relations (2.7) and (3.1) defining the deformation gradient **F** and the velocity **v**. The existence of the map $\mathbf{f}(\cdot,\cdot)$ requires then the following integrability conditions:

$$\frac{\partial \mathbf{F}}{\partial t}(\mathbf{X}, t) = \operatorname{Grad} \mathbf{v}(\mathbf{X}, t), \tag{4.39}$$

whose spatial counterpart is given by (3.21). It is useful to see the relation (4.39) as a **balance equation for the deformation gradient F**. Using the coordinate systems we have

$$\frac{\partial F^k{}_\alpha}{\partial t} = \frac{\partial v^k}{\partial X^\alpha} = \frac{\partial}{\partial X^\beta}\left(v^k \delta^\beta_\alpha\right). \tag{4.40}$$

Hence, the quantity $v^k \delta^\beta{}_\alpha \mathbf{e}_k \otimes \mathbf{e}_\alpha \otimes \mathbf{e}_\beta \equiv \mathbf{v} \otimes \mathbf{1}$ plays the role of the flux for **F**. Integrating (4.39) over an arbitrary subbody $\mathcal{P}$ we have

$$\frac{d}{dt}\int_{\mathcal{P}} \mathbf{F}\,d\mathrm{V} = \oint_{\partial\mathcal{P}} \mathbf{v} \otimes \mathbf{N}\,d\mathrm{S}. \tag{4.41}$$

This global balance law for **F** should hold, of course, also in the case of subbodies containing singular points of **F**. Consequently, we can write the following **jump condition** for **F**:

$$-\mathrm{U}[[\mathbf{F}]] = [[\mathbf{v}]] \otimes \mathbf{N}. \tag{4.42}$$

This equation is sometimes called the **kinematic compatibility condition** for a singular surface. It shows that for a given singular surface, the discontinuity of the velocity determines uniquely the discontinuity of the deformation gradient. This property is of fundamental importance for the theory of the so-called shock waves (strong discontinuity waves). We shall also use it frequently in this book.

For the material surface $\mathrm{U} \equiv 0$, the equation (4.42) implies

$$[[\mathbf{v}]] = 0. \tag{4.43}$$

Hence, the relation (4.38) reduces to:

$$[[\mathbf{Q}]] \cdot \mathbf{N} = 0, \tag{4.44}$$

i.e. the normal component of the heat flux must be continuous across the material surface. The relation (4.44) is called the **Fourier condition**, and it forms the basis for the boundary conditions in terms of the temperature gradient within the frame of the classical theory of heat conduction.

The jump condition for the energy $(4.34)_4$ plays a very important role in the thermodynamic theories of phase transitions. Let us denote

$$\langle \mathbf{PN} \rangle \equiv \tfrac{1}{2}\left(\mathbf{P}^+ + \mathbf{P}^-\right)\mathbf{N}. \tag{4.45}$$

Then the jump conditions yield the following relation for the jump of the internal energy:

$$\mathrm{U}\left(\rho_0[[\varepsilon]] - \langle \mathbf{PN} \rangle \cdot [[\mathbf{F}]]\mathbf{N}\right) = [[\mathbf{Q}]] \cdot \mathbf{N}. \tag{4.46}$$

This equation can be used as an equation of motion for an interface owing to the relation between the jump of the internal energy and the so-called latent heat of the phase transformation.

Exercise 4.3. Prove the relation (4.46)•

4.5 Spatial Form of Balance Equations

The global balance equation (4.17) can be transformed easily to the spatial (Eulerian) description. To this aim, the integration must be carried out over the spatial images of subbodies

$$\forall \mathcal{P} \in \mathcal{A}: \quad \mathcal{P}_t \equiv \mathbf{f}(\mathcal{P}, t), \tag{4.47}$$

at an arbitrary instant of time $t \in \mathcal{T}$, and we have to transform the volume and the surface measures into their spatial counterparts. Certainly, we have

$$d\mathrm{v} = J\,d\mathrm{V}, \qquad \mathbf{n}\,d\mathrm{s} = J\mathbf{F}^{-T}\mathbf{N}\,d\mathrm{S}, \tag{4.48}$$

where the considerations of the remark 2.1. were used.

Exercise 4.4. Let Ξ_a, a=1,2, be the parametrization of a material surface $\partial\mathcal{P}$ such that

$$\forall \mathbf{X} \in \partial\mathcal{P}: \ \mathbf{X} = \Phi(\Xi_1, \Xi_2), \qquad \mathbf{T}_a = \frac{\partial \Phi}{\partial \Xi_a}, \quad \mathbf{T}_a \cdot \mathbf{T}_b = \delta_{ab}.$$

Show that the following transformation rule for the oriented surface elements holds

$$\mathbf{N}\,d\mathrm{S} \equiv (\mathbf{T}_1 d\Xi_1) \times (\mathbf{T}_2 d\Xi_2) \Rightarrow \mathbf{n}\,d\mathrm{s} \equiv (\mathbf{t}_1 d\Xi_1) \times (\mathbf{t}_2 d\Xi_2) = J\mathbf{F}^{-T}\mathbf{N}\,d\mathrm{S},$$
$$\mathbf{t}_a \equiv \mathbf{F}\mathbf{T}_a \ \bullet$$

Bearing this in mind, we have

$$\forall \mathcal{P}_t \subset \Re^3 \ \text{s.t.}\ \mathbf{f}^{-1}(\mathcal{P}_t, t) \in \mathcal{A}:$$
$$\frac{d}{dt}\int_{\mathcal{P}_t} \varphi_t\,d\mathrm{v} = \oint_{\partial\mathcal{P}_t} \psi_t\,\mathbf{n}\,d\mathrm{s} + \int_{\mathcal{P}_t} \psi_{Vt}\,d\mathrm{v} + \int_{\mathcal{P}_t} \varphi_t^*\,d\mathrm{v}, \tag{4.49}$$

where

$$\varphi_t \equiv J^{-1}\varphi(\mathbf{X},t)\Big|_{\mathbf{X}=\mathbf{f}^{-1}(\mathbf{x},t)}, \quad \psi_{Vt} \equiv J^{-1}\psi_V(\mathbf{X},t)\Big|_{\mathbf{X}=\mathbf{f}^{-1}(\mathbf{x},t)},$$
$$\psi_t \equiv J^{-1}\psi\,\mathbf{F}^T(\mathbf{X},t)\Big|_{\mathbf{X}=\mathbf{f}^{-1}(\mathbf{x},t)}, \quad \varphi_t^* \equiv J^{-1}\varphi^*(\mathbf{X},t)\Big|_{\mathbf{X}=\mathbf{f}^{-1}(\mathbf{x},t)}. \tag{4.50}$$

In the case of positions $\mathbf{x}$ of regular points $\mathbf{X} \in \mathcal{B}_0$, we reduce the equation (4.49) to its local form in the following manner. The differentiation on the left-hand side yields

$$\int_{\mathcal{P}_t} \frac{\partial}{\partial t}(\varphi_t)\,d\mathrm{v} + \oint_{\partial\mathcal{P}_t} \varphi_t\,\mathbf{v}\cdot\mathbf{n}\,d\mathrm{s} = \oint_{\partial\mathcal{P}_t} \psi_t\,\mathbf{n}\,d\mathrm{s} + \int_{\mathcal{P}_t} \psi_{Vt}\,d\mathrm{v} + \int_{\mathcal{P}_t} \varphi_t^*\,d\mathrm{v}.$$

Now Stokes (divergence) theorem yields

$$\int_{\mathcal{P}_t} \left\{ \frac{\partial}{\partial t}(\varphi_t) + \mathrm{div}[\varphi_t \mathbf{v} - \psi_t] - \psi_{Vt} - \varphi_t^* \right\} d\mathrm{v} = 0.$$

This equation holds for the images of all subbodies for a given instant of time t. Hence, the argument of descending families of subbodies similar to that used in Sect. 4.3. leads to the local balance equation in a generic point $\mathbf{x} \in \mathcal{B}_t$,

$$\frac{\partial}{\partial t}(\varphi_t) + \mathrm{div}(\varphi_t \mathbf{v} - \psi_t) = \psi_{Vt} + \varphi_t^*. \tag{4.51}$$

It can be seen that the flux contains an additional part $\varphi_t \mathbf{v}$, which is called **convective**, whose appearance follows from the motion of the boundary $\partial\mathcal{P}_t$.

We proceed to discuss the form of the balance equation in points of discontinuity. Again, we assume that these points form a two-dimensional differentiable and orientable manifold $\mathcal{S}_t$ which is given by the equation

$$g(\mathbf{x},t)=0, \qquad \mathbf{x}\in\mathcal{B}_t. \tag{4.52}$$

Its normal velocity in a chosen reference system and the field of unit normal vectors are given by the relations

$$c=-\frac{\frac{\partial g}{\partial t}}{|\operatorname{grad} g|}, \quad \mathbf{n}=\frac{\operatorname{grad} g}{|\operatorname{grad} g|}, \tag{4.53}$$

grad denoting the gradient of g with respect to $\mathbf{x}$.

Again, we introduce a descending family of subbodies $\left\{\mathcal{P}_t^i\right\}_{i=1}^{\infty}$ at the instant of time t with the property [compare (4.18)]

$$\forall i,j:\ \mathcal{P}_t^i\cap\mathcal{S}_t=\mathcal{P}_t^j\cap\mathcal{S}_t, \quad \bigcap_{i=1}^{\infty}\mathcal{P}_t^i=\mathcal{P}_t^1\cap\mathcal{S}_t. \tag{4.54}$$

Then

$$\frac{d}{dt}\int_{\mathcal{P}_t^i}\varphi_t\,dv=\int_{\mathcal{P}_t^i}\frac{\partial}{\partial t}\varphi_t\,dv+\int_{\partial\mathcal{P}_t^{i-}}\varphi_t\mathbf{w}\cdot\mathbf{n}\,ds+\int_{\partial\mathcal{P}_t^{i+}}\varphi_t\mathbf{w}\cdot\mathbf{n}\,ds, \tag{4.55}$$

where $\mathcal{P}_t^{i-}, \mathcal{P}_t^{i+}$ denote the parts of $\mathcal{P}_t^i$ lying on the negative and positive sides of the surface $\mathcal{S}_t$, respectively. The velocity $\mathbf{w}$ of the points on the surfaces $\partial\mathcal{P}_t^{i-}$ and $\partial\mathcal{P}_t^{i+}$ are given by the relations

$$\mathbf{w}=\begin{cases}c\mathbf{n} \text{ for } \mathbf{x}\in\partial\mathcal{P}_t^{i-}\cap\mathcal{S}_t,\\ \mathbf{v} \text{ for } \mathbf{x}\in\partial\mathcal{P}_t^{i-}\setminus\mathcal{S}_t,\end{cases} \qquad \mathbf{w}=\begin{cases}-c\mathbf{n} \text{ for } \mathbf{x}\in\partial\mathcal{P}_t^{i+}\cap\mathcal{S}_t,\\ \mathbf{v} \text{ for } \mathbf{x}\in\partial\mathcal{P}_t^{i+}\setminus\mathcal{S}_t.\end{cases} \tag{4.56}$$

In these relations, $\mathbf{v}$ denotes the velocity of material points whose instantaneous position $\mathbf{x}$ belongs to $\partial\mathcal{P}_t$.

Substitution of (4.55) in (4.49) yields

$$\begin{aligned}\int_{\mathcal{P}_t^i}\left[\frac{\partial\varphi_t}{\partial t}-\psi_{Vt}-\varphi_t^*\right]dv= &\int_{\mathcal{P}_t^{i-}\setminus\mathcal{S}_t}(-\varphi_t\mathbf{v}\cdot\mathbf{n}+\psi\mathbf{n})\,ds+\\ &+\int_{\mathcal{P}_t^{i+}\setminus\mathcal{S}_t}(-\varphi_t\mathbf{v}\cdot\mathbf{n}+\psi\mathbf{n})\,ds+\int_{\mathcal{P}_t^i\cap\mathcal{S}_t}\left(\varphi_t^+c-\varphi_t^-c\right)ds,\end{aligned} \tag{4.57}$$

where φ_t^+, φ_t^- denote the limits of φ_t on $\mathcal{S}_t$.

Taking the limit over the descending family in (4.57) under the assumption of boundedness of the integrand in the volume integral of (4.57), we obtain

$$\int_{\mathcal{P}_t^1\cap\mathcal{S}_t}[[-\varphi_t(\mathbf{v}\cdot\mathbf{n}-c)+\psi\mathbf{n}]]\,ds=0, \tag{4.58}$$

where

$$[[\cdot]] \equiv (\cdot)^+ - (\cdot)^- . \tag{4.59}$$

For the continuous integrand on $\mathcal{S}_t$ we arrive at **Kotchine's condition** in the current configuration

$$\left[\left[\varphi_t (\mathbf{v} \cdot \mathbf{n} - c) - \psi\, \mathbf{n}\right]\right] = 0. \tag{4.60}$$

Let us notice that the quantity ($\mathbf{v}\cdot\mathbf{n}$-c) can be, in general, discontinuous. It describes the speed of material points instantaneously entering the singular surface, say $\mathbf{v}^-\cdot\mathbf{n}$-c, or leaving this surface, say $\mathbf{v}^+\cdot\mathbf{n}$-c, relative to the observer moving with this surface. Taken with the opposite sign, they are called the **local speeds of propagation** of the singular surface.

4.6 Spatial Form of Local Conservation Laws

Now we apply the results of the previous section to the quantities satisfying the conservation laws.

In the case of mass conservation we have

$$J^{-1}\rho_0\big|_{X=f^{-1}(x,t)} = \rho, \quad \psi_t \equiv 0, \quad \psi_{Vt} \equiv 0, \quad \varphi_t^* \equiv 0, \tag{4.61}$$

where $\rho=\rho(\mathbf{x},t)$ is mass density in the current configuration. Substitution in (4.51) yields

$$\frac{\partial \rho}{\partial t} + \operatorname{div}(\rho \mathbf{v}) = 0. \tag{4.62}$$

Exercise 4.5. Prove the identity

$$\dot{J} = J \operatorname{div} \mathbf{v}.$$

Answer:

$$\dot{J} = \operatorname{tr}\left(\frac{\partial J}{\partial \mathbf{F}} \dot{\mathbf{F}}^T\right) = \operatorname{tr}\left(J\mathbf{F}^{-T}\dot{\mathbf{F}}^T\right) = J \operatorname{tr} \mathbf{L} = J \operatorname{div} \mathbf{v} \;\bullet$$

Obviously, the relation $(4.61)_1$ can be considered as the general solution to the equation (4.62). It means that in the case of the deformation gradient $\mathbf{F}$ given as the solution to the other field equations describing the fields of the model, the mass density ρ follows immediately from (4.61), and the equation (4.62) does not have to be included in the set of the governing equations. Such a situation appears in the Lagrangian description of solids.

This is, however, not the case if we use the Eulerian description. Then the initial configuration is not the reference configuration. Consequently, neither $\mathbf{F}$ nor J follow from the field equations, and the equation (4.62) must be included in the set of the governing equations in order to specify ρ. We return to this problem later, discussing the examples of various field equations.

As the counterpart of the jump condition $(4.34)_1$ we obtain from Kotchine's condition (4.60) the following relation for the mass density ρ:

$$[[\rho(\mathbf{v}\cdot\mathbf{n}-c)]]=0. \tag{4.63}$$

This means that the mass flux (the amount of mass per unit surface and unit time) entering the singular surface must leave it on the other side, i.e. there is no mass sink on the surface. Taken with the opposite sign, the relation above describes the amount of mass swept through a running singular surface in relation to the motion of the body behind and ahead of the surface. According to (4.63), they must be equal.We denote this amount by m, and deduce

$$m\equiv\rho^{+}\left(c-\mathbf{v}^{+}\cdot\mathbf{n}\right)=\rho^{-}\left(c-\mathbf{v}^{-}\cdot\mathbf{n}\right). \tag{4.64}$$

In the particular case of a **material surface**

$$c=\mathbf{v}^{+}\cdot\mathbf{n}=\mathbf{v}^{-}\cdot\mathbf{n}. \tag{4.65}$$

The relation (4.63) is, of course, identically satisfied.

On the other hand, the continuity of mass density ρ yields

$$[[\rho]]=0 \quad\Rightarrow\quad [[\mathbf{v}]]\cdot\mathbf{n}=0. \tag{4.66}$$

This means that a jump of mass density is necessary for a discontinuity of the normal component of the velocity of material points. This conclusion has a particular bearing on the theory of shock waves. In such case we have

$$c-\langle\mathbf{v}\rangle\cdot\mathbf{n}=\frac{\langle\rho\rangle}{[[\rho]]}[[\mathbf{v}]]\cdot\mathbf{n},\qquad \langle\rho\rangle\equiv\tfrac{1}{2}\left(\rho^{+}+\rho^{-}\right),\quad \langle\mathbf{v}\rangle\equiv\tfrac{1}{2}\left(\mathbf{v}^{+}+\mathbf{v}^{-}\right), \tag{4.67}$$

i.e. the discontinuities of mass density and velocity determine the relative speed of propagation c of a singular surface with respect to the mean velocity of material points instantaneously located on the singular surface.

Now we turn our attention to the momentum balance. We have

$$\begin{aligned}\varphi_t&\equiv J^{-1}\rho_0\mathbf{v}\big|_{X=f^{-1}(x,t)}=\rho\mathbf{v}(\mathbf{x},t),\qquad \psi_t=J^{-1}\mathbf{P}\mathbf{F}^{T}\big|_{X=f^{-1}(x,t)}\equiv\mathbf{T},\\ \psi_{Vt}&\equiv J^{-1}\rho_0\mathbf{b}\big|_{X=f^{-1}(x,t)}=\rho\mathbf{b}(\mathbf{x},t),\qquad \varphi_t^{*}\equiv 0.\end{aligned} \tag{4.68}$$

Hence, according to (4.51)

$$\frac{\partial(\rho\mathbf{v})}{\partial t}+\operatorname{div}(\rho\mathbf{v}\otimes\mathbf{v}-\mathbf{T})=\rho\mathbf{b}. \tag{4.69}$$

The second rank tensor $\mathbf{T}$ related to the first Piola-Kirchhoff tensor $\mathbf{P}$ by the relation $(4.68)_2$ is called the **Cauchy stress tensor**. In contrast to $\mathbf{P}$, it defines the momentum flux per unit area of a surface in its *current* configuration rather than in the reference configuration. This flux

$$\mathbf{t}_n=\mathbf{T}\mathbf{n}, \tag{4.70}$$

is called the **traction** in the current configuration (Eulerian description).

According to the relation $(4.33)_3$, we have

$$\mathbf{T}=\mathbf{T}^{T}, \tag{4.71}$$

i.e. the Cauchy stress tensor is symmetric.

Exercise 4.6. Bearing in mind the global moment of momentum conservation law in the Eulerian description

$$\frac{d}{dt}\int_{\mathcal{P}_t}(\mathbf{x}-\mathbf{x}_0)\times(\rho\mathbf{v})\,dv=\oint_{\partial\mathcal{P}_t}(\mathbf{x}-\mathbf{x}_0)\times(\mathbf{Tn})\,ds+\int_{\mathcal{P}_t}(\mathbf{x}-\mathbf{x}_0)\times(\rho\mathbf{b})\,dv, \tag{4.72}$$

prove the symmetry condition (4.71)•

Example 4.1. *Stresses in the Homogeneous Extension of a Prism*
To see clearly the difference between the Piola-Kirchhoff and Cauchy stress tensors, let us consider again the prism in the homogeneous extension (compare Example 2.1.). If the stretch λ is not too big, we can assume that this deformation is sustained by the homogeneous tractions t applied on the upper and lower faces of the prism,

$$\mathbf{t}=\mathrm{t}\,\mathbf{n},\quad \mathbf{n}=\pm\mathbf{e}_3 \quad\Rightarrow\quad \mathbf{e}_3\cdot\mathbf{T}\mathbf{e}_3=\mathrm{t}, \tag{4.73}$$

where the implication follows from the continuity of the traction across the faces. We shall discuss this problem in the following.

On the other hand, the total resultant force acting on the prism in the $\mathbf{e}_3$-direction has the value

$$\mathrm{K}=\mathrm{A}\,\mathrm{t}, \tag{4.74}$$

where A is the current area of the loaded face. This area is related to its initial value A_0 by the relation [compare (2.46)]

$$\mathrm{A}_0=\left(\sqrt{\lambda}\right)^2\mathrm{A}. \tag{4.75}$$

Consequently,

$$\mathrm{K}=\frac{1}{\lambda}\mathrm{A}_0\mathrm{t} \tag{4.76}$$

and the traction referred to the initial configuration becomes

$$\mathbf{t}_N=\mathrm{t}_0\mathbf{e}_3=\frac{\mathrm{K}}{\mathrm{A}_0}\mathbf{e}_3=\frac{1}{\lambda}\mathrm{t}\,\mathbf{e}_3. \tag{4.77}$$

Hence,

$$\mathbf{e}_3\cdot\mathbf{P}\mathbf{e}_3=\mathbf{e}_3\cdot\mathbf{t}_N=\frac{1}{\lambda}\mathbf{e}_3\cdot\mathbf{T}\mathbf{e}_3,\qquad \mathbf{e}_3\equiv\mathbf{e}_3. \tag{4.78}$$

The component of the Cauchy stress, appearing on the right-hand side, is the quantity which is measured in the experiment because it refers to the current configuration. On the other hand, the component of the Piola-Kirchhoff stress refers to the geometry of the reference configuration, and in the case of extension (λ>1) it can be considerably smaller than the corresponding component of the Cauchy stress. Solely for small deformations, the stretch λ can be taken to be approximately equal to the one in the relation (4.78). In such cases, there is generally no difference between both stress tensors because the influence of the rotation tensor on **F** in the relation $(4.68)_2$ can also be neglected ■

On a singular surface Kotchine's condition yields the following relation for the momentum conservation

$$[[\rho \mathbf{v}(\mathbf{v}\cdot\mathbf{n}-c)]]-[[\mathbf{T}]]\mathbf{n}=0, \tag{4.79}$$

or bearing in mind the continuity of the mass flux (4.64),

$$m[[\mathbf{v}]]+[[\mathbf{T}]]\mathbf{n}=0. \tag{4.80}$$

Hence, on a non-material singular surface the jump of traction [[**T**]]**n** is necessary and sufficient for the appearance of a jump of velocity. For instance, in an ideal gas, the Cauchy stress tensor reduces to the pressure and the mass balance together with the momentum balance above on a singular surface yield the following relations for jumps of the velocity and mass density as functions of the pressure jump

$$\begin{gathered} m[[\mathbf{v}]]=[[p]]\mathbf{n} \quad\Rightarrow\quad m[[\mathbf{v}\cdot\mathbf{n}]]\equiv m[[\mathbf{v}\cdot\mathbf{n}-c]]\equiv m^2\left[\left[\frac{1}{\rho}\right]\right]=[[p]] \quad\Rightarrow \\ \Rightarrow\quad [[\rho]]=\frac{(\rho^+)^2[[p]]}{\rho^+[[p]]-m^2}, \qquad \mathbf{T}\equiv -p\mathbf{1}. \end{gathered} \tag{4.81}$$

These relations form the basis for the classical theory of shock waves in gases.

In the case of a material surface, the condition (4.80) reduces to the counterpart of Poisson's condition in the current configuration

$$c=\mathbf{v}\cdot\mathbf{n} \quad\Rightarrow\quad [[\mathbf{T}]]\mathbf{n}\equiv[[\mathbf{t}_n]]=0. \tag{4.82}$$

This relation determines two most important mechanical properties of the continuum. First of all, it shows that the mechanical interactions between two subbodies, say $\mathcal{P}^-$ and $\mathcal{P}^+$, which interact through a common part of the boundary surface (the surface of intersection of the body, see: Fig. 4.4.) are determined solely by the tractions on this surface. The tractions on these two parts of the body must have equal magnitude and opposite signs. Consequently, if we want the interactions to be determined uniquely – and this is certainly the case in classical local thermomechanics – the body forces in the momentum conservation law cannot contribute any more to the internal interactions. They must have an external character controllable by parameters independent of the body, and it is only the stress tensor which determines the surface contributions to the momentum flux. This problem was mentioned already at the beginning of this chapter, and it shall be discussed again, together with the constitutive relations defining a material.

Secondly, Poisson's condition yields the natural boundary conditions for stresses. Namely, in the case of a material surface coinciding with the external boundary of a body, the tractions on the positive side of the surface must be identified with the density of the external surface loading $\mathbf{t}_{ext}$ which is again controlled from the external world. We have

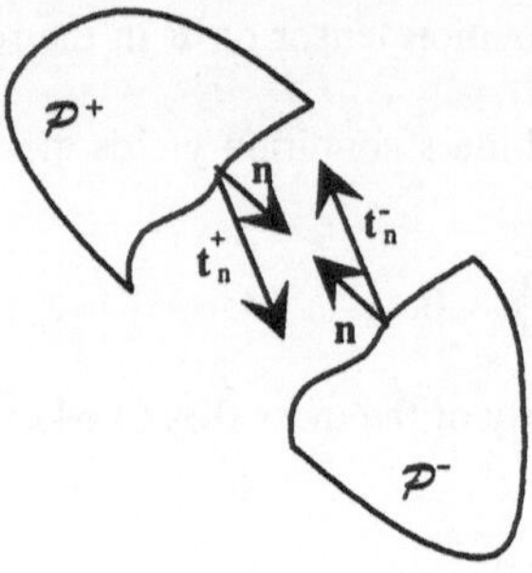

Fig. 4.4. *Mechanical interaction forces in an intersection of a body*

$$\mathbf{Tn}(\mathbf{x},t)\big|_{\mathbf{x}\in\partial\mathcal{B}_t} = \mathbf{t}_{\text{ext}}, \tag{4.83}$$

where the left-hand side is determined in the body, and the right-hand side is given as a control quantity. Owing to the uniqueness of the interaction forces, we cannot require any additional conditions for these forces at the same point of the boundary without violating the momentum conservation law. Such wrong boundary conditions, in which an additional control of the pressure in incompressible materials is required, appear sometimes in the literature.

It is easy to see that the jump condition for the moment of momentum is trivial.

We proceed to the local form of energy conservation. In this case

$$\begin{aligned} &J^{-1}\rho_0\left(\tfrac{1}{2}v^2+\varepsilon\right)\Big|_{X=f^{-1}(x,t)} = \rho\left(\tfrac{1}{2}v^2+\varepsilon\right)(\mathbf{x},t), \\ &\psi_t = \mathbf{v}\cdot\mathbf{Tn}-\mathbf{q}, \qquad \mathbf{q}=J^{-1}\mathbf{FQ}\big|_{X=f^{-1}(x,t)}, \\ &J^{-1}\rho_0(\mathbf{v}\cdot\mathbf{b}+r)\big|_{X=f^{-1}(x,t)} = \rho(\mathbf{v}\cdot\mathbf{b}+r), \quad \varphi_t^* \equiv 0, \end{aligned} \tag{4.84}$$

where $\mathbf{q}$ denotes the heat flux vector in the current configuration. Substitution in (4.17) yields

$$\frac{\partial}{\partial t}\left[\rho\left(\tfrac{1}{2}v^2+\varepsilon\right)\right]+\operatorname{div}\left[\rho\left(\tfrac{1}{2}v^2+\varepsilon\right)\mathbf{v}-\mathbf{Tv}+\mathbf{q}\right]=\rho(\mathbf{v}\cdot\mathbf{b}+r). \tag{4.85}$$

This is the conservation law of the total energy in its spatial form.

On a singular surface we obtain the following jump condition for the energy:

$$\left[\!\left[\rho\left(\tfrac{1}{2}v^2+\varepsilon\right)(\mathbf{v}\cdot\mathbf{n}-c)\right]\!\right]-[\![\mathbf{Tv}-\mathbf{q}]\!]\cdot\mathbf{n}=0, \tag{4.86}$$

or bearing in mind the continuity of the mass flux,

$$m\left[\!\left[\tfrac{1}{2}v^2+\varepsilon\right]\!\right]+[\![\mathbf{Tv}-\mathbf{q}]\!]\cdot\mathbf{n}=0. \tag{4.87}$$

On a non-material surface we can further simplify this relation using the condition (4.80). We have

$$[[\mathbf{T}\mathbf{v}]]\cdot\mathbf{n} = \langle\mathbf{v}\rangle\cdot[[\mathbf{T}]]\mathbf{n} + [[\mathbf{v}]]\cdot\langle\mathbf{T}\rangle\mathbf{n} = -m[[\tfrac{1}{2}v^2]] - \frac{1}{2m}[[(\mathbf{T}\mathbf{n})\cdot(\mathbf{T}\mathbf{n})]], \tag{4.88}$$

where the symmetry of the tensor **T** was used. Hence,

$$m[[\varepsilon]] - \frac{1}{2m}[[\mathbf{n}\cdot\mathbf{T}^2\mathbf{n}]] - [[\mathbf{q}]]\cdot\mathbf{n} = 0. \tag{4.89}$$

Some other equivalent forms of this jump condition for the internal energy appear in various applications. For instance, in the particular case of the spherical stress tensor, i.e. in the case when the stress tensor is given solely by the pressure [see:(4.81)] we have

$$m\left[\left[\varepsilon - \frac{p^2}{2m^2}\right]\right] = [[\mathbf{q}]]\cdot\mathbf{n}. \tag{4.90}$$

This condition is used to formulate the equation for the motion of an interface in the classical theory of evaporation.

Particularly important owing to its role in the conduction problems is the case of the material surface m≡0. We easily obtain the counterpart of the Fourier condition in the current configuration

$$[[\mathbf{q}]]\cdot\mathbf{n} = 0, \tag{4.91}$$

and on the boundary of the body $\mathcal{B}_t$

$$\mathbf{q}\cdot\mathbf{n}(\mathbf{x},t)\big|_{\mathbf{x}\in\partial\mathcal{B}_t} = q_{ext}, \tag{4.92}$$

where q_{ext} is the amount of heat flowing out of the body to the external world per unit surface and unit time. The role of Fourier's condition (4.91) in the determination of the energetic interactions through a surface of intersection is similar to that of Poisson's condition for tractions.

Let us summarize the results above. After easy manipulations, the conservation laws of mass, momentum, moment of momentum, and energy in their spatial representation and in the inertial reference frame can be written in the form presented in the table below.

(4.93) Balance laws in a regular point	(4.94) Balance laws in a point of a singular surface
$\dot{\rho} + \rho\,\mathrm{div}\,\mathbf{v} = 0,$	$m \equiv \rho^+(c - \mathbf{v}^+\cdot\mathbf{n}) = \rho^-(c - \mathbf{v}^-\cdot\mathbf{n}),$
$\rho\dot{\mathbf{v}} = \mathrm{div}\,\mathbf{T} + \rho\mathbf{b},$	$m[[\mathbf{v}]] + [[\mathbf{T}]]\mathbf{n} = 0,$
$\mathbf{T} = \mathbf{T}^T,$	identity,
$\rho\dot{\varepsilon} = \mathrm{tr}\,\mathbf{T}\mathbf{L}^T - \mathrm{div}\,\mathbf{q} + \rho r,$	$m[[\tfrac{1}{2}v^2 + \varepsilon]] + [[\mathbf{T}\mathbf{v} - \mathbf{q}]]\cdot\mathbf{n} = 0,$

where

$$\dot{\rho} \equiv \frac{\partial \rho}{\partial t} + (\mathbf{v} \cdot \mathrm{grad})\rho, \quad \dot{\mathbf{v}} \equiv \frac{\partial \mathbf{v}}{\partial t} + (\mathbf{v} \cdot \mathrm{grad})\mathbf{v}, \quad \dot{\varepsilon} \equiv \frac{\partial \varepsilon}{\partial t} + (\mathbf{v} \cdot \mathrm{grad})\varepsilon, \tag{4.95}$$

denote the material time derivatives (**X**=const.!) of the current mass density ρ, the velocity **v**, and the internal energy ε. The last equation $(4.93)_4$ is called the **balance law of the internal energy.** It is not a conservation law because of the contribution of the stress power tr $\mathbf{TL}^T$.

Exercise 4.7. Prove the relations (4.93)•

Exercise 4.8. An arbitrary Cauchy stress tensor **T** can be decomposed into **spherical** and **deviatoric (traceless)** parts

$$\mathbf{T} = -p\mathbf{1} + \mathbf{T}^D, \qquad p \equiv -\frac{1}{3}\mathrm{tr}\,\mathbf{T}, \quad \mathbf{T}^D \equiv \mathbf{T} + p\mathbf{1}, \tag{4.96}$$

where p is the **pressure.**

Prove **Bernoulli's Theorem**: Consider a flow with the stress **T** whose deviatoric part is identically zero and the body forces are **conservative**

$$\mathbf{b} = -\mathrm{grad}\,\Phi. \tag{4.97}$$

1. If the flow is **steady**, i.e. $\frac{\partial \mathbf{v}}{\partial t} = 0$, then

$$\mathbf{v} \cdot \mathrm{grad}\left(\frac{v^2}{2} + \Phi\right) + \frac{1}{\rho}\mathbf{v} \cdot \mathrm{grad}\,p = 0. \tag{4.98}$$

2. If the flow is steady and **irrotational**, i.e. curl **v**=0, then

$$\mathrm{grad}\left(\frac{v^2}{2} + \Phi\right) + \frac{1}{\rho}\mathrm{grad}\,p = 0 \bullet \tag{4.99}$$

Exercise 4.9. Suppose that the stress is in a hydrostatic state: **T**=-p **1** and **q**=0. Show that the following jump conditions hold

$$\left[\left[p + \frac{m^2}{\rho}\right]\right] = 0, \quad [[\mathbf{v} - (\mathbf{v} \cdot \mathbf{n})\mathbf{n}]] = 0, \quad \left[\left[\varepsilon + \frac{p}{\rho} + \frac{m^2}{2\rho^2}\right]\right] = 0 \bullet \tag{4.100}$$

4.7 Conservation Laws in a Non-Inertial Frame of Reference

The conservation laws discussed in the previous sections have been formulated in inertial reference frames.

Now we proceed to discuss their form in an arbitrary reference frame obtained by a Euclidean transformation (see: Sect. 3.3.). To this aim we need the transformation properties of quantities appearing in the conservation laws. We limit our attention to the spatial (Eulerian) description; the Lagrangian description follows easily after similar considerations.

Let $\Phi^* \in \Sigma$ be an arbitrary frame of reference. The transformation of the event from the inertial frame to this new frame is then given by the relations

$$\mathbf{x}^* = \mathbf{O}(t)(\mathbf{x} - \mathbf{x}^0) + \mathbf{c}(t), \quad t^* = t + a, \qquad \mathbf{O}^T = \mathbf{O}^{-1}, \quad a = \text{const.}, \tag{4.101}$$

[compare formula (3.46)].

The mass density ρ, the traction vector $\mathbf{t}_n$, the body force $\mathbf{b}$, the specific internal energy ε, the heat flux vector $\mathbf{q}$, and the energy supply r are all assumed to be objective, i.e.

$$\rho^* = \rho, \quad \mathbf{t}_n^* = \mathbf{O}\mathbf{t}_n, \quad \mathbf{b}^* = \mathbf{O}\mathbf{b}, \quad \varepsilon^* = \varepsilon, \quad \mathbf{q}^* = \mathbf{O}\mathbf{q}, \quad r^* = r, \tag{4.102}$$

whereas the velocity $\mathbf{v}$ and the acceleration $\mathbf{a}$ are not objective and transform according to the rules [see: (3.59), (3.61)]

$$\begin{aligned} \mathbf{O}\mathbf{v} &= \mathbf{v}^* - \Omega\left(\mathbf{x}^* - \mathbf{c}\right) - \dot{\mathbf{c}}, \\ \mathbf{O}\mathbf{a} &= \mathbf{a}^* - 2\Omega\left(\mathbf{v}^* - \dot{\mathbf{c}}\right) + \Omega^2\left(\mathbf{x}^* - \mathbf{c}\right) - \dot{\Omega}\left(\mathbf{x}^* - \mathbf{c}\right) - \ddot{\mathbf{c}}. \end{aligned} \tag{4.103}$$

The rule of transformation for the traction $\mathbf{t}_n$ yields easily the transformation rule for the Cauchy stress tensor

$$\mathbf{T}^* = \mathbf{O}\mathbf{T}\mathbf{O}^T, \tag{4.104}$$

i.e. the Cauchy stress tensor is objective as well.

Exercise 4.10. Prove the relation (4.104)•

It is also easy to see that the stress power is objective. Namely,

$$\operatorname{tr}\mathbf{T}^*\mathbf{L}^{*T} = \operatorname{tr}\mathbf{T}^*\mathbf{D}^* = \operatorname{tr}\mathbf{O}\mathbf{T}\mathbf{O}^T\mathbf{O}\mathbf{D}\mathbf{O}^T = \operatorname{tr}\mathbf{T}\mathbf{D}, \tag{4.105}$$

where the symmetry of the Cauchy stress tensor was used.

Inspection of the conservation laws (4.93) shows that the mass conservation and the internal energy conservation equations are objective whereas the momentum conservation equation requires a special structure of the body forces to be objective. Namely,

$$\begin{aligned} &\dot{\rho}^* + \rho^* \operatorname{div}^*\mathbf{v}^* = 0, \qquad \operatorname{div}^*\mathbf{v}^* \equiv \frac{\partial\, {}^*v^k}{\partial\, {}^*x^k}, \\ &\rho^*\mathbf{a}^* = \operatorname{div}^*\mathbf{T}^* + \rho^*\mathbf{b}^*_{app}, \quad \mathbf{b}^*_{app} \equiv \mathbf{b}^* + \mathbf{i}^*, \\ &\rho^*\dot{\varepsilon}^* = \operatorname{tr}\mathbf{T}^*\mathbf{D}^* - \operatorname{div}^*\mathbf{q}^* + \rho^* r^*, \end{aligned} \tag{4.106}$$

where

$$\mathbf{i}^* \equiv 2\Omega\left(\mathbf{v}^* - \dot{\mathbf{c}}\right) - \Omega^2\left(\mathbf{x}^* - \mathbf{c}\right) + \dot{\Omega}\left(\mathbf{x}^* - \mathbf{c}\right) + \ddot{\mathbf{c}}. \tag{4.107}$$

The term $\rho^*\mathbf{i}^*$ in the momentum conservation equation is called the **inertial body force** and it is, exactly in the same way as for the acceleration, not objective. The sum of the body force $\rho^*\mathbf{b}^*$ and the inertial body force is called the **apparent body force**.

With that notion, the form of the momentum conservation equation has the same form in inertial and non-inertial reference systems.

As indicated in connection with the mass conservation equation, the divergence div* is calculated with respect to the variables of the non-inertial frame.

Let us notice that in the case of Galilean transformations of inertial frames, i.e. for $\Phi^* \in \Gamma$, the inertial body force vanishes identically. Hence, for all transformations $\Phi^* \in \Sigma \backslash \Gamma$, the inertial body force $\rho^* \mathbf{i}^*$ does not depend on the choice of the inertial frame. Therefore $\rho^* \mathbf{i}^*$ depends only on the motion and on the transformation to the non-inertial frame. In other words, like the acceleration $\mathbf{a}^*$, the inertial body force is an observable quantity [see: (3.54)].

The relation $(4.106)_2$ yields, certainly,

$$\mathbf{i}^* = \mathbf{a}^* - \mathbf{O}\mathbf{a}, \tag{4.108}$$

and this means that the quantity $(\mathbf{a}^* - \mathbf{i}^*)$ is objective.

Exercise 4.11. Write down the momentum conservation law in a rotating frame with the angular velocity vector ω relative to an inertial frame•

5 Structure of Field Equations

5.1 Introductory Remarks

Let us remind ourselves of the aim of the ordinary thermodynamics of thermomechanical processes. The fields which are usually sought within such a description are either the map **f** describing the motion relative to a chosen reference configuration and the temperature T (Lagrangian description) or the mass density ρ, the velocity field **v**, and again the temperature T (Eulerian description). As we mentioned already, the Lagrangian description can be formulated equivalently for the deformation gradient **F** and the velocity field **v** instead of the map **f**. We show the list of those fields and the corresponding conservation laws in the juxtaposition in the table below.

Lagrangian description	**Eulerian description**
Variables: $\mathbf{X} \in \mathcal{B}_0,\ t \in \mathcal{T}$	Variables: $\mathbf{x} \in \mathbf{f}(\mathcal{B}_0, t),\ t \in \mathcal{T}$
Fields: **F**, **v**, T	Fields: ρ, **v**, T
Conservation laws:	
- integrability condition $\frac{\partial \mathbf{F}}{\partial t} = \operatorname{Grad} \mathbf{v},$	- mass conservation $\frac{\partial \rho}{\partial t} + \operatorname{div}(\rho \mathbf{v}) = 0,$
- momentum conservation $\rho_0 \frac{\partial \mathbf{v}}{\partial t} = \operatorname{Div} \mathbf{P} + \rho_0 \mathbf{b},$	- momentum conservation $\frac{\partial(\rho \mathbf{v})}{\partial t} + \operatorname{div}(\rho \mathbf{v} \otimes \mathbf{v} - \mathbf{T}) = \rho \mathbf{b},$
- energy conservation $\rho_0 \frac{\partial}{\partial t}\left(\frac{1}{2} v^2 + \varepsilon\right) = \operatorname{Div}\left(\mathbf{P}^T \mathbf{v} - \mathbf{Q}\right) + {} + \rho_0(\mathbf{v} \cdot \mathbf{b} + r),$	- energy conservation $\frac{\partial}{\partial t}\left[\rho\left(\frac{1}{2} v^2 + \varepsilon\right)\right] + \operatorname{div}\left[\rho\left(\frac{1}{2} v^2 + \varepsilon\right)\mathbf{v} - {} - \mathbf{T}\mathbf{v} + \mathbf{q}\right] = \rho(\mathbf{v} \cdot \mathbf{b} + r).$

Neither of the above sets of equations form a closed set of field equations for the fields appearing in this table. In the case of the Lagrangian description we have to add the relations connecting the first Piola-Kirchhoff stress tensor **P**, the internal energy ε, and the heat flux **Q** with the fields **F**, **v**, and T. In the case of the Eulerian description we have to add the relations connecting the Cauchy stress tensor **T**, the internal energy ε, and the heat flux **q** with the fields ρ, **v**, and T.

These relations, called the **constitutive equations**, turn the conservation laws into field equations. They specify the material of a system. The constitutive equations of the ordinary thermodynamics usually comprise the spatial and temporal derivatives of the fields. In the following two examples we will illustrate these properties.

Example 5.1. *Isothermal Processes in a Linearly Elastic Rod*
In order to recall the structure of a continuous model, let us first consider a one-dimensional problem of small deformations of a body described by the fields of deformation e(X,t) and velocity v(X,t), T=const., satisfying the equations

$$\frac{\partial e}{\partial t}=\frac{\partial v}{\partial X}, \qquad \rho_0\frac{\partial v}{\partial t}=\frac{\partial \sigma}{\partial X}, \qquad \rho_0=\text{const.}, \quad |e|<<1, \tag{5.1}$$

where

$$\sigma = Ee, \qquad E=\text{const.} \tag{5.2}$$

The stress σ describes the $P^{1}{}_{1}$-component of the first Piola-Kirchhoff stress tensor as well as the T^{11}-component of the Cauchy stress tensor because they are identical for small deformations. The latter are given by the $E^{1}{}_{1}$-component of the Green-St.Venant deformation tensor which is, in turn, identical for a small deformation with the e^{11}-component of the Almansi-Hamel deformation tensor, and is denoted by e. The $F^{1}{}_{1}$-component of the deformation gradient has in this case the form 1+e. The constant E in (5.2) denotes the **Young modulus.**

Instead of reducing the above system of equations to a single higher-order partial differential equation for a single unknown field, for instance the displacement, we write this system in the form of a set of first order equations

$$\frac{\partial e}{\partial t}-\frac{\partial v}{\partial X}=0, \qquad \frac{\partial v}{\partial t}-\frac{1}{\rho_0}\frac{\partial \sigma}{\partial X}=0, \qquad \frac{\partial \sigma}{\partial t}-E\frac{\partial v}{\partial X}=0, \tag{5.3}$$

which constitutes, certainly, the system of field equations for the fields e, v, and σ. The corresponding eigenvalue problem (see: Appendix C)

$$\begin{vmatrix} -\mu & -1 & 0 \\ 0 & -\mu & -\dfrac{1}{\rho_0} \\ 0 & -E & -\mu \end{vmatrix} = -\mu\left(\mu^2-\frac{E}{\rho_0}\right)=0, \tag{5.4}$$

yields the eigenvalues

$$\mu_1=0, \qquad \mu_{2,3}=\pm\sqrt{\frac{E}{\rho_0}}, \tag{5.5}$$

which are real, provided the following condition

$$E>0, \tag{5.6}$$

is fulfilled. We shall see further that this condition is connected with the so-called stability of the thermodynamical equilibrium state. The components of the eigenvectors (not normalized!) in the (e,v,σ)-field space can be written in the form of the following matrices

$$\mathbf{k}^{(1)}=(1,0,0)^T, \quad \mathbf{k}^{(2)}=\left(-\sqrt{\frac{\rho_0}{E}},1,-\rho_0\sqrt{\frac{E}{\rho_0}}\right)^T, \quad \mathbf{k}^{(3)}=\left(\sqrt{\frac{\rho_0}{E}},1,\rho_0\sqrt{\frac{E}{\rho_0}}\right)^T. \tag{5.7}$$

They are linearly independent, i.e. they span the space of solutions of the system (5.3). Hence, it follows in this case that the field equations form a **hyperbolic system** of partial differential equations (compare Appendix C). This means mathematically that the Cauchy initial value problems are well-posed. The eigenvalues μ describe the characteristic speeds $\pm\sqrt{\frac{E}{\rho_0}}$ of the propagation of disturbances along the characteristics, i.e. the sound waves in a body ■

Example 5.2. *The Heat Conducting Linearly Elastic Rod*
Again, we consider the one-dimensional problem of small deformations of a body described by the fields of deformation e(X,t), the velocity v(X,t), and the temperature T(X,t). They satisfy the following field equations

$$\frac{\partial e}{\partial t}=\frac{\partial v}{\partial X},\qquad \rho_0\frac{\partial v}{\partial t}=\frac{\partial \sigma}{\partial X},\qquad \rho_0\frac{\partial \varepsilon}{\partial t}=\sigma\frac{\partial e}{\partial t}-\frac{\partial q}{\partial X},$$
$$\rho_0=\text{const.},\quad |e|<<1 \tag{5.8}$$

where

$$\sigma=Ee,\qquad q=-K\frac{\partial T}{\partial X},\qquad \varepsilon=\varepsilon(e,T),\qquad E,K>0-\text{const.} \tag{5.9}$$

This is the simplest case of a thermoelastic material with a constant **thermal conductivity** K.

This set of equations can again be written as a set of first-order partial differential equations which are

$$\frac{\partial e}{\partial t}-\frac{\partial v}{\partial X}=0,\qquad \frac{\partial v}{\partial t}-\frac{1}{\rho_0}\frac{\partial \sigma}{\partial X}=0,\qquad \frac{\partial \sigma}{\partial t}-E\frac{\partial v}{\partial X}=0,$$
$$\frac{\partial T}{\partial t}+\frac{1}{c_V}\frac{\partial v}{\partial X}\left(\frac{\partial \varepsilon}{\partial e}-\frac{\sigma}{\rho_0}\right)+\frac{1}{\rho_0 c_V}\frac{\partial q}{\partial X}=0,\qquad K\frac{\partial T}{\partial X}+q=0,\qquad c_V\equiv\frac{\partial \varepsilon}{\partial T}, \tag{5.10}$$

where e, v, σ, T, and q are unknown fields. It is obvious that the above system is not hyperbolic owing to the singularity of the matrix of coefficients of the time derivatives. It is known that the thermal disturbances, as described by such a model, propagate with infinite speed. This can be seen easily from the last two equations in the particular case of the homogeneous velocity field. Then the combination of those two equations yields

$$\frac{\partial T}{\partial t}=\frac{K}{\rho_0 c_V}\frac{\partial^2 T}{\partial X^2}, \tag{5.11}$$

which is the simplest **parabolic heat conduction equation** ■

The above two examples show those important features of the construction of models within the continuum thermodynamics which motivate further considerations of this book.

First of all, in contrast to the ordinary thermodynamics of continua, we have constructed the above field equations in such a way that they appear as a system of first-order PDE. Within the framework of the Lagrangian description, this can be achieved by replacing the unknown field **f** by the deformation gradient **F** and the velocity field **v**, in our examples by e and v.

Secondly, we **extended** the number of unknown fields by adding σ and q to the set of fields. This was done by writing the constitutive relations (5.2), (5.9) in the form of additional differential equations. Let us notice that these equations have the form of **balance equations**: the equations $(5.3)_3$ and $(5.10)_3$ without the source term and the equation $(5.10)_5$ with the source term equal to (-q), even though they are degenerated owing to the absence of a term proportional to the proper time derivative $\left(\frac{\partial q}{\partial t}\right)$. In this way, the constitutive relations completing the set of balance equations, do not involve a dependence on the spatial gradients of the unknown fields which, in turn, allows us to write the set of field equations as the set of first-order PDE.

The above considerations expose important properties of the continuous models which are not easily visible in the usual formulation of the continuum thermodynamics. Physically, they show the existence of certain relaxation times connected with the sources in the balance equations. This is closely related to the dissipation in the macroscopical systems, and simultaneously, indicates the ways in which the models should be extended in order to account for such effects if they are not present in the theoretical description. Mathematically, this form of field equations simplifies considerably general considerations, such as the existence and stability of solutions of initial-boundary value problems, the properties of wave solutions, and the like, in particular in the case of hyperbolic systems of field equations. In addition, the system of first-order partial differential equations admits simple methods of numerical analysis for problems of practical importance. This has been demonstrated, for instance, in the analysis of one-dimensional problems of dynamical viscoplasticity [e.g.: N. CRISTESCU, I. SULICIU (1982)].

The strategy of constructing continuous models above described, in particular the extension of the set of unknown fields, is commonly known as the **EXTENDED CONTINUUM THERMODYNAMICS**. The above examples do not contain all elements of extended thermodynamics because we have not made any use of the second law of thermodynamics.

In this chapter, we discuss the procedure of constructing the field equations for the extended set of unknown fields, leaving the problem of the second law to Chap. 6 of this book.

5.2 Isotropic Functions

The examples that we presented in the previous section show that the construction of the field equations requires the constitutive relations, and these are functions on various tensor spaces. For physical reasons, such functions are supposed to satisfy certain invariance conditions, which we shall discuss further on in this book. Some properties of these functions will be needed, however, in the next section. Therefore, we present below a brief survey of the mathematical results without going into detail about the problem of their physical background at this stage.

Let us limit our attention to scalar, vector, and symmetric tensor functions ψ, $\mathbf{q}$, and $\mathbf{T}$, respectively. We assume they are objective, i.e. for any proper orthogonal tensor $\mathbf{O}(t)$ they are supposed to satisfy the following transformation rules

$$\psi^* = \psi, \quad \mathbf{q}^* = \mathbf{O}\mathbf{q}, \quad \mathbf{T}^* = \mathbf{O}\mathbf{T}\mathbf{O}^T, \tag{5.12}$$

(see: Sect. 3.3. of this book).

We start by considering the functions defined on the vector space $\mathcal{V}^3$. We say that these functions are **isotropic** if they satisfy the following conditions [see: (3.55)]

$$\forall \mathbf{w} \in \mathcal{V}^3, \quad \mathbf{w}^* = \mathbf{O}\mathbf{w} \in \mathcal{V}^3:$$
$$\psi(\mathbf{w}^*) = \psi(\mathbf{w}), \quad \mathbf{q}(\mathbf{w}^*) = \mathbf{O}\mathbf{q}(\mathbf{w}), \quad \mathbf{T}(\mathbf{w}^*) = \mathbf{O}\mathbf{T}(\mathbf{w})\mathbf{O}^{\mathrm{T}}. \tag{5.13}$$

Theorem 5.1. (*on the representation of isotropic functions of a vector variable*)
Let ψ, $\mathbf{q}$, $\mathbf{T}$ be isotropic scalar-, vector-, and symmetric tensor-valued functions of a vector variable $\mathbf{w}$. Then, it is necessary and sufficient that they have the following representations:

$$\begin{aligned} &1.\ \ \psi(\mathbf{w}) = \psi(\mathrm{w}^2), \\ &2.\ \ \mathbf{q}(\mathbf{w}) = \mathrm{q}(\mathrm{w}^2)\,\mathbf{w}, \\ &3.\ \ \mathbf{T}(\mathbf{w}) = \tau_0(\mathrm{w}^2)\,\mathbf{1} + \tau_1(\mathrm{w}^2)\,\mathbf{w}\otimes\mathbf{w}, \end{aligned} \tag{5.14}$$

where ψ, q, τ_0, τ_1 are arbitrary scalar functions.

Proof: The sufficiency is trivial. Now we show the necessity.
In order to prove 1. we show that

$$\forall \mathbf{w},\mathbf{u} \in \mathcal{V}^3: \quad \mathbf{w}\cdot\mathbf{w} = \mathbf{u}\cdot\mathbf{u} \;\Rightarrow\; \psi(\mathbf{w}) = \psi(\mathbf{u}). \tag{5.15}$$

According to the requirement, the vectors $\mathbf{u}$ and $\mathbf{w}$ are of the same length, which means that there exists a rotation described by an orthogonal tensor $\mathbf{O}$ such that $\mathbf{u}=\mathbf{O}\mathbf{w}$. Hence

$$\psi(\mathbf{u}) = \psi(\mathbf{O}\mathbf{w}) = \psi(\mathbf{w}), \tag{5.16}$$

where the definition $(5.13)_1$ of the isotropic scalar function was used.
For the isotropic vector function $\mathbf{q}(\mathbf{w})$ we have

$$\forall \mathbf{O}: \mathbf{w} = 0 \;\Rightarrow\; \mathbf{q}(0) = \mathbf{O}\mathbf{q}(0) \;\Rightarrow\; \mathbf{q}(0) = 0, \tag{5.17}$$

and 2. is satisfied. Let us assume that $\mathbf{w}\neq 0$. Then

$$\mathbf{q}(\mathbf{w}) = \left[\frac{1}{\mathrm{w}^2}\mathbf{q}(\mathbf{w})\cdot\mathbf{w}\right]\mathbf{w} + \left[\mathbf{q}(\mathbf{w}) - \left(\frac{1}{\mathrm{w}^2}\mathbf{q}(\mathbf{w})\cdot\mathbf{w}\right)\mathbf{w}\right], \tag{5.18}$$

the second vector on the right-hand side being orthogonal to $\mathbf{w}$. Let us write (5.18) in a (*)-reference system. For isotropic functions, we have

$$\begin{aligned} \mathbf{q}(\mathbf{w}^*) &= \left[\frac{1}{\mathrm{w}^2}\mathbf{q}(\mathbf{O}\mathbf{w})\cdot\mathbf{O}\mathbf{w}\right]\mathbf{O}\mathbf{w} + \left[\mathbf{q}(\mathbf{O}\mathbf{w}) - \left(\frac{1}{\mathrm{w}^2}\mathbf{q}(\mathbf{O}\mathbf{w})\cdot\mathbf{O}\mathbf{w}\right)\mathbf{O}\mathbf{w}\right] = \\ &= \left[\frac{1}{\mathrm{w}^2}\mathbf{q}(\mathbf{w})\cdot\mathbf{w}\right]\mathbf{O}\mathbf{w} + \mathbf{O}\left[\mathbf{q}(\mathbf{w}) - \left(\frac{1}{\mathrm{w}^2}\mathbf{q}(\mathbf{w})\cdot\mathbf{w}\right)\mathbf{w}\right], \end{aligned} \tag{5.19}$$

where $(5.13)_2$ has been used. Now we consider the rotation by 180^0 about the vector $\mathbf{w}$, i.e. $\mathbf{O}\,\mathbf{w}=\mathbf{w}$. The relation (5.19) yields then

$$\mathbf{q}(\mathbf{w}) = \left[\frac{1}{\mathrm{w}^2}\mathbf{q}(\mathbf{w})\cdot\mathbf{w}\right]\mathbf{w} - \left[\mathbf{q}(\mathbf{w}) - \left(\frac{1}{\mathrm{w}^2}\mathbf{q}(\mathbf{w})\cdot\mathbf{w}\right)\mathbf{w}\right], \tag{5.20}$$

where the orthogonality of the second term in (5.18) to $\mathbf{w}$ has been accounted for. Comparison of (5.18) and (5.20) shows that the second term must vanish, hence

$$\mathbf{q}(\mathbf{w}) = \left(\frac{1}{\mathrm{w}^2}\mathbf{q}(\mathbf{w})\cdot\mathbf{w}\right)\mathbf{w}. \tag{5.21}$$

The coefficient of **w** is, certainly, an objective scalar. Hence, bearing 1. in mind, we obtain 2..

In order to prove 3. we introduce the following vector function:

$$\mathbf{h}(\mathbf{w}) = \mathbf{T}(\mathbf{w})\mathbf{w}. \tag{5.22}$$

It is an isotropic function. Namely,

$$\mathbf{h}(\mathbf{O}\mathbf{w}) = \mathbf{T}(\mathbf{O}\mathbf{w})\mathbf{O}\mathbf{w} = \mathbf{O}\mathbf{T}(\mathbf{w})\mathbf{O}^T\mathbf{O}\mathbf{w} = \mathbf{O}\mathbf{h}(\mathbf{w}). \tag{5.23}$$

Therefore, there exists an isotropic scalar function h such that

$$\mathbf{h}(\mathbf{w}) = \mathrm{h}(\mathrm{w}^2)\mathbf{w} \quad \Rightarrow \quad \left[\mathbf{T}(\mathbf{w}) - \mathrm{h}(\mathrm{w}^2)\mathbf{1}\right]\mathbf{w} = 0. \tag{5.24}$$

Hence, **w** is an eigenvector of **T**. Since **T** is a symmetric tensor, we can write **T** according to the spectral theorem in the following form:

$$\mathbf{T}(\mathbf{w}) = \alpha_1(\mathbf{w})\mathbf{w}\otimes\mathbf{w} + \alpha_2(\mathbf{w})\mathbf{u}^{(1)}\otimes\mathbf{u}^{(1)} + \alpha_3(\mathbf{w})\mathbf{u}^{(2)}\otimes\mathbf{u}^{(2)}, \tag{5.25}$$

where $\mathbf{u}^{(1)}$ and $\mathbf{u}^{(2)}$ are the two remaining unit eigenvectors of **T**. The three vectors $\{\mathbf{w}, \mathbf{u}^{(1)}, \mathbf{u}^{(2)}\}$ are, of course, orthogonal to each other. Let us choose a particular rotation about the vector **w** which satisfies the conditions

$$\mathbf{O}\mathbf{w} = \mathbf{w}, \quad \mathbf{O}\mathbf{u}^{(1)} = \mathbf{u}^{(2)}, \quad \mathbf{O}\mathbf{u}^{(2)} = \mathbf{u}^{(1)}. \tag{5.26}$$

Then the relation (5.25) implies for this **O**

$$\mathbf{T}(\mathbf{w}) = \alpha_1(\mathbf{w})\mathbf{w}\otimes\mathbf{w} + \alpha_2(\mathbf{w})\mathbf{u}^{(2)}\otimes\mathbf{u}^{(2)} + \alpha_3(\mathbf{w})\mathbf{u}^{(1)}\otimes\mathbf{u}^{(1)},$$

and hence

$$\alpha_2(\mathbf{w}) = \alpha_3(\mathbf{w}). \tag{5.27}$$

Therefore, we can write (5.25) as

$$\mathbf{T}(\mathbf{w}) = \tau_0(\mathbf{w})\mathbf{1} + \tau_1(\mathbf{w})\mathbf{w}\otimes\mathbf{w}, \qquad \tau_1(\mathbf{w}) \equiv \alpha_1(\mathbf{w}) - \frac{1}{\mathrm{w}^2}\alpha_2(\mathbf{w}). \tag{5.28}$$

Moreover, according to $(5.13)_3$, the coefficients τ_0 and τ_1 must be scalar isotropic functions. This completes the proof ■

In the case of functions defined on the space $\mathcal{S}_2^{sym}$ of symmetric tensors of the second rank, we say that the functions (5.12) are **isotropic** if they satisfy the following conditions:

$$\begin{gathered}\forall \mathbf{A}\in\mathcal{S}_2^{sym}, \quad \mathbf{A}^* = \mathbf{O}\mathbf{A}\mathbf{O}^T \in \mathcal{S}_2^{sym}: \\ \psi(\mathbf{A}^*) = \psi(\mathbf{A}), \quad \mathbf{q}(\mathbf{A}^*) = \mathbf{O}\mathbf{q}(\mathbf{A}), \quad \mathbf{T}(\mathbf{A}^*) = \mathbf{O}\mathbf{T}(\mathbf{A})\mathbf{O}^T.\end{gathered} \tag{5.29}$$

Then

Theorem 5.2. (RIVLIN, ERICKSEN; *on the representation of isotropic functions of a symmetric tensor variable*)

Let ψ, $\mathbf{q}$, $\mathbf{T}$ be isotropic scalar-, vector-, and symmetric tensor-valued functions of a symmetric tensor variable $\mathbf{A}$. Then it is necessary and sufficient that they have the following representations:

$$\begin{aligned} &1.\ \psi(\mathbf{A})=\psi(a^{(1)}, a^{(2)}, a^{(3)}),\\ &2.\ \mathbf{q}(\mathbf{A})=0,\\ &3.\ \mathbf{T}(\mathbf{A})=\tau_0\mathbf{1}+\tau_1\mathbf{A}+\tau_2\mathbf{A}^2, \end{aligned} \tag{5.30}$$

where ψ, τ_r, r=0,1,2 are scalar functions of the three eigenvalues $\{a^{(i)}\}$, i=1,2,3 of $\mathbf{A}$.

Proof: The sufficiency is trivial again. Next we prove the necessity. Let us start with the vector functions. According to $(5.29)_2$ we have for all orthogonal tensors

$$\mathbf{q}\left(\mathbf{OAO}^{\mathrm{T}}\right)=\mathbf{Oq}(\mathbf{A}). \tag{5.31}$$

Taking $\mathbf{O}=-\mathbf{1}$ we obtain $\mathbf{q}(\mathbf{A})=-\mathbf{q}(\mathbf{A})$ and $(5.30)_2$ follows.

For the scalar functions we have to show that whenever two tensors $\mathbf{A}$ and $\mathbf{B}$ have the same eigenvalues, say $\{a^{(i)}\}$, i=1,2,3, then $\psi(\mathbf{A})=\psi(\mathbf{B})$. Using the spectral theorem we obtain for $\mathbf{A}$ and $\mathbf{B}$

$$\mathbf{A}=\sum_{i=1}^{3}a^{(i)}\mathbf{u}^{(i)}\otimes\mathbf{u}^{(i)},\quad \mathbf{B}=\sum_{i=1}^{3}a^{(i)}\mathbf{v}^{(i)}\otimes\mathbf{v}^{(i)}, \tag{5.32}$$

where $\mathbf{u}^{(i)}$ and $\mathbf{v}^{(i)}$ are (orthonormal) eigenvectors of $\mathbf{A}$ and $\mathbf{B}$, respectively. Let us choose a transformation $\mathbf{O}$ in such a way that

$$\mathbf{u}^{(i)}=\mathbf{Ov}^{(i)}. \tag{5.33}$$

Then the relations (5.32) yield

$$\mathbf{A}=\mathbf{OBO}^{\mathrm{T}}. \tag{5.34}$$

Hence

$$\psi(\mathbf{A})=\psi\left(\mathbf{OBO}^{\mathrm{T}}\right), \tag{5.35}$$

and the isotropy condition $(5.29)_1$ proves the assertion.

In order to prove the necessity of 3., we have to show first, that any eigenvector of $\mathbf{A}$ is also an eigenvector of $\mathbf{T}(\mathbf{A})$. Let us choose $\mathbf{O}$ as a rotation by 180^0 about the eigenvector $\mathbf{u}^{(1)}$ of $\mathbf{A}$. Then

$$\mathbf{Ou}^{(1)}=\mathbf{u}^{(1)},\quad \mathbf{Ou}^{(2)}=-\mathbf{u}^{(2)},\quad \mathbf{Ou}^{(3)}=-\mathbf{u}^{(3)}. \tag{5.36}$$

Since $\mathbf{T}(\mathbf{A})$ is isotropic, we have

$$\mathbf{OT}(\mathbf{A})\mathbf{O}^{\mathrm{T}}=\mathbf{T}\left(\mathbf{OAO}^{\mathrm{T}}\right)\ \Rightarrow\ \mathbf{OT}(\mathbf{A})=\mathbf{T}(\mathbf{A})\mathbf{O}^{\mathrm{T}}, \tag{5.37}$$

where we used the obvious relation $\mathbf{A}=\mathbf{OAO}^{\mathrm{T}}$. Hence,

$$\mathbf{OT}(\mathbf{A})\mathbf{u}^{(1)} = \mathbf{T}(\mathbf{A})\mathbf{O}\mathbf{u}^{(1)} = \mathbf{T}(\mathbf{A})\mathbf{u}^{(1)}. \tag{5.38}$$

For the above chosen transformation **O** (5.38) can only be satisfied if $\mathbf{T}(\mathbf{A})\mathbf{u}^{(1)}$ points in the direction of $\mathbf{u}^{(1)}$. Hence, $\mathbf{u}^{(1)}$ is an eigenvector of **T**. For the remaining two eigenvectors the proof is similar.

Now by means of the above result we can write

$$\mathbf{T}(\mathbf{A}) = \sum_{i=1}^{3} b^{(i)}\mathbf{u}^{(i)} \otimes \mathbf{u}^{(i)}, \tag{5.39}$$

where $b^{(i)}$ are functions of **A**. We show that the formula (5.39) implies $(5.30)_3$. Firstly we suppose that the eigenvalues $a^{(i)}$ of **A** are distinct. We consider the set of equations

$$\tau_0 + a^{(i)}\tau_1 + \left(a^{(i)}\right)^2 \tau_2 = b^{(i)}, \quad i = 1,2,3. \tag{5.40}$$

Since the determinant

$$\begin{vmatrix} 1 & a^{(1)} & \left(a^{(1)}\right)^2 \\ 1 & a^{(2)} & \left(a^{(2)}\right)^2 \\ 1 & a^{(3)} & \left(a^{(3)}\right)^2 \end{vmatrix} = \left(a^{(1)} - a^{(2)}\right)\left(a^{(2)} - a^{(3)}\right)\left(a^{(3)} - a^{(1)}\right), \tag{5.41}$$

does not vanish we can solve (5.40) for τ_0, τ_1, τ_2. Substitution of (5.40) in (5.39) yields

$$\begin{aligned} \mathbf{T}(\mathbf{A}) &= \tau_0 \sum_{i=1}^{3} \mathbf{u}^{(i)} \otimes \mathbf{u}^{(i)} + \tau_1 \sum_{i=1}^{3} a^{(i)}\mathbf{u}^{(i)} \otimes \mathbf{u}^{(i)} + \tau_2 \sum_{i=1}^{3} \left(a^{(i)}\right)^2 \mathbf{u}^{(i)} \otimes \mathbf{u}^{(i)} = \\ &= \tau_0(\mathbf{A})\mathbf{1} + \tau_1(\mathbf{A})\mathbf{A} + \tau_2(\mathbf{A})\mathbf{A}^2. \end{aligned} \tag{5.42}$$

According to the *Cayley-Hamilton theorem* for tensors generated by a three-dimensional vector space, we have

$$\mathbf{A}^3 - \mathrm{I}^A \mathbf{A}^2 + \mathrm{II}^A \mathbf{A} - \mathrm{III}^A \mathbf{1} = 0, \tag{5.43}$$

[compare with formula (2.45)], where I^A, II^A, III^A are invariants of **A**. The relation (5.43) shows that the set $\{\mathbf{1}, \mathbf{A}, \mathbf{A}^2\}$ is a basis of the three linearly independent tensors $\mathbf{1}, \mathbf{A}, \mathbf{A}^2$ for the space $\mathfrak{S}_2^{sym}$. Hence, the isotropy of **T** implies that the coefficients in (5.42) are isotropic scalars

$$\tau_r\left(\mathbf{OAO}^T\right) = \tau_r(\mathbf{A}), \quad r = 0,1,2. \tag{5.44}$$

This proves 3. for distinct eigenvalues of **A**.

In a similar way, we can show for two distinct eigenvalues that 3. follows provided τ_2=0. In the case of a single eigenvalue $\mathbf{A}=a\mathbf{1}$, every vector is an eigenvector and then, according to (5.39), we have $\mathbf{T}(\mathbf{A})=\tau_0\mathbf{1}$ which is, certainly, a special case of 3. This completes the proof of the theorem ■

Exercise 5.1. Prove the relation (5.44)•

Since the eigenvalues $a^{(i)}$ are roots of the characteristic equation for **A**

$$\det(a\mathbf{1}-\mathbf{A}) = a^3 - I^A a^2 + II^A a - III^A = 0, \tag{5.45}$$

where [see:(2.45)]

$$\begin{aligned} I^A &= \operatorname{tr}\mathbf{A} = \sum_{i=1}^{3} a^{(i)}, \quad II^A = \tfrac{1}{2}\left[(\operatorname{tr}\mathbf{A})^2 - \operatorname{tr}\mathbf{A}^2\right] = a^{(1)}a^{(2)} + a^{(1)}a^{(3)} + a^{(2)}a^{(3)}, \\ III^A &= \det\mathbf{A} = \tfrac{1}{6}\left[(\operatorname{tr}\mathbf{A})^3 - 3(\operatorname{tr}\mathbf{A})\operatorname{tr}\mathbf{A}^2 + 2\operatorname{tr}\mathbf{A}^3\right] = a^{(1)}a^{(2)}a^{(3)}, \end{aligned} \tag{5.46}$$

the three sets $\{a^{(i)}\}_{i=1,2,3}$, $\{I^A, II^A, III^A\}$, and $\{\operatorname{tr}\mathbf{A}, \operatorname{tr}\mathbf{A}^2, \operatorname{tr}\mathbf{A}^3\}$ are equivalent in the sense that each of them determines uniquely the other two. Hence, we deduce from Theorem 5.2. a **Corollary** that any isotropic functions ψ, $\mathbf{q}$, $\mathbf{T}$ of a second order symmetric tensor **A** can be written in the form

$$\begin{aligned} &\psi(\mathbf{A}) = \psi\left(I^A, II^A, III^A\right), \\ &\mathbf{q}(\mathbf{A}) = 0, \\ &\mathbf{T}(\mathbf{A}) = \tau_0\mathbf{1} + \tau_1\mathbf{A} + \tau_2\mathbf{A}^2, \qquad \tau_r = \tau_r\left(I^A, II^A, III^A\right), \quad r = 0,1,2. \end{aligned} \tag{5.47}$$

In most cases of practical interest, the constitutive functions appearing in the field equations depend on more than one variable. The problem of representations of such functions usually becomes much more involved than in the two cases which we considered above.

Now we proceed to reproduce the basic results for such functions. As the domain of the functions we consider the Cartesian product of m vector spaces $\mathcal{V}^3$ and n spaces $\mathcal{S}_2^{sym}$ of symmetric tensors of the second rank

$$\mathcal{D} \equiv \underbrace{\mathcal{V}^3 \times \ldots \times \mathcal{V}^3}_{m} \times \underbrace{\mathcal{S}_2^{sym} \times \ldots \times \mathcal{S}_2^{sym}}_{n}. \tag{5.48}$$

Scalar-, vector-, and symmetric tensor-valued functions ψ, $\mathbf{q}$, $\mathbf{T}$ on $\mathcal{D}$ are called **scalar**, **vector**, and **tensor invariants** relative to a subgroup $\mathcal{G}$ of the orthogonal group $\mathcal{O}$ if

$$\begin{aligned} &\forall \vec{v} \equiv (\mathbf{v}_1, \ldots, \mathbf{v}_m) \in \underbrace{\mathcal{V}^3 \times \ldots \times \mathcal{V}^3}_{m}, \\ &\vec{A} \equiv (\mathbf{A}_1, \ldots, \mathbf{A}_n) \in \underbrace{\mathcal{S}_2^{sym} \times \ldots \times \mathcal{S}_2^{sym}}_{n}, \qquad \mathbf{G} \in \mathcal{G}: \\ &\qquad \psi\left(\mathbf{G}\vec{v}, \mathbf{G}\vec{A}\mathbf{G}^T\right) = \psi\left(\vec{v}, \vec{A}\right), \\ &\qquad \mathbf{q}\left(\mathbf{G}\vec{v}, \mathbf{G}\vec{A}\mathbf{G}^T\right) = \mathbf{G}\mathbf{q}\left(\vec{v}, \vec{A}\right), \\ &\qquad \mathbf{T}\left(\mathbf{G}\vec{v}, \mathbf{G}\vec{A}\mathbf{G}^T\right) = \mathbf{G}\mathbf{T}\left(\vec{v}, \vec{A}\right)\mathbf{G}^T, \end{aligned} \tag{5.49}$$

where

$$G\bar{v} \equiv (Gv_1, \ldots, Gv_m), \qquad G\bar{A}G^T \equiv (GA_1G^T, \ldots, GA_nG^T). \tag{5.50}$$

If $\mathcal{G}$ coincides with the orthogonal group $\mathcal{O}$, the invariants are called **isotropic** or **isotropic functions**. Otherwise, they are called **anisotropic invariants**.

It is easy to see that for arbitrary $\mathbf{u}, \mathbf{v} \in \mathcal{V}^3$, $\mathbf{A}, \mathbf{B} \in \mathcal{S}_2^{sym}$ the following functions are isotropic:

- scalar invariants

$$\mathbf{v} \cdot \mathbf{u}, \quad \det \mathbf{A}, \quad \mathrm{tr}\left(\mathbf{A}^p \mathbf{B}^q\right), \quad \left(\mathbf{A}^p \mathbf{v}\right) \cdot \left(\mathbf{B}^q \mathbf{u}\right), \tag{5.51}$$

- vector invariants

$$\mathbf{A}^p \mathbf{v}, \quad \mathbf{A}^p \mathbf{B}^q \mathbf{v}, \tag{5.52}$$

- tensor invariants

$$\mathbf{A}^p, \quad \mathbf{A}^p \mathbf{v} \otimes \mathbf{B}^q \mathbf{v}. \tag{5.53}$$

Let us denote by ς_s, ς_v, ς_t, the sets of scalar, vector, and tensor invariants, respectively. Obviously, for arbitrary scalar functions ψ, ψ_a, ψ_b, the following functions are isotropic:

$$\psi = \psi(\varsigma_s), \qquad \mathbf{q} = \sum_{\mathbf{u}_a \in \varsigma_v} \psi_a(\varsigma_s)\mathbf{u}_a, \qquad \mathbf{T} = \sum_{\mathbf{T}_b \in \varsigma_t} \psi_b(\varsigma_s)\mathbf{T}_b. \tag{5.54}$$

The representation problem is to find the smallest sets of invariants ς_s, ς_v, ς_t required to describe isotropic functions in the form (5.54). Such a ς_s-set is called the set of **basic invariants** and ς_v- and ς_t-sets are called the **generating sets (sets of generators)** for the isotropic vector- and tensor-valued functions.

A set of basic invariants ς_s is called the **integrity basis** if its elements are polynomially independent, i.e. no element of the integrity basis can be expressed as a polynomial of the other elements. Similarly the sets of ς_v and ς_t are called the integrity bases if no element of each of these sets can be expressed as a polynomial of the other elements of the same set with coefficient functions being scalar invariants.

A set of basic invariants or generators is called a **functional basis** if its elements are functionally independent. Certainly, such a set cannot be bigger than the integrity basis.

The above sets are also called polynomially or functionally **irreducible** sets.

The following four tables contain the results for the isotropic invariants in terms of functional bases*). Obviously, the representation theorems proved earlier in this section are just the simplest particular cases of those results.

In the tables below, **v** denotes a vector, **A** – a symmetric tensor of the second rank, and **W** a skew–symmetric tensor of the second rank.

---

*) These results have been obtained in a painstaking analysis over many years of work. Here we quote only a few standard reference papers on this subject.
For the functional bases: C.-C. WANG (1970).
For the integrity bases: G. F. SMITH (1965), J. M. SPENCER (1976).
For the anisotropic invariants: I-SHIH LIU (1982), G. F. SMITH (1985).

Example 5.3. EULER; *the Basic Invariants and Generators for* m *Vector Variables*
For $\mathcal{D}=\{(\mathbf{v}_1,\ldots,\mathbf{v}_m)\in \underbrace{\mathcal{V}^3\times\ldots\times\mathcal{V}^3}_{m}\}$ we have the following functional bases

$$\varsigma_s=\{\mathbf{v}_1\cdot\mathbf{v}_1,\mathbf{v}_1\cdot\mathbf{v}_2,\ldots,\mathbf{v}_{m\text{-}1}\cdot\mathbf{v}_m,\mathbf{v}_m\cdot\mathbf{v}_m\},\qquad \varsigma_v=\{\mathbf{v}_1,\ldots,\mathbf{v}_m\},$$
$$\varsigma_t=\{\mathbf{1},\mathbf{v}_1\otimes\mathbf{v}_1,\ldots,\mathbf{v}_m\otimes\mathbf{v}_m,(\mathbf{v}_1\otimes\mathbf{v}_2+\mathbf{v}_2\otimes\mathbf{v}_1),\ldots$$
$$\ldots,(\mathbf{v}_{m-1}\otimes\mathbf{v}_m+\mathbf{v}_m\otimes\mathbf{v}_{m-1})\},\qquad\blacksquare\quad(5.55)$$

Example 5.4. *The Basic Invariants and Generators for one Vector and one Symmetric Tensor Variable*

$$\varsigma_s=\{\operatorname{tr}\mathbf{A},\operatorname{tr}\mathbf{A}^2,\operatorname{tr}\mathbf{A}^3,\mathbf{v}\cdot\mathbf{v},\mathbf{v}\cdot\mathbf{A}\mathbf{v},\mathbf{v}\cdot\mathbf{A}^2\mathbf{v}\},$$
$$\varsigma_v=\{\mathbf{v},\mathbf{A}\mathbf{v},\mathbf{A}^2\mathbf{v}\},\qquad\blacksquare\quad(5.56)$$
$$\varsigma_t=\{\mathbf{1},\mathbf{A},\mathbf{A}^2,\mathbf{v}\otimes\mathbf{v},(\mathbf{A}\mathbf{v}\otimes\mathbf{v}+\mathbf{v}\otimes\mathbf{A}\mathbf{v}),\mathbf{A}\mathbf{v}\otimes\mathbf{A}\mathbf{v}\},$$

Table 5.1. *Isotropic scalar invariants*

	Invariant elements
One variable:	
$\mathbf{v}$	$\mathbf{v}\cdot\mathbf{v}$ compare: (5.14)
$\mathbf{A}$	$\operatorname{tr}\mathbf{A}$, $\operatorname{tr}\mathbf{A}^2$, $\operatorname{tr}\mathbf{A}^3$ compare: (5.46)
$\mathbf{W}$	$\operatorname{tr}\mathbf{W}^2$
Two variables:	
$\mathbf{v}_1$, $\mathbf{v}_2$	$\mathbf{v}_1\cdot\mathbf{v}_2$
$\mathbf{v}$, $\mathbf{A}$	$\mathbf{v}\cdot\mathbf{A}\mathbf{v}$, $\mathbf{v}\cdot\mathbf{A}^2\mathbf{v}$
$\mathbf{v}$, $\mathbf{W}$	$\mathbf{v}\cdot\mathbf{W}^2\mathbf{v}$
$\mathbf{A}_1$, $\mathbf{A}_2$	$\operatorname{tr}\mathbf{A}_1\mathbf{A}_2$, $\operatorname{tr}\mathbf{A}_1\mathbf{A}_2^2$, $\operatorname{tr}\mathbf{A}_1^2\mathbf{A}_2$, $\operatorname{tr}\mathbf{A}_1^2\mathbf{A}_2^2$
$\mathbf{W}_1$, $\mathbf{W}_2$	$\operatorname{tr}\mathbf{W}_1\mathbf{W}_2$
Three variables:	
$\mathbf{v}_1$, $\mathbf{v}_2$, $\mathbf{A}$	$\mathbf{v}_1\cdot\mathbf{A}\mathbf{v}_2$, $\mathbf{v}_1\cdot\mathbf{A}^2\mathbf{v}_2$
$\mathbf{v}_1$, $\mathbf{v}_2$, $\mathbf{W}$	$\mathbf{v}_1\cdot\mathbf{W}\mathbf{v}_2$, $\mathbf{v}_1\cdot\mathbf{W}^2\mathbf{v}_2$
$\mathbf{v}$, $\mathbf{A}_1$, $\mathbf{A}_2$	$\mathbf{v}\cdot\mathbf{A}_1\mathbf{A}_2\mathbf{v}$
$\mathbf{v}$, $\mathbf{W}_1$, $\mathbf{W}_2$	$\mathbf{v}\cdot\mathbf{W}_1\mathbf{W}_2\mathbf{v}$, $\mathbf{v}\cdot\mathbf{W}_1\mathbf{W}_2^2\mathbf{v}$, $\mathbf{v}\cdot\mathbf{W}_2\mathbf{W}_1^2\mathbf{v}$
$\mathbf{v}$, $\mathbf{A}$, $\mathbf{W}$	$\mathbf{v}\cdot\mathbf{W}\mathbf{A}\mathbf{v}$, $\mathbf{v}\cdot\mathbf{W}\mathbf{A}^2\mathbf{v}$, $\mathbf{v}\cdot\mathbf{W}\mathbf{A}\mathbf{W}^2\mathbf{v}$
$\mathbf{A}_1$, $\mathbf{A}_2$, $\mathbf{A}_3$	$\operatorname{tr}\mathbf{A}_1\mathbf{A}_2\mathbf{A}_3$
$\mathbf{W}_1$, $\mathbf{W}_2$, $\mathbf{W}_3$	$\operatorname{tr}\mathbf{W}_1\mathbf{W}_2\mathbf{W}_3$
$\mathbf{A}_1$, $\mathbf{A}_2$, $\mathbf{W}$	$\operatorname{tr}\mathbf{A}_1\mathbf{A}_2\mathbf{W}$, $\operatorname{tr}\mathbf{A}_1\mathbf{A}_2^2\mathbf{W}$, $\operatorname{tr}\mathbf{A}_2\mathbf{A}_1^2\mathbf{W}$, $\operatorname{tr}\mathbf{A}_1\mathbf{W}\mathbf{A}_2\mathbf{W}^2$
$\mathbf{A}$, $\mathbf{W}_1$, $\mathbf{W}_2$	$\operatorname{tr}\mathbf{A}_1\mathbf{W}_1\mathbf{W}_2$, $\operatorname{tr}\mathbf{A}\mathbf{W}_1\mathbf{W}_2^2$, $\operatorname{tr}\mathbf{A}\mathbf{W}_2\mathbf{W}_1^2$
Four variables	
$\mathbf{v}_1$, $\mathbf{v}_2$, $\mathbf{A}_1$, $\mathbf{A}_2$	$\mathbf{v}_1\cdot\mathbf{A}_1\mathbf{A}_2\mathbf{v}_2$, $\mathbf{v}_1\cdot\mathbf{A}_2\mathbf{A}_1\mathbf{v}_2$
$\mathbf{v}_1$, $\mathbf{v}_2$, $\mathbf{W}_1$, $\mathbf{W}_2$	$\mathbf{v}_1\cdot\mathbf{W}_1\mathbf{W}_2\mathbf{v}_2$, $\mathbf{v}_1\cdot\mathbf{W}_2\mathbf{W}_1\mathbf{v}_2$
$\mathbf{v}_1$, $\mathbf{v}_2$, $\mathbf{A}$, $\mathbf{W}$	$\mathbf{v}_1\cdot\mathbf{A}\mathbf{W}\mathbf{v}_2$, $\mathbf{v}_1\cdot\mathbf{W}\mathbf{A}\mathbf{v}_2$

Table 5.2. *Isotropic vector invariants*

	Generator elements
one variable	
$\mathbf{v}$	$\mathbf{v}$
$\mathbf{A}$ or $\mathbf{W}$	0
two variables	
$\mathbf{v}, \mathbf{A}$	$\mathbf{Av}, \mathbf{A}^2\mathbf{v}$
$\mathbf{v}, \mathbf{W}$	$\mathbf{Wv}, \mathbf{W}^2\mathbf{v}$
three variables	
$\mathbf{v}, \mathbf{A}_1, \mathbf{A}_2$	$\mathbf{A}_1\mathbf{A}_2\mathbf{v}, \mathbf{A}_2\mathbf{A}_1\mathbf{v}$
$\mathbf{v}, \mathbf{W}_1, \mathbf{W}_2$	$\mathbf{W}_1\mathbf{W}_2\mathbf{v}, \mathbf{W}_2\mathbf{W}_1\mathbf{v}$
$\mathbf{v}, \mathbf{A}, \mathbf{W}$	$\mathbf{AWv}, \mathbf{WAv}$

Table 5.3. *Isotropic symmetric tensor invariants*

	Generator elements
No variable	
0	$\mathbf{1}$
One variable	
$\mathbf{v}$	$\mathbf{v}\otimes\mathbf{v}$
$\mathbf{A}$	$\mathbf{A}, \mathbf{A}^2$
$\mathbf{W}$	$\mathbf{W}^2$
Two variables	
$\mathbf{v}_1, \mathbf{v}_2$	$\mathbf{v}_1\otimes\mathbf{v}_2+\mathbf{v}_2\otimes\mathbf{v}_1$
$\mathbf{v}, \mathbf{A}$	$\mathbf{v}\otimes\mathbf{Av}+\mathbf{Av}\otimes\mathbf{v}, \mathbf{Av}\otimes\mathbf{Av}$
$\mathbf{v}, \mathbf{W}$	$\mathbf{v}\otimes\mathbf{Wv}+\mathbf{Wv}\otimes\mathbf{v}, \mathbf{Wv}\otimes\mathbf{Wv},$ $\mathbf{Wv}\otimes\mathbf{W}^2\mathbf{v}+\mathbf{W}^2\mathbf{v}\otimes\mathbf{Wv}$
$\mathbf{A}_1, \mathbf{A}_2$	$\mathbf{A}_1\mathbf{A}_2+\mathbf{A}_2\mathbf{A}_1, \mathbf{A}_1\mathbf{A}_2\mathbf{A}_1, \mathbf{A}_2\mathbf{A}_1\mathbf{A}_2$
$\mathbf{W}_1, \mathbf{W}_2$	$\mathbf{W}_1\mathbf{W}_2+\mathbf{W}_2\mathbf{W}_1, \mathbf{W}_1\mathbf{W}_2^2-\mathbf{W}_2^2\mathbf{W}_1, \mathbf{W}_1^2\mathbf{W}_2-\mathbf{W}_2\mathbf{W}_1^2$
$\mathbf{A}, \mathbf{W}$	$\mathbf{AW}-\mathbf{WA}, \mathbf{WAW}, \mathbf{A}^2\mathbf{W}-\mathbf{WA}^2, \mathbf{WAW}^2-\mathbf{W}^2\mathbf{AW}$

Table 5.4. *Isotropic skew-symmetric tensor invariants*

	Generator elements
One variable	
$\mathbf{v}$ or $\mathbf{A}$	0
$\mathbf{W}$	$\mathbf{W}$
Two variables	
$\mathbf{v}_1, \mathbf{v}_2$	$\mathbf{v}_1\otimes\mathbf{v}_2-\mathbf{v}_2\otimes\mathbf{v}_1$
$\mathbf{v}, \mathbf{A}$	$\mathbf{v}\otimes\mathbf{Av}-\mathbf{Av}\otimes\mathbf{v}$, $\mathbf{v}\otimes\mathbf{A}^2\mathbf{v}-\mathbf{A}^2\mathbf{v}\otimes\mathbf{v}$, $\mathbf{Av}\otimes\mathbf{A}^2\mathbf{v}-\mathbf{A}^2\mathbf{v}\otimes\mathbf{Av}$
$\mathbf{v}, \mathbf{W}$	$\mathbf{v}\otimes\mathbf{Wv}-\mathbf{Wv}\otimes\mathbf{v}$, $\mathbf{v}\otimes\mathbf{W}^2\mathbf{v}-\mathbf{W}^2\mathbf{v}\otimes\mathbf{v}$
$\mathbf{A}_1, \mathbf{A}_2$	$\mathbf{A}_1\mathbf{A}_2-\mathbf{A}_2\mathbf{A}_1$, $\mathbf{A}_1\mathbf{A}_2^2-\mathbf{A}_2^2\mathbf{A}_1$, $\mathbf{A}_1^2\mathbf{A}_2-\mathbf{A}_2\mathbf{A}_1^2$, $\mathbf{A}_1\mathbf{A}_2\mathbf{A}_1^2-\mathbf{A}_1^2\mathbf{A}_2\mathbf{A}_1$, $\mathbf{A}_2\mathbf{A}_1\mathbf{A}_2^2-\mathbf{A}_2^2\mathbf{A}_1\mathbf{A}_2$
$\mathbf{W}_1, \mathbf{W}_2$	$\mathbf{W}_1\mathbf{W}_2-\mathbf{W}_2\mathbf{W}_1$
$\mathbf{A}, \mathbf{W}$	$\mathbf{AW}+\mathbf{WA}$, $\mathbf{AW}^2-\mathbf{W}^2\mathbf{A}$
Three variables	
$\mathbf{v}_1, \mathbf{v}_2, \mathbf{A}$	$\mathbf{v}_1\otimes\mathbf{Av}_2-\mathbf{Av}_2\otimes\mathbf{v}_1$, $\mathbf{v}_2\otimes\mathbf{Av}_1-\mathbf{Av}_1\otimes\mathbf{v}_2$
$\mathbf{v}_1, \mathbf{v}_2, \mathbf{W}$	$\mathbf{v}_1\otimes\mathbf{Wv}_2-\mathbf{Wv}_2\otimes\mathbf{v}_1$, $\mathbf{v}_2\otimes\mathbf{Wv}_1-\mathbf{Wv}_1\otimes\mathbf{v}_2$
$\mathbf{v}, \mathbf{A}_1, \mathbf{A}_2$	$\mathbf{A}_1\mathbf{v}\otimes\mathbf{A}_2\mathbf{v}-\mathbf{A}_2\mathbf{v}\otimes\mathbf{A}_1\mathbf{v}$, $\mathbf{A}_1\mathbf{A}_2\mathbf{v}\otimes\mathbf{v}-\mathbf{v}\otimes\mathbf{A}_1\mathbf{A}_2\mathbf{v}$, $\mathbf{A}_2\mathbf{A}_1\mathbf{v}\otimes\mathbf{v}-\mathbf{v}\otimes\mathbf{A}_2\mathbf{A}_1\mathbf{v}$
$\mathbf{A}_1, \mathbf{A}_2, \mathbf{A}_3$	$\mathbf{A}_1\mathbf{A}_2\mathbf{A}_3-\mathbf{A}_3\mathbf{A}_2\mathbf{A}_1$, $\mathbf{A}_2\mathbf{A}_3\mathbf{A}_1-\mathbf{A}_1\mathbf{A}_3\mathbf{A}_2$, $\mathbf{A}_3\mathbf{A}_1\mathbf{A}_2-\mathbf{A}_2\mathbf{A}_1\mathbf{A}_3$

In many cases of practical interest we need only certain linear or quadratic representations. Further on in this book, we shall discuss examples of such representations. The following example shows such representations for the functions ψ, $\mathbf{q}$, $\mathbf{T}$ defined on $\mathcal{D}=\{(\mathbf{v},\mathbf{A})\}$.

Example 5.5. *The Linearization for one Vector and one Symmetric Tensor Variable*

1) The functions linearized in $\mathbf{A}$:

$$\begin{aligned}
&\psi = \psi_1 + \psi_2 \operatorname{tr}\mathbf{A} + \psi_3\, \mathbf{v}\cdot\mathbf{Av},\\
&\mathbf{q} = [q_1 + q_2 \operatorname{tr}\mathbf{A} + q_3 \mathbf{v}\cdot\mathbf{Av}]\mathbf{v} + q_4\mathbf{Av},\\
&\mathbf{T} = [\tau_1 + \tau_2 \operatorname{tr}\mathbf{A} + \tau_3\mathbf{v}\cdot\mathbf{Av}]\mathbf{1} + \tau_4\mathbf{A} + [\tau_5 + \tau_6 \operatorname{tr}\mathbf{A} + \tau_7\mathbf{v}\cdot\mathbf{Av}]\mathbf{v}\otimes\mathbf{v} +\\
&\qquad +\tau_8(\mathbf{v}\otimes\mathbf{Av} + \mathbf{Av}\otimes\mathbf{v}),
\end{aligned} \tag{5.57}$$

where the coefficients $\psi_1,\ldots,q_1,\ldots,\tau_1,\ldots,\tau_8$ are scalar functions of v^2.

2) The functions linearized in $\mathbf{v}$ and $\mathbf{A}$:

$$\begin{aligned}
&\psi = \psi_1 + \psi_2 \operatorname{tr}\mathbf{A},\\
&\mathbf{q} = q_1\mathbf{v},\\
&\mathbf{T} = (\tau_1 + \tau_2 \operatorname{tr}\mathbf{A})\mathbf{1} + \tau_3\mathbf{A},
\end{aligned} \tag{5.58}$$

where $\psi_1,\ldots,\tau_3$ are independent of **v** and **A** ■

Exercise 5.2. Consider the variables $(\mathbf{v}_1, \mathbf{v}_2, \mathbf{A})$.

1) Find the general representation of the scalar, vector, and isotropic symmetric tensor functions.

2) Find their representations which are linearized in the variables $\mathbf{v}_2$ and **A**•

Exercise 5.3. Suppose that **M**(**v**) is a third rank tensor isotropic function of a vector variable, i.e. for any orthogonal tensor **O**

$$M_{ijk}(\mathbf{O}\mathbf{v}) = O_i{}^l O_j{}^m O_k{}^n M_{lmn}(\mathbf{v}),$$

relative to a Cartesian coordinate system. Find a representation formula for **M**. [Note that if **S**(**v**,**u**) is defined by $S_{ij}=M_{ijk}u_k$, then **S** is a second rank tensor isotropic function of two vector variables]•

5.3 Galilean Invariance of Field Equations

In the following, we will focus on the Eulerian description of the motion of a body. For this reason, the remaining part of this chapter will be formulated in this description. We return to the Lagrangian description in the last chapter of this book.

As we saw in the fourth chapter, the conservation laws of mass, momentum, moment of momentum, and energy are Galilean invariant. The extension of the set of fields, as carried out in the examples of Sect. 5.1., requires additional balance equations which are usually not in the form of conservation laws any longer. These additional equations contain non-trivial source terms whose structure must be investigated. In many cases of practical bearing, the structure of these terms is motivated by microscopical considerations, for instance, based on kinetic theories. However, the main purpose of these notes is a phenomenological macroscopical approach, and such microscopical arguments cannot be presented to any considerable extent even when they are available, which is not always the case. Furthermore, we will present two brief remarks in which we expose some kinetic aspects following from the Boltzmann equation for ideal fluids and from the Boltzmann-Peierls equation for phonons. These should not be considered, however, as an immanent part of the extended thermodynamics discussed in this book.

The considerations of this section are based on a paper by T. RUGGERI (1989). Parts of the results were also published as the sixth chapter of a book by I. MÜLLER and T. RUGGERI (1993).

Let us denote the elements of the extended set of fields by $\mathbf{w} \equiv (\mathbf{v},\mathbf{u}) \in \Re^N$, where **v** denotes the velocity field and **u** denotes such fields as the mass density ρ, the Cauchy stress tensor **T**, the temperature T, the heat flux vector **q**, etc. We shall specify **u** whenever we discuss a particular model of a body. The velocity field is exposed particularly in this notation owing to its role in the invariance considerations. We have already seen that it causes particular difficulties in such invariance problems as the rotation of the observer.

For simplicity, we limit our attention to the inertial reference frames. The balance equations of mass (4.62), momentum (4.69), and energy (4.85) possess an explicit dependence on the velocity field. Namely, the mass balance contains a linear term with respect to the velocity, the momentum balance a quadratic term, the energy balance a cubic term owing to the contribution of the kinetic energy. In the extended thermodynamics all remaining quantities in these equations with the exception of the internal energy ε are fields. If we assume that the internal energy is independent of the velocity, which we will see follows from the so-called material objectivity, then the explicit dependence on the velocity determines the order of this dependence uniquely. There seems to be a certain regularity of this dependence owing to the growing tensorial rank of equations if the energy balance equation was the trace of a tensor equation of second rank. This property is indeed indicated by the kinetic theory of gases. Now we proceed to investigate this property for the general case of the balance equation. The results will be used in the equations which must be additionally formulated in models with the extended set of fields.

The general system of the **local balance laws** for the **density** $\boldsymbol{F}^0$ can then be written in the following compact form:

$$\frac{\partial \boldsymbol{F}^0}{\partial t}+\frac{\partial \boldsymbol{F}^k}{\partial x^k}=\boldsymbol{f}, \qquad \boldsymbol{F}^k \equiv \boldsymbol{F}^0 v^k + \boldsymbol{G}^k. \tag{5.59}$$

The product of the density $\boldsymbol{F}^0$ and the velocity $\mathbf{v}$ is called the **convective** part of the **flux** $\boldsymbol{F}^k$. The other part $\boldsymbol{G}^k$ is called **non-convective.**

If the objects appearing in these equations are given as functions of $\boldsymbol{w}$=(**v**,**u**)

$$\boldsymbol{F}^0=\boldsymbol{F}^0(\mathbf{v},\mathbf{u}), \quad \boldsymbol{G}^k=\boldsymbol{G}^k(\mathbf{v},\mathbf{u}), \qquad \boldsymbol{f}=\boldsymbol{f}(\mathbf{v},\mathbf{u}), \tag{5.60}$$

then the set (5.59) is supposed to form a **closed set of field equations** for (**v**,**u**). This is the famous **closure problem** of field equations and the above functions are, certainly, the **constitutive relations** defining the material. We shall extensively discuss the possible forms of these relations further on in this book. In this section it is solely the dependence on the velocity field **v** which is limited by the principle of the Galilean invariance.

Assuming that we deal with N independent components of fields

$$\mathbf{v}\in\Re^3, \quad \mathbf{u}\in\Re^{N-3}, \quad \boldsymbol{w}=\left(v^1,v^2,v^3,u^1,\dots,u^{N-3}\right), \tag{5.61}$$

we then have for any **determined** system of field equations

$$\boldsymbol{F}^0\in\Re^N, \quad \forall k: \boldsymbol{F}^k\in\Re^N, \boldsymbol{G}^k\in\Re^N, \boldsymbol{f}\in\Re^N. \tag{5.62}$$

In these notes, we skip almost entirely the problem of overdetermined systems. The only exception will be made in the section on materials with internal constraints.

Now we proceed to investigate the assumption that an **arbitrary system of field equations must be Galilean invariant.** We choose a Galilean transformation in which **O=1** and **c**(t)=**V**t as the time-independent rotations and translations have no essential influence on the structure of field equations. This means that we consider the transformation [compare (3.63)]

$$\mathbf{x}^*=\mathbf{x}+\mathbf{V}t, \quad \mathbf{V}=\text{const.}, \qquad t^*=t, \qquad \mathbf{v}^*=\mathbf{v}+\mathbf{V}. \tag{5.63}$$

According to the above assumption, we impose the requirement that the field equations

$$\frac{\partial}{\partial t}\boldsymbol{F}^0(\mathbf{v},\mathbf{u})+\frac{\partial}{\partial x^k}\left[\boldsymbol{F}^0(\mathbf{v},\mathbf{u})v^k+\boldsymbol{G}^k(\mathbf{v},\mathbf{u})\right]=\boldsymbol{f}(\mathbf{v},\mathbf{u}) \tag{5.64}$$

must have the same form in the new (*)-frame

$$\frac{\partial}{\partial t^*}F^0(v^*,u)+\frac{\partial}{\partial x^k}\left[F^0(v^*,u)v^{*k}+G^k(v^*,u)\right]=f(v^*,u),$$
$$\frac{\partial}{\partial t^*}=\frac{\partial}{\partial t}-V^k\frac{\partial}{\partial x^k},\quad \frac{\partial}{\partial x^{*k}}\equiv\frac{\partial}{\partial x^k},\quad u^*\equiv u. \tag{5.65}$$

We prove the following

Lemma. If the field equations (5.64) are invariant with respect to the Galilean transformation (5.63), then there exists a linear operator $\mathbf{x}(\mathbf{v})$ such that

$$\begin{aligned}F^0(v,u)&=\mathbf{x}(v)F^0(0,u),\\ G^k(v,u)&=\mathbf{x}(v)G^k(0,u),\\ f(v,u)&=\mathbf{x}(v)f(0,u),\end{aligned} \tag{5.66}$$

with the following properties:

$$\forall\, \mathbf{a},\mathbf{b}\in\Re^3:\quad \mathbf{x}(\mathbf{a}+\mathbf{b})=\mathbf{x}(\mathbf{a})\mathbf{x}(\mathbf{b}),\quad \mathbf{x}(0)=\mathbf{1}, \tag{5.67}$$

i.e. $\mathbf{x}$ is an exponential operator

$$\mathbf{x}(v)=\exp\left(A^k v_k\right)=\mathbf{1}+A^k v_k+\tfrac{1}{2}A^k A^l v_k v_l+\ldots, \tag{5.68}$$

A^k being the constant (N×N)-matrices such that

$$\forall\, k,l:\quad A^k A^l=A^l A^k. \tag{5.69}$$

Proof: Under the transformation (5.63), the equation (5.65) becomes

$$\frac{\partial}{\partial t}F^0(v+V,u)+\frac{\partial}{\partial x^k}\left[F^0(v+V,u)v^k+G^k(v+V,u)\right]=f(v+V,u), \tag{5.70}$$

and this must be equivalent to (5.64). This means that the terms appearing in (5.70) must be the linear combinations of the corresponding terms appearing in (5.64) with coefficients which may depend on **V**. Hence, we conclude that there exists a matrix $\mathbf{x}(\mathbf{V})$ with $\mathbf{x}(0)=\mathbf{1}$ such that

$$\begin{aligned}F^0(v+V,u)&=\mathbf{x}(V)F^0(v,u),\\ G^k(v+V,u)&=\mathbf{x}(V)G^k(v,u),\\ f(v+V,u)&=\mathbf{x}(V)f(v,u).\end{aligned} \tag{5.71}$$

Taking **v**=0, we obtain (5.66) and choosing **v**=0 and **V**=**a**+**b** in (5.71) we obtain (5.67).

The conditions (5.67) are equivalent to the Cauchy problem

$$\frac{\partial\mathbf{x}}{\partial v^k}=A^k\mathbf{x}(v)=\mathbf{x}(v)A^k,\quad A^k\equiv\frac{\partial\mathbf{x}}{\partial v^k}(0),\quad \mathbf{x}(0)=\mathbf{1}. \tag{5.72}$$

The solution to this problem gives (5.68), and the differentiation with respect to v_l in (5.72) proves (5.69) ■

Substitution of the above results in the general balance equation (5.59) yields the following Galilean invariant form:

$$\left(\hat{\boldsymbol{F}}^0\right)^{\bullet}+\frac{\partial}{\partial x^k}\hat{\boldsymbol{G}}^k+\left(v_k\boldsymbol{A}^k+\mathbf{1}\,\mathrm{div}\,\mathbf{v}\right)\hat{\boldsymbol{F}}^0+\frac{\partial v_k}{\partial x^m}\boldsymbol{A}^k\hat{\boldsymbol{G}}^m=\hat{\boldsymbol{f}}, \tag{5.73}$$

the dot denoting the material time derivative

$$(\cdot)^{\bullet}\equiv\frac{\partial}{\partial t}(\cdot)+\left(\mathbf{v}\cdot\mathrm{grad}\right)(\cdot), \tag{5.74}$$

and the hat indicates that the function should be taken for $\mathbf{v}=0$.

Exercise 5.4. Prove the relation (5.73)•

So far we have dealt with fields, fluxes, and production terms without specifying their tensorial character. To proceed further, we have to assume a certain **hierarchical** structure of the field equations (5.59). We will show in the remark (5.1) that such a structure follows, in some cases, from an appropriate kinetic theory. Within the frame of extended thermodynamics the following cases are considered:

- *five moments* with

$$\boldsymbol{F}^0=\left(\mathrm{F},\mathrm{F}_k,\mathrm{F}^m{}_m\right)^T, \tag{5.75}$$

F being the mass density ρ, F_k the momentum ρv_k and $\frac{1}{2}F^m{}_m$ the specific energy $\rho(\frac{1}{2}v^2+\varepsilon)$,

- *thirteen moments* with

$$\boldsymbol{F}^0=\left(\mathrm{F},\mathrm{F}_k,\mathrm{F}_{km},\mathrm{F}_{km}{}^m\right)^T, \tag{5.76}$$

F being the mass density ρ, F_k the momentum density ρv_k, $F_{km}=\frac{1}{3}F^p{}_p\delta_{km}+F_{<km>}$, $\frac{1}{2}F^p{}_p$ the specific energy $\rho(\frac{1}{2}v^2+\varepsilon)$, $F_{<km>}$ being connected with the deviatoric part of the Cauchy stresses and $F_{km}{}^m$ with the heat flux,

- *fourteen moments* with

$$\boldsymbol{F}^0=\left(\mathrm{F},\mathrm{F}_k,\mathrm{F}_{km},\mathrm{F}_{km}{}^m,\mathrm{F}^k{}_k{}^m{}_m\right)^T, \tag{5.77}$$

with a similar interpretation of the fields as before, and the trace $F^k{}_k{}^m{}_m$ corresponding to the trace of stresses (pressure).

Furthermore, we shall consider these models in some detail. At this stage it is important to notice that they contain fields of growing tensorial rank.

The components of $\boldsymbol{F}^0$, $\boldsymbol{G}^k$, $\boldsymbol{f}$ are scalar, vectors, symmetric tensors of second rank etc., i.e.

$$\boldsymbol{F}^0 = \begin{pmatrix} F \\ F_{k_1} \\ F_{k_1k_2} \\ \vdots \\ F_{k_1\ldots k_n} \end{pmatrix}, \qquad \boldsymbol{G}^k = \begin{pmatrix} G^k \\ G^k_{k_1} \\ G^k_{k_1k_2} \\ \vdots \\ G^k_{k_1\ldots k_n} \end{pmatrix}, \qquad \boldsymbol{f} = \begin{pmatrix} f \\ f_{k_1} \\ f_{k_1k_2} \\ \vdots \\ f_{k_1\ldots k_n} \end{pmatrix}. \tag{5.78}$$

The number n is called the **tensorial rank of the system**. The examples mentioned above correspond to n=2 for 5 moments, n=3 for 13 moments, and n=4 for 14 moments.

Each block of the tensorial field equations of rank j governs the evolution of the field $F_{k_1,k_2,\ldots,k_j}$. According to the kinetic considerations [e.g. see: I. MÜLLER, T. RUGGERI (1993)] it seems natural to **assume** that

the covariance of the block of tensorial rank j **is not influenced by the equations of tensorial character greater than** j.

This means that the matrix $\boldsymbol{x}(\mathbf{v})$ should be **subtriangular** [compare (5.71)]

$$\boldsymbol{x}(\mathbf{v}) = \begin{pmatrix} \chi & & & \\ \chi_{k_1} & \chi^{m_1}_{k_1} & & \\ \chi_{k_1k_2} & \chi^{m_1}_{k_1k_2} & \chi^{m_1m_2}_{k_1k_2} & \\ \vdots & & & \\ \chi_{k_1\ldots k_n} & \chi^{m_1}_{k_1\ldots k_n} & \ldots & \ldots\chi^{m_1\ldots m_n}_{k_1\ldots k_n} \end{pmatrix}, \tag{5.79}$$

with the block elements being fully symmetric with respect to the indices $m_1, m_2, \ldots,$ and $k_1, k_2, \ldots$.

Lemma. The matrices $\boldsymbol{A}^k$ are the subtriangular block matrices with blocks of zeros on the main diagonal. If the number of field equations is finite, they are nilpotent of degree n+1, i.e.

$$\forall k_j = 1,2,3,\ j = 1,\ldots,n+1:\quad \boldsymbol{A}^{k_1}\boldsymbol{A}^{k_2}\ldots\boldsymbol{A}^{k_{n+1}} = 0. \tag{5.80}$$

The matrix $\boldsymbol{x}(\mathbf{v})$ is then a polynomial matrix in $\mathbf{v}$ of order n.

Proof: Let us denote by $\mathcal{T}$ the set of subtriangular block matrices, and by $\mathcal{T}_\mu$ the set of triangular matrices such that the block $a_{ij}=0$ if $i<j+\mu$, $\mu\geq 0$. Then we have:

1. $\mathcal{T}_0=\mathcal{T}$,

2. $\mathcal{T}_1$ is a set of triangular block matrices with blocks of zero on the main diagonal,

3. let $\mathcal{T}_\mu\mathcal{T}_\nu \equiv \{\boldsymbol{AB} | \boldsymbol{A}\in\mathcal{T}_\mu \text{ and } \boldsymbol{B}\in\mathcal{T}_\nu\}$; then $\mathcal{T}_\mu\mathcal{T}_\nu \subset \mathcal{T}_{\mu+\nu}$,

4. a necessary and sufficient condition for a triangular matrix $\boldsymbol{A}$ to be nilpotent is that $\boldsymbol{A}\in\mathcal{T}_1$.

According to the assumption (5.79), $\boldsymbol{x}(\mathbf{v})\in\mathcal{T}$. Therefore, the relations (5.72) imply $\boldsymbol{A}^k\in\mathcal{T}$, i.e.

$$A^k = \begin{pmatrix} a^k & & & \\ a^k_{k_1} & a^{km_1}_{k_1} & & \\ a^k_{k_1k_2} & a^{km_1}_{k_1k_2} & & \\ \vdots & & & \\ a^k_{k_1k_2\ldots k_n} & a^{km_1}_{k_1k_2\ldots k_n} & \cdots & a^{km_1\ldots m_n}_{k_1k_2\ldots k_n} \end{pmatrix}. \tag{5.81}$$

Bearing (5.71) in mind, we see that $\boldsymbol{x}(\mathbf{v})$ are objective tensors, i.e. for each proper orthogonal transformation $\mathbf{O}$ we have

$$\boldsymbol{x}^* = \mathbf{O}\,\boldsymbol{x}. \tag{5.82}$$

This means that A^k must be isotropic functions.

It follows that the elements $a^{km_1\ldots m_q}_{k_1\ldots k_p}$ of A^k are zero for p+q being even numbers and a combination of Kronecker symbols otherwise. In particular, the block elements of the diagonal are zero. Bearing the definition of the sets $\mathcal{T}_\mu$ in mind, we obtain $A^{k_1} \in \mathcal{T}_1$, $A^{k_1}A^{k_2} \in \mathcal{T}_2$, etc.

We proceed to show that the matrices A^k are completely determined, and hence, we can split uniquely F^0, G^k, and f into convective (velocity-dependent) and non-convective (objective) quantities.

Since $\boldsymbol{x}(\mathbf{v}) \in \mathcal{T}$ and $A^k \in \mathcal{T}_1$ we have from (5.72) that $\partial \boldsymbol{x}(\mathbf{v}) / \partial \mathrm{v}_k \in \mathcal{T}_1$. Simultaneously $\mathcal{T}(0)=\mathbf{1}$ and, therefore, the elements of the diagonal of $\mathcal{T}(\mathbf{v})$ are blocks of Kronecker symbols

$$\chi = 1, \quad \chi^{m_1}_{k_1} = \delta^{m_1}_{k_1}, \ldots, \chi^{m_1\ldots m_n}_{k_1\ldots k_n} = \delta^{m_1}_{k_1}\ldots\delta^{m_n}_{k_n}, \tag{5.83}$$

and the remaining elements $\chi^{m_1\ldots m_q}_{k_1\ldots k_p}$, $0 \le p < q \le n$, are homogeneous polynomials of degree q-p.

Non-zero elements $a^{km_1\ldots m_q}_{k_1\ldots k_p}$ of the matrices A^k are linear combinations of Kronecker symbols with arbitrary coefficients so far. These are constrained by the symmetry of those matrices and by the commutation condition (5.69). It can be shown that they are determined uniquely by these conditions. To this aim one has to integrate the system of differential equations

$$\begin{aligned} &\frac{\partial F^0}{\partial \mathrm{v}_k} = A^k F^0, \quad F^0(0) = \hat{F}^0, \\ &\frac{\partial G^m}{\partial \mathrm{v}_k} = A^k G^m, \quad G^m(0) = \hat{G}^m, \\ &\frac{\partial f}{\partial \mathrm{v}_k} = A^k f, \quad f(0) = \hat{f}, \end{aligned} \tag{5.84}$$

whose integrabilty is guaranteed by (5.69) ■

We limit our attention to a few particular cases that are of importance for further consideration in this book.

1. In the case of a single scalar equation (n=0, N=1 in (5.62)), we have

$$\boldsymbol{A}^k = \left(0^k\right), \quad \boldsymbol{x}(\mathbf{v}) = (1) \quad \Rightarrow \quad \mathrm{F} = \hat{\mathrm{F}}, \quad \mathrm{G}^k = \hat{\mathrm{G}}^k, \quad \mathrm{f} = \hat{\mathrm{f}}. \tag{5.85}$$

In applications to thermomechanics, $\hat{\mathrm{F}}$ corresponds to the mass density and $\hat{\mathrm{G}}^k, \hat{\mathrm{f}}$ are zero. Thus,

$$\frac{\partial \hat{\mathrm{F}}}{\partial t} + \frac{\partial}{\partial x^k}\left(\hat{\mathrm{F}} v^k\right) = 0, \tag{5.86}$$

which is the mass balance equation (4.58).

2. For a system of one scalar and one vector equation (n=2, N=4) we have

$$\boldsymbol{A}^k = \begin{pmatrix} 0^k & \\ c\delta^k_{k_1} & 0^{k m_1}_{k_1} \end{pmatrix}, \quad \boldsymbol{A}^k \boldsymbol{A}^m \equiv 0, \tag{5.87}$$

and

$$\mathrm{F}_{k_1} = c v_{k_1} \hat{\mathrm{F}} + \hat{\mathrm{F}}_{k_1}, \quad \mathrm{G}^k_{k_1} = \hat{\mathrm{G}}^k_{k_1}, \quad \mathrm{f}_{k_1} = \hat{\mathrm{f}}_{k_1}, \tag{5.88}$$

where c is an arbitrary constant which can be set equal to 1. Otherwise it is only necessary to redefine the objective parts (with hats). According to the above relations, the scalar equation has the form (5.86) and the vector equation looks as follows:

$$\frac{\partial}{\partial t}\left(\hat{\mathrm{F}} v_{k_1}\right) + \frac{\partial}{\partial x^k}\left(\hat{\mathrm{F}} v_{k_1} v^k + \hat{\mathrm{G}}^k_{k_1}\right) = \hat{\mathrm{f}}_{k_1}. \tag{5.89}$$

We recognize the momentum balance equation (4.70) with $\hat{\mathrm{F}}$ being the mass density, the non-convective part of momentum $\hat{\mathrm{F}}_{k_1}$ vanishing in the case of the classical continuum, $\hat{\mathrm{G}}^k_{k_1}$ the Cauchy stress tensor with the minus sign and $\hat{\mathrm{f}}_k$ body forces.

3. In the case of an additional equation of second rank, i.e. for n=2 (N=10) we have

$$\boldsymbol{A}^k = \begin{pmatrix} 0^k & & \\ \delta^k_{k_1} & 0^{k m_1}_{k_1} & \\ 0^k_{k_1 k_2} & a^{k m_1}_{k_1 k_2} & 0^{k m_1 m_2}_{k_1 k_2} \end{pmatrix}, \tag{5.90}$$

and the symmetry implies

$$a^{k m_1}_{k_1 k_2} = a\delta_{k_1 k_2}\delta^{k m_1} + b\left(\delta^k_{k_1}\delta^{m_1}_{k_2} + \delta^k_{k_2}\delta^{m_1}_{k_1}\right), \tag{5.91}$$

with the commutation condition (5.69) satisfied identically. Hence,

$$
\begin{aligned}
F_{k_1k_2} &= \tfrac{1}{2}\hat{F}\left(av^2\delta_{k_1k_2} + 2bv_{k_1}v_{k_2}\right) + \hat{F}_{k_1k_2}, \\
G^k_{k_1k_2} &= a\hat{G}^k_{m_1}v^{m_1}\delta_{k_1k_2} + b\left(v_{k_1}\hat{G}^k_{k_2} + v_{k_2}\hat{G}^k_{k_1}\right) + \hat{G}^k_{k_1k_2}, \\
f_{k_1k_2} &= a\hat{f}_{m_1}v^{m_1}\delta_{k_1k_2} + b\left(v_{k_1}\hat{f}_{k_2} + v_{k_2}\hat{f}_{k_1}\right) + \hat{f}_{k_1k_2}.
\end{aligned}
\tag{5.92}
$$

The two arbitrary constants, a and b, are chosen conveniently by rescaling the equations. Usually these equations are written separately for the trace and the deviatoric parts. We choose $\frac{1}{2}(3a+2b)=1$ and $b=1$ by an appropriate change of the objective terms.

Exercise 5.5. Specify the trace and the deviatoric parts of the equations for n=2•

It follows that

$$
\begin{aligned}
F_{k_1k_2} &= \hat{F}v_{k_1}v_{k_2} + \hat{F}_{k_1k_2}, \\
G^k_{k_1k_2} &= v_{k_1}\hat{G}^k_{k_2} + v_{k_2}\hat{G}^k_{k_1} + \hat{G}^k_{k_1k_2}, \\
f_{k_1k_2} &= v_{k_1}\hat{f}_{k_2} + v_{k_2}\hat{f}_{k_1} + \hat{f}_{k_1k_2}.
\end{aligned}
\tag{5.93}
$$

Now the tensorial equation of second rank has the form

$$
\begin{aligned}
&\frac{\partial}{\partial t}\left(\hat{F}v_{k_1}v_{k_2} + \hat{F}_{k_1k_2}\right) + \frac{\partial}{\partial x^k}\Big[\left(\hat{F}v_{k_1}v_{k_2} + \hat{F}_{k_1k_2}\right)v^k + \\
&+ v_{k_1}\hat{G}^k_{k_2} + v_{k_2}\hat{G}^k_{k_1} + \hat{G}^k_{k_1k_2}\Big] = v_{k_1}\hat{f}_{k_2} + v_{k_2}\hat{f}_{k_1} + \hat{f}_{k_1k_2}.
\end{aligned}
\tag{5.94}
$$

The trace of this equation gives the energy balance equation (4.88) if we identify

$$
\begin{aligned}
&\hat{F} = \rho, \quad \hat{F}_{k_1}{}^{k_1} = 2\rho\varepsilon, \quad \hat{G}^k_{k_1} = -\sigma^k{}_{k_1}, \quad \mathbf{T} \equiv \sigma^k{}_{k_1}\mathbf{e}_k \otimes \mathbf{e}_{k_1}, \\
&\hat{G}^k_{k_1}{}^{k_1} = 2q^k, \quad \mathbf{q} \equiv q^k\mathbf{e}_k, \quad \hat{f}_{k_1} = \rho b_{k_1}, \quad \hat{f}_{k_1}{}^{k_1} = 2\rho r, \quad \hat{f}_{\langle k_1k_2\rangle} = 0,
\end{aligned}
\tag{5.95}
$$

$\hat{f}_{k_1}{}^{k_1}$ being the energy **supply** in contrast to the deviatoric part $\hat{f}_{\langle k_1k_2\rangle}$, which physically describes the **production** of the quantity $\hat{F}_{\langle k_1k_2\rangle}$. We will return later to the discussion of this problem.

The equation (5.94) illustrates the *purpose* of the considerations of this section. It is obvious that the balance equations, as discussed in Chap. 4, cannot predict the way in which the velocity field enters the balance equations of quantities which are not the classical conserved quantities of mass density, momentum density, moment of momentum density, and energy density. The deviatoric part of the tensor $\hat{F}_{k_1k_2}$ is such a new quantity appearing in the extended thermodynamics.

As we will show in the remark 5.1. below, the structure of the equation (5.94) can be motivated by the kinetic theory of ideal gases, but it cannot be derived from such a theory.

4. The case n=3 (N=20) requires the application of the condition (5.84). We have

$$A^k = \begin{pmatrix} 0^k & & & \\ \delta^k_{k_1} & 0^{km_1}_{k_1} & & \\ 0^k_{k_1k_2} & 2\delta^k_{(k_1}\delta^{m_1}_{k_2)} & 0^{km_1m_2}_{k_1k_2} & \\ a^k_{k_1k_2k_3} & 0^{km_1}_{k_1k_2k_3} & a^{km_1m_2}_{k_1k_2k_3} & 0^{km_1m_2}_{k_1k_2k_3} \end{pmatrix}, \tag{5.96}$$

where the parenthesis denotes the symmetric part of the tensor with respect to the corresponding indices.

According to (5.84) for the elements of the tensorial equation of third rank, we have

$$\begin{aligned} \frac{\partial F_{k_1k_2k_3}}{\partial v_k} &= \underbrace{\hat{F} a^k_{k_1k_2k_3}} + \left(\hat{F} v_{m_1} v_{m_2} + \hat{F}_{m_1m_2}\right) a^{km_1m_2}_{k_1k_2k_3}, \\ \frac{\partial G^m_{k_1k_2k_3}}{\partial v_k} &= \left(v_{m_1}\hat{G}^m_{m_2} + v_{m_2}\hat{G}^m_{m_1} + \hat{G}^m_{m_1m_2}\right) a^{km_1m_2}_{k_1k_2k_3}, \\ \frac{\partial f_{k_1k_2k_3}}{\partial v_k} &= \left(v_{m_1}\hat{f}_{m_2} + v_{m_2}\hat{f}_{m_1} + \hat{f}_{m_1m_2}\right) a^{km_1m_2}_{k_1k_2k_3}, \end{aligned} \tag{5.97}$$

and the commutation condition (5.69) is of the form

$$a^{km_1m_2}_{k_1k_2k_3} = a^{m_1km_2}_{k_1k_2k_3}. \tag{5.98}$$

Let us notice that the marked coefficient $a^k_{k_1k_2k_3}$ appears only in the first equation. Therefore, if we integrate we obtain a term proportional to $\hat{F}v_k$, which can be eliminated by means of a vectorial equation (momentum balance). Hence, we can rescale the objective terms eliminating this coefficient entirely. For this reason, without loss of generality, we can assume this coefficient to be zero.

On the other hand, bearing (5.98) in mind, and proceeding in the same manner as for n=2, we obtain

$$a^{km_1m_2}_{k_1k_2k_3} = \delta^k_{k_1}\delta^{m_1}_{k_2}\delta^{m_2}_{k_3} + \delta^k_{k_2}\delta^{m_1}_{k_3}\delta^{m_2}_{k_1} + \delta^k_{k_3}\delta^{m_1}_{k_1}\delta^{m_2}_{k_2}. \tag{5.99}$$

Finally,

$$\begin{aligned} F_{k_1k_2k_3} &= \hat{F} v_{k_1} v_{k_2} v_{k_3} + v_{k_1}\hat{F}_{k_2k_3} + v_{k_2}\hat{F}_{k_3k_1} + v_{k_3}\hat{F}_{k_1k_2} + \hat{F}_{k_1k_2k_3}, \\ G^k_{k_1k_2k_3} &= v_{k_1}v_{k_2}\hat{G}^k_{k_3} + v_{k_2}v_{k_3}\hat{G}^k_{k_1} + v_{k_3}v_{k_1}\hat{G}^k_{k_2} + \\ &\quad + v_{k_1}\hat{G}^k_{k_2k_3} + v_{k_2}\hat{G}^k_{k_3k_1} + v_{k_3}\hat{G}^k_{k_1k_2} + \hat{G}^k_{k_1k_2k_3}, \\ f_{k_1k_2k_3} &= v_{k_1}v_{k_2}\hat{f}_{k_3} + v_{k_2}v_{k_3}\hat{f}_{k_1} + v_{k_3}v_{k_1}\hat{f}_{k_2} + \hat{f}_{k_1k_2k_3}, \end{aligned} \tag{5.100}$$

and the form of the balance equation of third tensorial rank follows:

$$\frac{\partial}{\partial t}\left(F_{k_1k_2k_3}\right) + \frac{\partial}{\partial x^k}\left(F_{k_1k_2k_3}v^k + G^k_{k_1k_2k_3}\right) = f_{k_1k_2k_3}, \tag{5.101}$$

with the terms given by (5.100).

The equations of higher tensorial rank n can be derived in a similar manner. The general result can be found in the quoted paper by T. RUGGERI (1989).

Example 5.6. *The Field Equations for Thirteen Fields*
The above considerations yield in the case of 13 fields, vis.

$$\left(\rho, v^k, \sigma^{km}, q^k\right), \tag{5.102}$$

the following set of equations in a Cartesian inertial frame of reference:

$$\frac{\partial \rho}{\partial t}+\frac{\partial}{\partial x^k}\left(\rho v^k\right)=0,$$

$$\frac{\partial}{\partial t}\left(\rho v^k\right)+\frac{\partial}{\partial x^m}\left(\rho v^k v^m-\sigma^{km}\right)=\rho b^k, \tag{5.103}$$

$$\frac{\partial}{\partial t}\left(\tfrac{1}{2}\rho v^2+\rho\varepsilon\right)+\frac{\partial}{\partial x^k}\left[\left(\tfrac{1}{2}\rho v^2+\rho\varepsilon\right)v^k-v_m\sigma^{km}+q^k\right]=\rho v_k b^k+\rho r,$$

$$\frac{\partial}{\partial t}\left(\rho v_{\langle k}v_{l\rangle}+\hat{F}_{\langle kl\rangle}\right)+\frac{\partial}{\partial x^m}\left[\left(\rho v_{\langle k}v_{l\rangle}+\hat{F}_{\langle kl\rangle}\right)v^m-2v_{\langle k}\sigma^m_{l\rangle}+\hat{G}^m_{\langle kl\rangle}\right]=2\rho v_{\langle k}b_{l\rangle}+\hat{f}_{\langle kl\rangle},$$

$$\frac{\partial}{\partial t}\left[v_k\left(\tfrac{1}{2}\rho v^2+\rho\varepsilon\right)+v_m\hat{F}^m_k+\tfrac{1}{2}\hat{F}^m_{km}\right]+\frac{\partial}{\partial x^m}\left[v^n\hat{G}^m_{\langle kn\rangle}-v_k v^n\sigma^m_n-\tfrac{1}{2}v^2\sigma^m_k-\right.$$

$$\left.+2v_k q^m+\tfrac{1}{2}\hat{G}^{mn}_{kn}\right]=\rho v_k v^m b_m+\tfrac{1}{2}\rho v^2 b_k+3\rho v_k r+v^n\hat{f}_{\langle kn\rangle}+\tfrac{1}{2}\hat{f}^m_{km}. \tag{5.104}$$

Here we used the standard notation

$$\rho=\hat{F},\quad \varepsilon=\frac{1}{2\rho}\hat{F}^m{}_m,\quad \sigma^k_m=-\hat{G}^k_m,\quad q^k=\tfrac{1}{2}\hat{G}^{km}_m,\quad \rho b^k=\hat{f}^k. \tag{5.105}$$

The <·>-brackets denote the traceless (deviatoric) part of a tensor, e.g. $v_{\langle k}v_{l\rangle}=v_k v_l-\frac{1}{3}v^2\delta_{kl}$. The equation (5.94) was split into the trace part $(5.103)_3$ and the deviatoric part $(5.104)_1$. From the equation (5.101) we used only the contracted part with respect to two indices.

It is obvious that the above set becomes a set of field equations for the fields (5.102) if we specify the constitutive relations for the following quantities:

$$\varepsilon,\ \hat{F}_{\langle km\rangle},\ \hat{F}_{km}{}^m,\ \hat{G}^m_{\langle kn\rangle},\ \hat{G}^{mn}_{kn},\ \hat{f}_{\langle km\rangle},\ \hat{f}_{kn}{}^n. \tag{5.106}$$

We will return to the discussion of this problem for some specific materials after the presentation of the second law of thermodynamics, which is the subject of Chap. 6 ■

Remark 5.1. *On the Moments of the Boltzmann Equation*
The hierarchical character of the field equations of extended thermodynamics is motivated by a statistical model of an ideal gas. Such a gas is considered to be a collection of N particles of equal mass m whose microscopical state is described by the positions $\mathbf{y}_\alpha$ and the velocities $\mathbf{c}_\alpha$ of all particles, $\alpha=1,\ldots N$. The particles are assumed to move freely without any interactions except very brief encounters with each other and collisions with the walls of the container. On the microscopical level, the motion of such a system is described by a set of N Newton's equations of motion.

Boltzmann (1844–1906) introduced a density function f(**y**,**c**,t), the so-called **phase function**, which determines the number of particles at the instant of time t in a unit volume of the phase space $\mu=\{(\mathbf{y},\mathbf{c})\}$ with the centre at a chosen generic point (**y**,**c**). The phase space itself is the Cartesian product of the $\Re^3$-space of positions **y** and the $\mathcal{V}^3$-space of velocities **c**. Divided by the number of particles N, the phase function f can be considered as the **probability density** for the appearance of given positions and velocities in the system at a chosen instant of time. If the phase volume enclosing the whole system of particles at the time t is $\Omega(t)$, then we have

$$N = \int_{\Omega(t)} f(\mathbf{y}, \mathbf{c}, t)\, d\, v(\mu), \tag{5.107}$$

where $v(\mu)$ is the volume measure of the phase space μ.

It was proved by Boltzmann that the phase function f must satisfy the differential equation

$$\frac{\partial f}{\partial t} + c^k \frac{\partial f}{\partial y^k} + k^k \frac{\partial f}{\partial c^k} = \mathcal{C}(f), \tag{5.108}$$

where **k** is the density of external forces – apparent if is used a non-inertial reference frame and $\mathcal{C}(f)$ is the so-called **collision operator**. The following integrals over the space of velocities of the Boltzmann collision operator vanish identically:

$$\int_{\mathcal{V}^3} \mathcal{C}(f)\, d\mathbf{c} = 0, \quad \int_{\mathcal{V}^3} \mathbf{c}\, \mathcal{C}(f)\, d\mathbf{c} = 0, \quad \int_{\mathcal{V}^3} c^2 \mathcal{C}(f)\, d\mathbf{c} = 0. \tag{5.109}$$

We do not have to go into another details of the structure of this operator.

Multiplying the equation (5.108) by m and integrating with respect to the velocity we obtain

$$\frac{\partial}{\partial t}\left(m \int_{\mathcal{V}^3} f\, d\mathbf{c} \right) + \frac{\partial}{\partial x^k}\left(m \int_{\mathcal{V}^3} c^k f\, d\mathbf{c} \right) = 0, \tag{5.110}$$

where we have used the assumption that $\lim_{|\mathbf{c}| \to \infty} f(\mathbf{y}, \mathbf{c}, t) = 0$, i.e. the particles must have finite velocities.

We have also identified the coordinates y^k of the space of positions **y** with the Eulerian coordinates x^k.

The equation (5.110) reminds the mass conservation law. Indeed, the definition of the phase function should yield the following mass density in the current configuration:

$$\rho(\mathbf{x}, t) = m \int_{\mathcal{V}^3} f\, d\mathbf{c} \;\Rightarrow\; \int_{\Re^3} \rho\, dV = mN. \tag{5.111}$$

Hence the macroscopic momentum density $\rho\mathbf{v}$ should be defined by the following phase average:

$$\rho v^k = m \int_{\mathcal{V}^3} c^k f\, d\mathbf{c}. \tag{5.112}$$

The above considerations show that the mass conservation equation follows as the **zeroth moment** of the Boltzmann equation, i.e. obtained by integration with respect to the velocity **c** in the equation (5.108) multiplied by the velocity **c** in the zeroth power.

Now let us perform similar manipulations in the case of the Boltzmann equation multiplied by m and by the velocity **c** in the first power. To this aim it is convenient to introduce the so-called **peculiar velocity**

$$\mathbf{C} \equiv \mathbf{c} - \mathbf{v}. \tag{5.113}$$

Bearing in mind $(5.109)_2$ and (5.111), (5.112), we obtain

$$\frac{\partial}{\partial t}\left(\rho v^k\right) + \frac{\partial}{\partial x^m}\left(v^k v^m - \sigma^{km}\right) = \rho k^k, \tag{5.114}$$

where

$$\sigma^{km} \equiv -m \int_{\mathcal{V}^3} C^k C^m f \, d\mathbf{c}. \tag{5.115}$$

Obviously, the equation (5.114) coincides with the momentum conservation equation (4.70) if we identify $\mathbf{T}=\sigma^{km}\mathbf{e}_k\otimes\mathbf{e}_m$ with the Cauchy stress tensor and $\mathbf{k}=k^k\mathbf{e}_k$ with the body forces **b**. Hence, the **first moment** of the Boltzmann equation yields the momentum conservation law with the Cauchy stress given by (5.115).

Again, the multiplication of (5.108) by m and c^2 and integration with respect to the velocity **c** yields

$$\frac{\partial}{\partial t}\left[\rho\left(\tfrac{1}{2}v^2 + \varepsilon\right)\right] + \frac{\partial}{\partial x^k}\left[\rho\left(\tfrac{1}{2}v^2 + \varepsilon\right)v^k - \sigma^k{}_m v^m + q^k\right] = \rho b_k v^k, \tag{5.116}$$

i.e. the energy conservation law (4.88). Here we introduced the notation

$$\rho\varepsilon \equiv \frac{m}{2}\int_{\mathcal{V}^3} C^2 f \, d\mathbf{c}, \quad q^k \equiv \frac{m}{2}\int_{\mathcal{V}^3} C^2 C^k f \, d\mathbf{c}, \tag{5.117}$$

and used the definitions (5.111), (5.112), and (5.115) as well as the property (5.109) of the collision operator. Hence, the energy conservation law follows as the trace of the **second moment** of the Boltzmann equation.

Let us notice, that according to (5.115), the pressure in this approach is given by

$$p = -\tfrac{1}{3}\operatorname{tr}\mathbf{T} = \frac{m}{3}\int_{\mathcal{V}^3} C^2 f \, d\mathbf{c}. \tag{5.118}$$

Therefore, bearing $(5.117)_1$ in mind, we obtain

$$p = \frac{2}{3}\rho\varepsilon. \tag{5.119}$$

This property is characteristic, solely, for an ideal gas. This means that the Boltzmann equation and its moments, obtained by the above described **Maxwell iteration**, can produce a continuum model **appropriate only for ideal gases**. Consequently, the extended thermodynamics as a strategy for constructing general continuous models may use those kinetic results only as hints for the structure of field equations without, however, some peculiar symmetries of those results such as (5.119).

There is also another reason for disregarding a kinetic approach in a general macroscopical modelling procedure. Namely, except for those moments which we introduced above, all other quantities would require a calculation of integrals involving the collision operator $\mathcal{C}$. This would yield **production terms** in the corresponding balance equations whose structure can be found from such kinetic considerations only in few exceptional cases.

For instance, if we multiply the Boltzmann equation by the velocity in the second power **c**⊗**c** rather than only by its trace c^2, then apart from the equation (5.116) we would obtain the following equation for the stress **T**:

$$\frac{\partial}{\partial t}\left(\rho v^k v^m - \sigma^{km}\right) + \frac{\partial}{\partial x^n}\left(\rho v^k v^m v^n - 3\sigma^{(km} v^{n)} + m^{kmn}\right) = P^{km}, \tag{5.120}$$

where, for simplicity, we neglected body forces and used an inertial reference frame. In the above equation

$$m^{kmn} = \int_{\mathcal{V}^3} C^k C^m C^n f \, d\mathbf{c},$$
$$P^{km} == \int_{\mathcal{V}^3} C^k C^m \mathcal{C}(f) d\mathbf{c}, \quad P^k{}_k \equiv 0, \tag{5.121}$$

and

$$\sigma^{(km} v^{n)} = \frac{1}{3!}\left(\sigma^{km} v^n + \sigma^{kn} v^m + \sigma^{nm} v^k\right), \tag{5.122}$$

i.e. the round brackets denote the full symmetrization. Obviously, the trace in the equation (5.120) together with the relation (5.119) yields the energy conservation law (5.116). The traceless part of the equation (5.120) contains the production **P** which shows that it is not a conservation law any longer.

The part of the **third moment** of the Boltzmann equation obtained by the multiplication by $c^2 c^k$ would yield a balance equation for m^{nnk}, i.e. for the heat flux q^k.

Other moments would produce the balance equations for quantities of higher tensorial rank, i.e. a tensorial hierarchy of equations, which, apart from two exceptions, the flux of the heat flux and the non-equilibrium pressure appearing in the so-called 14 moments theory, would not have a clear physical interpretation in terms of macroscopical quantities.

Even though rigorous proofs are not available, it is claimed that the higher moments yield better approximations of nonequilibrium thermodynamic processes [see: I. MÜLLER, T. RUGGERI (1993)]. We discuss this problem further on in this book within the framework of extended thermodynamics. However, it should be mentioned that the presence of production terms in the balance laws is the reason for the macroscopical irreversibility in such a model. This is due to the fact that collisions in the model of N non-interacting particles create the only mechanism for approaching an equilibrium state which cannot be left spontaneously by the system without interaction with the external world•

Remark 5.2. *On the Material Objectivity*
The Galilean invariance of the field equations demonstrates one of the most important reasons for the construction of extended thermodynamics. Namely, it shows the status of the so-called **material objectivity** principle introduced in the continuous theories in the 1960s as a general requirement for all constitutive quantities. In 1966 INGO MÜLLER

showed in his PhD Thesis („*Zur Ausbreitungsgeschwindigkeit von Störungen in kontinuierlichen Medien*", Dissertation TH Aachen, Germany) that the kinetic theory of gases leads to the non-objective macroscopical constitutive laws for such quantities as the heat flux*). This started a vehement discussion concerning this principle and its universality.

Within the framework of the extended thermodynamics, this problem becomes immaterial. Certainly the field equations cannot be frame-indifferent; the Galilean invariance is the only natural invariance requirement imposed on those equations. We may require the isotropic dependence of the constitutive quantities on the fields as we do further on in this book, but this requirement does not imply the material objectivity. Numerous examples show that material objectivity follows only as an **approximation** within this theory, exactly in the same way as it follows within the macroscopical theories derived by, for instance, Maxwell iterations from kinetic theories.

To demonstrate the influence of the reference system on the quantities which are considered to be material frame-indifferent (materially objective) in the usual formulation of continuum thermodynamics, we consider an example of linear equations describing the isotropic rigid heat conductor. This problem will be investigated in Chap. 9 in its full generality.

The **rigid heat conductor** is defined within the frame of extended thermodynamics as a material whose processes are described by the fields of temperature T and the heat flux **q.** The appropriate field equations follow from the balance equations $(5.103)_3$ and $(5.104)_2$ with the vanishing field of velocity **v** and the constant mass density ρ, i.e.

$$\rho\frac{\partial \varepsilon}{\partial t}+\frac{\partial q^k}{\partial x^k}=0,\qquad \frac{\partial}{\partial t}\left(\tfrac{1}{2}\hat{F}_{km}{}^{m}\right)+\frac{\partial}{\partial x^n}\left(\tfrac{1}{2}\hat{G}^{nm}_{km}\right)=\tfrac{1}{2}\hat{f}_{km}{}^{m}, \tag{5.123}$$

where the energy supply r has been neglected for simplicity.

In the most general case the **isotropic** constitutive relations for the fields, appearing in the above set of equations, have the form (see: the Tables of Sect. 5.1.-5.4.)

$$\begin{aligned}
&\varepsilon=\varepsilon\left(T,q^2\right),\quad q^2\equiv q_k q^k,\\
&\tfrac{1}{2}\hat{F}_{km}{}^{m}=\alpha\left(T,q^2\right)q_k,\\
&\tfrac{1}{2}\hat{G}^{nm}_{km}=\beta_1\left(T,q^2\right)\delta^n_k+\beta_2\left(T,q^2\right)q_k q^n,\\
&\tfrac{1}{2}\hat{f}_{km}{}^{m}=\gamma\left(T,q^2\right)q_k.
\end{aligned} \tag{5.124}$$

Substitution in (5.123) yields

$$\begin{aligned}
&\underline{\rho c_V\frac{\partial T}{\partial t}+\frac{\partial q^k}{\partial x^k}}+\rho\frac{\partial \varepsilon}{\partial q^2}2q^k\frac{\partial q^k}{\partial t}=0,\\
&\underline{\tau\frac{\partial q^k}{\partial t}+K\frac{\partial T}{\partial x^k}+q^k}-\frac{1}{\gamma}\frac{\partial \alpha}{\partial t}q^k-\frac{1}{\gamma}\frac{\partial \beta_1}{\partial q^2}2q_m\frac{\partial q^m}{\partial x^k}-\frac{1}{\gamma}\frac{\partial}{\partial x^n}\left(\beta_2 q_k q^n\right)=0,
\end{aligned} \tag{5.125}$$

where

$$c_V\equiv\frac{\partial \varepsilon}{\partial T},\quad \tau\equiv-\frac{\alpha}{\gamma},\quad K\equiv-\frac{1}{\gamma}\frac{\partial \beta_1}{\partial T}. \tag{5.126}$$

*) e.g. see: I. MÜLLER (1972), (1976).

In a linear model we neglect higher-order terms in **q** and derivatives of the fields T and **q**. Hence, in the linear field equations there remain only the underlined terms of the equations (5.125) with the coefficients c_V, τ, and K which may solely depend on temperature. Thus

$$\rho c_V \frac{\partial T}{\partial t} + \operatorname{div} \mathbf{q} = 0, \quad \tau \frac{\partial \mathbf{q}}{\partial t} + K \operatorname{grad} T + \mathbf{q} = 0. \tag{5.127}$$

The second equation was proposed by C. CATTANEO (1948), and it is used commonly in models of the so-called second sound. As already mentioned, in Chap. 9 we will present a brief account of this problem.

It is obvious that the set (5.127) is Galilean invariant but not Euclidean frame-indifferent. Namely, if we change the observer according to the formula [see: (3.46)]

$$\mathbf{x}^* = \mathbf{O}(t)\mathbf{x}, \quad t^* = t, \tag{5.128}$$

then the field equations become

$$\rho c_V \frac{\partial T}{\partial t} + \operatorname{div}^* \mathbf{q}^* = 0, \quad \tau \frac{\partial \mathbf{q}^*}{\partial t} + K \operatorname{grad}^* T + \mathbf{q}^* + \tau \Omega \mathbf{q}^* = 0, \tag{5.129}$$

with

$$T = T^*, \quad \mathbf{q}^* = \mathbf{O}\mathbf{q}, \quad \Omega = \frac{\partial \mathbf{O}}{\partial t} \mathbf{O}^T. \tag{5.130}$$

The presence of a tensor of the relative angular velocities Ω in the second equation makes **q** frame-dependent. For instance, in the case of stationary problems we have

$$\mathbf{q}^* = -K(\mathbf{1} + \tau\Omega)^{-1} \operatorname{grad}^* T, \tag{5.131}$$

which replaces the classical Fourier law in the case of a non-inertial frame of reference. Obviously, in the case of small relaxation times τ, the relation (5.131) can be replaced **approximately** by the usual Fourier law. According to the experimental results, τ is of the order 10^{-6}–10^{-8} sec., which means that we would have to rotate the material relative to the observer with an angular velocity approx. 1–100 MHz to see the influence of the additional term. This is not very realistic under normal conditions.

We shall see further, that this lack of objectivity within the framework of extended thermodynamics and the moment equations of the kinetic theories is connected with some **unsolved** constitutive problems. In Chap. 8 we demonstrate a model of the so-called Maxwellian fluid, in which the above-described structure of the field equations seems to yield the wrong signs of certain material coefficients. A similar problem arises in the construction of a model of viscoelastic solids•

6 Entropy Principle

6.1 Preliminary Remarks

One of the most controversial problems connected with the construction of macroscopical continuous models is the formulation of a selection criterion for solutions of the field equations which are admissible in nature. Of course, we shall not debate all such criteria, known under the name of the **second law of thermodynamics**. However, in order to formulate the premises for the second law, we must discuss in this section the notion of **irreversibility**.

The question of how the macroscopical irreversibility of processes arises from microscopical phenomena, which seem to be reversible (invariant with respect to the **time reversal**), was focused on in particular by the publication of BOLTZMANN's famous paper on the H-Theorem (1868). From the equations of Newtonian mechanics for N colliding and otherwise non-interacting mass particles, Boltzmann derived a **kinetic equation** (5.108) whose direct consequence is the irreversibility of processes (i.e. lack of invariance with respect to the time reversal) described by this equation and reflected in the H-Theorem. The H-function as defined by Boltzmann is:

$$H(t) = \int_{\mathcal{V}^3} f \ln f \, d\mathbf{c}. \tag{6.1}$$

It is the prototype of the entropy function, and, according to this theorem, it must be non-increasing: $dH/dt \leq 0$.

These results have been vehemently objected to by Zermelo, Loschmidt, and many others who have constructed „paradoxes" of the Boltzmann's model. For instance, Zermelo showed that on the basis of Poincare's Recurrence Theorem, a closed dynamical system must, unless it happens to start from an exceptional initial state, eventually come back, after the recurrence time of the Poincare's cycle, arbitrarily near to its initial state. Thus, Boltzmann's H-function describing the irreversibility and being simultaneously a purely dynamical quantity could not always decrease.

In order to clarify these contradictions P. and T. EHRENFEST (1907) constructed an extremely simple and one of the most instructive models of physics. It is sometimes referred to as the *dog-flea model.*

Following M. KAC (1957), let us consider a simple system exposing the importance of the large number of microstates of the system in the macroscopic irreversibility. Let us imagine there are two containers in which there are a total of N balls, labelled from 1 to N. The **macroscopical state** of this system is defined by the number of balls in one of the containers, say N_1. Recall that the other container has $N_2=N-N_1$ balls. Let us assume, for simplicity, that initially all N balls are in one container: $N_1(t=0)=N$, $N_2(t=0)=0$. Now, we change the state of the system by drawing a number between 1 and N at random, and transferring the ball with the number drawn from one container to another. It is obvious that the „flux" must start from the first (full) container in the direction of the second (empty) container. It may happen that we draw the same number again and the ball then returns to the first container. However, for a very large N, say $N=10^{23}$ to be near the Avogadro number, it is much more probable that we get a different number and another ball moves from the first container to the second. Hence, despite small fluctuations, an average flux of balls has a definite direction until we reach the macrostate in which both containers have an almost equal number of balls $N_1 \approx N_2 \approx N/2$. Although it cannot be excluded that we draw the sequence of numbers which would make the second container empty again, it is clear that such a process is extremely exceptional. This is connected to the fact that the macroscopical state: $N_1=N_2=N/2$ corresponds to, approximately, 2^N microstates, each having balls with different combinations of numbers in a given container but realising the same macrostate. For very large N, it is almost impossible that within the time of the physical observation we will ever observe the initial state again. Hence, the macroscopical „irreversibility" of the system follows from the large number of microstates seen as the same at the macroscopical level of observations*).

This trivial example illustrates the fundamental source of paradoxes connected with Boltzmann's H-Theorem. The times of physical observations are, in practice, so short when compared with the recurrence time of the Poincare's cycle that we are unable to check the macroscopical reversibility even if it was a feature of processes in nature. Easy estimates show that the recurrence times of macroscopical systems exceed the time of existence of the Universe by many orders of magnitude. To Zermelo's criticism that a system should, after the recurrence time, return to the initial state, Boltzmann supposedly replied: „You should wait that long!"

However, it is also clear that Boltzmann's derivation of the kinetic equation and, consequently, the H-Theorem, as well as any other construction of a macroscopical model, must contain an assumption which destroys the initial reversibility of the microscopical model. This assumption, although justified by the nature of macroscopical observations, cannot be a part of the initial reversible model of microstates. In the case of Boltzmann's theory, this assumption is called **„Stoßzahlansatz"**, and we illustrate its meaning using a simple example designed by Ehrenfest.

Let us imagine N small circles distributed with the equal distance, Δt, on the circumference of the large circle (Fig. 6.1.). The small circles may have two different colours, say black and white. The dynamics in this model is due to the rotation of the wheel. The wheel has a set, S, of M spokes, $1 \ll M \ll N$ located at random in different middle positions between small circles. During the rotation, whenever a spoke passes a small circle, the circle changes colour, from white to black or vice versa. Hence, the instantaneous number of white and black circles is described by the following „equation of motion" („balance law"):

$$N_w(t+\Delta t) = N_w(t) + N_b(S,t) - N_w(S,t), \tag{6.2}$$

with

$$\forall t: \quad N_w(t) + N_b(t) = N, \qquad N_w(S,t) + N_b(S,t) = M. \tag{6.3}$$

*) The detailed discussion of probabilistic problems connected with both models presented in this section can be found in two books of M. KAC (1956), (1959). In the first, a manuscript of his lectures at Magnolia Oil Society in Dallas, Texas, the models are presented so that a reader with no mathematical background can understand them.

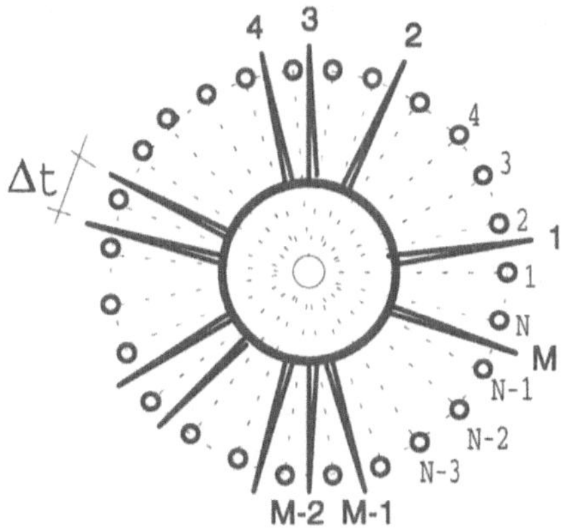

Fig. 6.1. *To demonstrate the Ehrenfest model illustrating the meaning of Stoßzahlansatz, t=0*

We denoted the number of white circles which change colour owing to the passage of the spoke in the step $t \to t+\Delta t$ by $N_w(S,t)$ and the number of black circles changing colour in this step by $N_b(S,t)$. Of course, $N_w(t)$ and $N_b(t)$ denote the number of white and black circles at the instant of time t. Obviously, in this model time is discrete with the elementary time step equal to Δt.

The equation (6.2) constitutes the counterpart of the **Liouville's equation** for N particles and describes reversible processes on the microlevel of our model. If we start, for instance, from the state $N_w(t=0)=N$, $N_b(t=0)=0$, then after two full rotations of the wheel with spokes the system returns to its initial state. Thus in this model the recurrence time of the Poincare's cycle is equal to $t_P = 2\,N\,\Delta t$.

The change in the surplus of white circles at the instant of time $t+\Delta t$ is, according to (6.2), given by the following equation

$$[N_w(t+\Delta t)-N_b(t+\Delta t)] = [N_w(t)-N_b(t)] - 2[N_w(S,t)-N_b(S,t)] \tag{6.4}$$

Now, we make the reasonable assumption that, owing to the randomness of S, after sufficiently many time steps the black and white circles will be regularly distributed, i.e.

$$N_w(S,t) \approx \frac{M}{N} N_w(t), \qquad N_b(S,t) \approx \frac{M}{N} N_b(t). \tag{6.5}$$

Substitution of this assumption into (6.4) yields the following difference equation

$$N_w(t+\Delta t)-N_b(t+\Delta t) = \left(1-2\frac{M}{N}\right)[N_w(t)-N_b(t)] \tag{6.6}$$

This difference equation can be solved immediately and we obtain the solution

$$N_w(t)-N_b(t) = N\left(1-2\frac{M}{N}\right)^t, \tag{6.7}$$

i.e.

$$\frac{N_w(t)}{N} = \frac{1}{2} + \frac{1}{2}\left(1 - 2\frac{M}{N}\right)^t . \tag{6.8}$$

Hence, we have

$$\frac{N_w(t=0)}{N} = 1, \qquad \lim_{t\to\infty} \frac{N_w(t)}{N} = \frac{1}{2}, \tag{6.9}$$

in other words, after a long time the fraction of circles of the different colours will be almost equal.

The most important property of the solution (6.8) is its irreversibility. We started with a simple initial state, but whatever this state is we will always obtain the same asymptotic end state, i.e. containing equal numbers of white and black circles. There is no trace left of the microscopical periodicity of processes. The reason for this result is hidden in the assumption (6.5), which corresponds to the Stoßzahlansatz of Boltzmann's theory. We see that the conditions under which the solution (6.8) makes sense are the randomness of the set S, and long, but not too long times of observations ($1 \ll t \ll t_P$). Furthermore, the above example shows that the evolution described by a macroscopic model terminates at the end state ($N_w(\infty)=N_b(\infty)=N/2$).

6.2 Entropy Inequality

According to the considerations in the previous section, processes in systems with a large number of particles must appear as irreversible in all macroscopical observations, such as all those which can be described by means of a continuum theory. In the case of thermodynamical processes described by the field equations (5.59), this means that the choice of the constitutive relations (5.60) cannot be arbitrary. Conversely, we expect to be limited to functions which deliver solutions to (5.59), and these solutions should have a particular order of appearance of the values of fields $\boldsymbol{w}(.,t)$ in subsequent instants of time.

This condition is usually formulated as the **entropy inequality:**

any solution $\boldsymbol{w}$ of the field equations

$$\frac{\partial \boldsymbol{F}^0}{\partial t} + \frac{\partial \boldsymbol{F}^k}{\partial x^k} = \boldsymbol{f}, \tag{6.10}$$

must identically satisfy the inequality

$$\frac{\partial \rho\eta}{\partial t} + \frac{\partial h^k}{\partial x^k} \geq 0, \tag{6.11}$$

where

$$\eta = \eta(\boldsymbol{w}), \qquad h^k = \rho\eta v^k + \mathrm{h}^k(\mathbf{u}), \tag{6.12}$$

are called the **specific entropy** and the **entropy flux**, respectively.

Here we used the Eulerian description of the deformation of the body.

Exercise 6.1. Transform the inequality (6.11) to the Lagrangian description•

In the above formulation of the **second law of thermodynamics,** as well as in all further considerations, we omit the body forces, the energy and the entropy supplies. In Chaps. 9 and 10, we consider the cases in which the mass, momentum, and energy balance laws are not conservation laws. This non-conservation may only happen when the body consists of more than one component (mixture), and the source terms, in the balance equations for each component, describe the local volume interactions with other components of the mixture. Otherwise, the body forces and supplies may appear solely as the inertial body forces caused by the non-inertial reference frames. This simplification is immaterial in most practical applications of the thermodynamic theory of constitutive relations. Furthermore, it eliminates certain unsolved problems of continuum thermodynamics.

In these notes, we shall not discuss problems connected with the form of inequality (6.11), its relation to the other formulations of the second law of thermodynamics or considerations from statistical mechanics, which justify its appearance. Certain aspects of these problems can be found in the literature.*)

It is convenient to have a global form of the inequality (6.11) [compare: (4.49)]

$$\begin{aligned} &\forall \mathcal{P}_t \subset \Re^3 \quad \text{s.t.} \quad \mathbf{f}^{-1}(\mathcal{P}_t, t) \in \mathcal{A}: \\ &\frac{d}{dt}\int_{\mathcal{P}_t} \rho\eta \, dv + \oint_{\partial\mathcal{P}_t} \mathbf{h}\cdot\mathbf{n} \, ds = \int_{\mathcal{P}_t} \rho s^* \, dv \geq 0, \end{aligned} \tag{6.13}$$

where s* denotes the density of the **entropy production**. Certainly, the inequality (6.13) is equivalent to (6.11) for all regular points of the body. On the singular surface, $\mathcal{S}_t$, we obtain from Kotchine's condition (4.60)

$$[[\rho\eta(\mathbf{v}\cdot\mathbf{n} - c)]] + [[\mathbf{h}\cdot\mathbf{n}]] \geq 0, \tag{6.14}$$

provided the production s* contains the singular part concentrated on a singular surface, $\mathcal{S}_t$. Otherwise, the jump condition for the entropy becomes a continuity condition. We discuss this problem in the following.

Let us notice that the inequality (6.11) jointly with the relations (6.12) is Galilean invariant.

The consequences of the entropy inequality in the theory of field equations can be derived in many different ways. However, the most appealing method is based on LIU's Theorem, which we proceed to formulate.

For the differentiable functions $\rho\eta$ and $\mathbf{h}$ using the chain rule of differentiation the entropy inequality can be written in the following explicit form:

$$\mathfrak{a}\cdot\mathcal{X} \geq 0, \qquad \mathfrak{a}, \mathcal{X} \in \Re^{4N}, \tag{6.15}$$

where

*) e.g. I. MÜLLER (1985), K. WILMANSKI (1992).

$$\mathfrak{a} \equiv \left(\frac{\partial \rho\eta}{\partial \omega^1}, \ldots, \frac{\partial \rho\eta}{\partial \omega^N}, \frac{\partial h^1}{\partial \omega^1}, \frac{\partial h^1}{\partial \omega^2}, \ldots, \frac{\partial h^3}{\partial \omega^N} \right), \quad \boldsymbol{w} \equiv \left(\omega^1, \ldots, \omega^N \right)^T,$$
$$\mathcal{X} \equiv \left(\frac{\partial \omega^1}{\partial t}, \ldots, \frac{\partial \omega^N}{\partial t}, \frac{\partial \omega^1}{\partial x^1}, \frac{\partial \omega^2}{\partial x^1}, \ldots, \frac{\partial \omega^N}{\partial x^3} \right)^T. \tag{6.16}$$

Now we consider the problem of solutions $\boldsymbol{w}$ of the inequality (6.15) which satisfy identically the field equations (6.10). The latter can be written conveniently in the form

$$\mathbf{A}\mathcal{X} - f = 0, \tag{6.17}$$

where the (N×4N)-matrix, $\mathbf{A}$, is defined as

$$\mathbf{A} = \begin{pmatrix} \dfrac{\partial F^{01}}{\partial \omega^1} & \dfrac{\partial F^{01}}{\partial \omega^2} & \cdots & \dfrac{\partial F^{31}}{\partial \omega^N} \\ \dfrac{\partial F^{02}}{\partial \omega^1} & \dfrac{\partial F^{02}}{\partial \omega^2} & \cdots & \dfrac{\partial F^{32}}{\partial \omega^1} \\ \vdots & \vdots & & \vdots \\ \dfrac{\partial F^{0N}}{\partial \omega^1} & \dfrac{\partial F^{0N}}{\partial \omega^2} & \cdots & \dfrac{\partial F^{3N}}{\partial \omega^N} \end{pmatrix}, \tag{6.18}$$

provided $\boldsymbol{F}^0$, $\boldsymbol{F}^k$ are differentiable.

It is obvious that the solutions to this problem are identical to those of the entropy inequality.

The constraints imposed by (6.17) on $\boldsymbol{w}$, the solutions of the inequality (6.15), can be eliminated in the same manner in which the constraints are eliminated in the variational problems of classical mechanics. Namely, one can introduce **Lagrange multipliers** whose existence is guaranteed by the following theorem.

Theorem (I-SHIH LIU; *on the existence of Lagrange multipliers*). Let $\mathbf{A}$ be given by (6.18) and $\mathcal{X}$ - by $(6.16)_3$ and

$$\mathbf{S} \equiv \left\{ \mathcal{X} \in \Re^{4N} \middle| \mathbf{A}\mathcal{X} - f = 0 \right\} \neq \varnothing. \tag{6.19}$$

Then, the following conditions are equivalent:

$$\text{i/} \quad \forall \mathcal{X} \in \mathbf{S}: \quad \mathfrak{a} \cdot \mathcal{X} \geq 0, \tag{6.20}$$

where $\mathfrak{a}$ is given by $(6.16)_1$,

$$\text{ii/} \quad \exists \Lambda \in \Re^N, \Lambda \neq 0 \ \forall \mathcal{X} \in \Re^{4N}: \quad \mathfrak{a} \cdot \mathcal{X} - \Lambda \cdot \left(\mathbf{A}\mathcal{X} - f \right) \geq 0, \tag{6.21}$$

$$\text{iii/} \quad \exists \Lambda \in \Re^N, \Lambda \neq 0: \quad \mathfrak{a} - \mathbf{A}^T \Lambda = 0, \quad \Lambda \cdot f \geq 0. \tag{6.22}$$

Proof. We prove the following implications:

i/ $\Leftarrow$ ii/ $\Leftrightarrow$ iii/ $\Leftarrow$ i/.

1. The implication ii/$\Rightarrow$i/ is immediate.

2. We shall prove the equivalence of ii/ and iii/. We can write (6.21) in the form

$$\forall \mathcal{X} \in \Re^{4N}: \quad \left(\mathbf{a} - \mathbf{A}^T \Lambda\right) \cdot \mathcal{X} + \Lambda \cdot f \geq 0.$$

Since this inequality holds for arbitrary $\mathcal{X}$, it follows necessarily that

$$\left(\mathbf{a} - \mathbf{A}^T \Lambda\right) = 0 \quad \Rightarrow \quad \Lambda \cdot f \geq 0.$$

3. It remains to prove the implication i/⇒iii/. With this aim, we define the following sets:

$$\begin{aligned}
&\mathbf{H} \equiv \left\{\mathcal{X} \in \Re^{4N} \middle| \mathbf{a} \cdot \mathcal{X} \geq 0\right\}, \\
&\mathbf{H}_0 \equiv \left\{\mathcal{X} \in \Re^{4N} \middle| \mathbf{a} \cdot \mathcal{X} = 0\right\}, \quad \mathbf{S}_0 \equiv \left\{\mathcal{X} \in \Re^{4N} \middle| \mathbf{A}\mathcal{X} = 0\right\}, \\
&\mathbf{H}_0^\perp \equiv \left\{\mathcal{Y} \in \Re^{4N} \middle| \forall \mathcal{X} \in \mathbf{H}_0 : \mathcal{Y} \cdot \mathcal{X} = 0\right\}, \\
&\mathbf{S}_0^\perp \equiv \left\{\mathcal{Y} \in \Re^{4N} \middle| \forall \mathcal{X} \in \mathbf{S}_0 : \mathcal{Y} \cdot \mathcal{X} = 0\right\}.
\end{aligned}$$

To prove the assertion we show first that

$$\mathbf{H}_0^\perp \subset \mathbf{S}_0^\perp.$$

It is easy to notice that $\mathbf{H}_0, \mathbf{H}_0^\perp, \mathbf{S}_0^\perp$ are subspaces of $\Re^{4N}$. Simultaneously, i/ implies that $\mathbf{S} \subset \mathbf{H}$; we obtain $\mathbf{S}_0 \subset \mathbf{H}_0$.

Suppose that the above relation does not hold. Then

$$\exists \mathcal{Y} \in \mathbf{S}_0: \quad \mathbf{a} \cdot \mathcal{Y} \neq 0.$$

However, $\mathbf{S}_0$ is the linear subspace of $\Re^{4N}$. Therefore,

$$\forall \mathrm{a} \in \Re: \quad \mathrm{a}\mathcal{Y} \in \mathbf{S}_0 \quad \Rightarrow \quad \forall \mathcal{Z} \in \mathbf{S}: \quad \mathcal{Z} + \mathrm{a}\mathcal{Y} \in \mathbf{S}.$$

On the other hand,

$$\mathbf{a} \cdot \left(\mathcal{Z} + \mathrm{a}\mathcal{Y}\right) = \mathrm{a}\,\mathbf{a} \cdot \mathcal{Y} + \mathbf{a} \cdot \mathcal{Z}.$$

Hence,

$$\exists \mathrm{a} \in \Re: \quad \mathrm{a}\,\mathbf{a} \cdot \mathcal{Y} \leq -\mathbf{a} \cdot \mathcal{Z} \quad \Rightarrow \quad \mathcal{Z} + \mathrm{a}\mathcal{Y} \notin \mathbf{H},$$

which is a contradiction. Hence, $\mathbf{S}_0 \subset \mathbf{H}_0$.

We proceed to prove the implication i/⇒iii/. By definition, we have $\mathbf{a} \in \mathbf{H}_0^\perp$ and consequently, $\mathbf{a} \in \mathbf{S}_0^\perp$. Let us construct, from the matrix $\mathbf{A}$, the sequence of vectors $\{\mathbf{A}_1, \ldots, \mathbf{A}_N\}$ whose coordinates coincide with the rows of $\mathbf{A}$. Then

$$\forall 1 \leq \Delta \leq N: \quad \mathbf{A}_\Delta \in \mathbf{S}_0^\perp.$$

On the other hand we have

$$\dim \mathbf{S}_0^\perp = \operatorname{rank} \mathbf{A},$$

and consequently, the vectors $\{\mathbf{A}_\Delta\}_{\Delta=1}^{N}$ span the space $\mathbf{S}_0^\perp$. It follows that

$$\exists \Lambda \in \Re^N, \Lambda \neq 0: \quad \boldsymbol{a} = \sum_{\Delta=1}^{N} \mathbf{A}_\Delta \Lambda^\Delta \equiv \mathbf{A}^T \Lambda.$$

Finally,

$$\forall \mathcal{X} \in \mathbf{S}: \quad \boldsymbol{a} \cdot \mathcal{X} = \left(\mathbf{A}^T \Lambda\right) \cdot \mathcal{X} = \Lambda \cdot \left(\mathbf{A} \mathcal{X}\right) = \Lambda \cdot f;$$

since $\mathbf{S} \subset \mathbf{H}$, i.e. $\boldsymbol{a} \cdot \mathcal{X} \geq 0$, we obtain

$$\Lambda \cdot f \geq 0,$$

which completes the proof ■

Clearly, if the rank of the matrix $\mathbf{A}$ satisfies the following condition:

$$\operatorname{rank} \mathbf{A} \geq N, \tag{6.23}$$

then the system of the algebraic equations (6.22) can be solved with respect to Λ. The multipliers, Λ, are auxiliary quantities in these considerations and, hence, such a solution does not contribute anything to the structure of field equations. However, the system (6.22) usually consists of more than N linearly independent equations. The remaining conditions, as well as the inequality (6.22) impose certain constraints on $\boldsymbol{F}^0$ and $\boldsymbol{F}^k$, which are called the **thermodynamical admissibility conditions**. We shall consider these conditions for the hyperbolic systems of field equations in the subsequent chapters. To demonstrate the type of conditions which we can expect, let us first present two simple examples.

Example 6.1. *Eulerian Heat Conducting Fluid*

Let us consider the fluid whose thermomechanical processes are described by the following fields:

$$\boldsymbol{w} = (\mathbf{v}, \mathbf{u}), \quad \mathbf{u} \equiv (\rho, \mathrm{T}, \mathbf{q}), \tag{6.24}$$

T being the absolute temperature (see: Appendix A). The field equations for these fields follow from the conservation laws (4.93)

$$\begin{aligned} &\dot{\rho} + \rho \operatorname{div} \mathbf{v} = 0, \\ &\rho \dot{\mathbf{v}} - \operatorname{div} \mathbf{T} = 0, \\ &\rho \dot{\varepsilon} + \operatorname{div} \mathbf{q} - \operatorname{tr} \mathbf{T} \mathbf{L}^T = 0, \quad \mathbf{L} = \operatorname{grad} \mathbf{v}, \end{aligned} \tag{6.25}$$

supplemented by the following constitutive relations:

$$\mathbf{T} = \mathbf{T}(\mathbf{u}), \quad \varepsilon = \varepsilon(\mathbf{u}). \tag{6.26}$$

Obviously, we are missing a vector equation in order to close the set of field equations. In the following chapters, we shall show how such an equation can be constructed within the frame of extended thermodynamics. For the purpose of our example, we assume the gradient of temperature to be a constitutive quantity

$$\operatorname{grad} \mathrm{T} = \mathbf{g}(\mathbf{u}). \tag{6.27}$$

This relation reminds the classical Fourier law for the heat conduction.

According to our considerations on the Galilean invariance, the velocity field does not enter the above constitutive relations.

The second law of thermodynamics has the following form:

$$\rho\dot{\eta} + \operatorname{div}\mathbf{h} \geq 0, \quad \eta = \eta(\mathbf{u}), \quad \mathbf{h} = \mathbf{h}(\mathbf{u}), \tag{6.28}$$

for all solutions $\boldsymbol{w}$ of the field equations (6.25-27).

We make the additional assumption on the **frame-indifference** [see: (3.55)] of the above constitutive quantities, i.e. $\mathbf{T}$, ε, η, $\mathbf{g}$, $\mathbf{h}$ are supposed to be isotropic functions of $\mathbf{u}$. Then, according to the tables in Sect. 5.2., we have

$$\begin{aligned}
&\mathbf{T} = -p\left(\rho, T, q^2\right)\mathbf{1} + s\left(\rho, T, q^2\right)\mathbf{q}\otimes\mathbf{q},\\
&\varepsilon = \varepsilon\left(\rho, T, q^2\right), \qquad \eta = \eta\left(\rho, T, q^2\right),\\
&\mathbf{g} = -\left[K\left(\rho, T, q^2\right)\right]^{-1}\mathbf{q}, \qquad \mathbf{h} = H\left(\rho, T, q^2\right)\mathbf{q}.
\end{aligned} \tag{6.29}$$

Certainly, in the above relations p is the candidate for the pressure, and K - for the heat conductivity.

We proceed to solve the entropy inequality. According to LIU's Theorem, there exist Lagrange multipliers $\Lambda^\rho \in \Re$, $\Lambda^v \in \Re^3$, $\Lambda^\varepsilon \in \Re$, $\Lambda^T \in \Re^3$, such that for **all fields**, $\boldsymbol{w}$, the following inequality must be satisfied:

$$\begin{aligned}
&\rho\dot{\eta} + \operatorname{div}\mathbf{h} - \Lambda^\rho\left(\dot{\rho} + \rho\operatorname{div}\mathbf{v}\right) - \Lambda^v\cdot\left(\rho\dot{\mathbf{v}} - \operatorname{div}\mathbf{T}\right) -\\
&-\Lambda^\varepsilon\left(\rho\dot{\varepsilon} + \operatorname{div}\mathbf{q} - \operatorname{tr}\mathbf{T}\mathbf{L}^T\right) - \Lambda^T\cdot\left(\mathbf{q} - K\operatorname{grad}T\right) \geq 0.
\end{aligned} \tag{6.30}$$

The chain rule of differentiation and the above constitutive relations yield the following form of the vector $\mathcal{X} \in \Re^{32}$:

$$\mathcal{X} = \left(\frac{\partial\mathbf{v}}{\partial t}, \frac{\partial\rho}{\partial t}, \frac{\partial T}{\partial t}, \frac{\partial\mathbf{q}}{\partial t}, \operatorname{grad}\mathbf{v}, \operatorname{grad}\rho, \operatorname{grad}T, \operatorname{grad}\mathbf{q}\right)^T. \tag{6.31}$$

After simple manipulations we arrive at the following explicit form of the relations (6.22):

$$\begin{aligned}
&\Lambda^v = 0,\\
&\frac{\partial\eta}{\partial\rho} - \frac{1}{\rho}\Lambda^\rho - \Lambda^\varepsilon\frac{\partial\varepsilon}{\partial\rho} = 0, \qquad \frac{\partial\eta}{\partial T} - \Lambda^\varepsilon\frac{\partial\varepsilon}{\partial T} = 0, \qquad \frac{\partial\eta}{\partial q^2} - \Lambda^\varepsilon\frac{\partial\varepsilon}{\partial q^2} = 0,\\
&\left(\rho\Lambda^\rho + p\Lambda^\varepsilon\right)\mathbf{1} - s\Lambda^\varepsilon\mathbf{q}\otimes\mathbf{q} = 0,\\
&\frac{\partial H}{\partial\rho} = 0, \qquad \frac{\partial H}{\partial T}\mathbf{q} - K\Lambda^T = 0, \qquad \left(H - \Lambda^\varepsilon\right)\mathbf{1} + 2\frac{\partial H}{\partial q^2}\mathbf{q}\otimes\mathbf{q} = 0,\\
&\Lambda^T\cdot\mathbf{q} \leq 0.
\end{aligned} \tag{6.32}$$

Exercise 6.2. Prove the relations (6.32)•

The deviatoric part of the identity $(6.32)_5$ yields immediately

$$s = 0. \tag{6.33}$$

Consequently, the stress tensor **T** reduces to the pressure. Similarly, the deviatoric part of $(6.32)_8$ leads to

$$\frac{\partial H}{\partial q^2} = 0. \tag{6.34}$$

Hence, according to $(6.32)_6$ and the trace part of $(6.32)_8$ we have

$$H = \Lambda^\varepsilon = \Lambda^\varepsilon(T). \tag{6.35}$$

This result together with $(6.32)_4$ shows that the function

$$\psi \equiv \varepsilon - \Lambda^{-1}\eta, \tag{6.36}$$

does not depend on q^2. Simultaneously, as a result of $(6.32)_2$, and $(6.32)_3$

$$\begin{aligned} \frac{\partial\psi}{\partial T} &= \frac{\partial\varepsilon}{\partial T} - \Lambda^{\varepsilon-1}\frac{\partial\eta}{\partial T} - \frac{d}{dT}\left(\frac{1}{\Lambda^\varepsilon}\right)\eta = -\frac{d}{dT}\left(\frac{1}{\Lambda^\varepsilon}\right)\eta, \\ \frac{\partial\psi}{\partial\rho} &= \frac{\partial\varepsilon}{\partial\rho} - \Lambda^{\varepsilon-1}\frac{\partial\eta}{\partial\rho} = -\frac{1}{\rho}\frac{\Lambda^\rho}{\Lambda^\varepsilon} = \frac{p}{\rho^2} \quad\Rightarrow\quad p = p(\rho, T), \end{aligned} \tag{6.37}$$

where the trace in the relation $(6.32)_5$ was used.

The above relations can be written in the compact form

$$d\psi = -\frac{d}{dT}\left(\frac{1}{\Lambda^\varepsilon}\right)\eta\, dT + \frac{p}{\rho^2}d\rho. \tag{6.38}$$

Comparison of this result with the Gibbs equation for the ideal fluids in the thermodynamical equilibrium (see: Appendix A) yields

$$\Lambda^\varepsilon = \frac{1}{T}, \tag{6.39}$$

i.e. the Lagrange multiplier is the **coldness** and the function ψ is the **Helmholtz free energy function** for the Eulerian fluid.

The formula (6.39) yields simultaneously the following relation between the heat flux, **q**, and the entropy flux, **h**, (see: $(6.29)_5$)

$$\mathbf{h} = \frac{1}{T}\mathbf{q}, \tag{6.40}$$

which is usually assumed within classical thermodynamics.

It remains to exploit the relations $(6.32)_7$ and $(6.32)_9$. We have

$$\Lambda^T = -\frac{1}{KT^2}\mathbf{q} \quad \Rightarrow \quad K \geq 0, \tag{6.41}$$

i.e. the classical condition for heat conductivity, which is even sometimes mistakenly identified with the second law of thermodynamics itself („the heat flows from the hotter to the colder region"). Let us notice that the above considerations do not otherwise limit the dependence of K on the mass density, ρ, temperature, T, and the modulus of the heat flux $\sqrt{q^2}$. Hence, according to our thermodynamical arguments, the relation (6.29) admits the generalization of the classical Fourier law

$$\mathbf{q} = -K \operatorname{grad} T, \qquad K = K(\rho, T, q^2) \geq 0. \tag{6.42}$$

Example 6.2. *Thermo-Viscoelastic Rod*

For the second example, we consider a material whose processes are described by the one-dimensional fields of velocity, v, deformation, e, stress, σ, temperature, T, and the heat flux, q, (compare the Example 5.2.). In the Lagrangian description, these fields are assumed to satisfy the following set of field equations:

$$\begin{aligned}
&\frac{\partial v}{\partial t} - \frac{1}{\rho_0}\frac{\partial \sigma}{\partial X} = 0, && \text{(momentum conservation,} \quad \rho_0 = \text{const.)}\\
&\frac{\partial e}{\partial t} - \frac{\partial v}{\partial X} = 0, && \text{(compatibility condition)}\\
&\frac{\partial \sigma}{\partial t} - E\frac{\partial v}{\partial X} = G, && \text{(Maxwell equation)}\\
&\frac{\partial \varepsilon}{\partial t} + \frac{1}{\rho_0}\frac{\partial q}{\partial X} - \frac{\sigma}{\rho_0}\frac{\partial v}{\partial X} = 0, && \text{(energy conservation)}\\
&\frac{\partial q}{\partial t} + \frac{K}{\tau}\frac{\partial T}{\partial X} = -\frac{1}{\tau}q, && \text{(Cattaneo equation,} \quad \tau = \text{const.)}
\end{aligned} \tag{6.43}$$

with the following constitutive relations:

$$\varepsilon = \varepsilon(\mathbf{u}), \quad G = G(\mathbf{u}), \quad \mathbf{u} \equiv (\sigma, T, e, q)^T. \tag{6.44}$$

The constant, τ, appearing in the Cattaneo equation $(6.43)_5$ has an obvious interpretation as the relaxation time of thermal disturbances. In the limit case, τ=0, we recover the Fourier law of heat conduction. For τ≠0, the Cattaneo equation yields a very important effect of the second sound which we discuss in more depth in Chap. 9.

In the case of isothermal processes: T=const., q≡0, the above set of equations reduces to the system which, owing to its practical importance, appears very extensively in the literature, particularly in the case of the function G describing the plastic behaviour of the material [see, for instance, the earlier quoted book of N. CRISTESCU, I. SULICIU (1982)].

As we see further, this example, in spite of its practical importance, contains certain faults connected with its one-dimensional character. For instance, we cannot investigate the isotropy properties of the constitutive relations (6.44) and, consequently, a certain caution is required in the transition to the linear models. Simultaneously, the evolution equations $(6.43)_{3,5}$ should contain certain non-linear terms which follow naturally within the frame of the three-dimensional model of extended thermodynamics and which are missing in the above equations.

In spite of these draw-backs, the above set of equations reflects a number of properties desired in the case of a model of thermo-viscoelastic materials. It is also relatively easy to handle in the demonstration of the consequences of the second law of thermodynamics.

Before we exploit the entropy inequality, we rewrite the set (6.43) in the normal form of PDE for the fields (v,e,σ,T,q). We have

$$\frac{\partial \boldsymbol{w}}{\partial t}+\mathbf{a}\frac{\partial \boldsymbol{w}}{\partial X}=\boldsymbol{f}, \qquad \boldsymbol{w}\equiv(v,\sigma,T,e,q)^T,$$

$$\mathbf{a}\equiv\begin{pmatrix} 0 & 0 & -\frac{1}{\rho_0} & 0 & 0 \\ -1 & 0 & 0 & 0 & 0 \\ -E & 0 & 0 & 0 & 0 \\ \varepsilon_T^{-1}\left(\varepsilon_e+E\varepsilon_\sigma-\frac{\sigma}{\rho_0}\right) & 0 & 0 & -\varepsilon_T^{-1}\varepsilon_q\frac{K}{\tau} & \varepsilon_T^{-1}\frac{1}{\rho_0} \\ 0 & 0 & 0 & \frac{K}{\tau} & 0 \end{pmatrix}, \tag{6.45}$$

$$\boldsymbol{f}\equiv\left(0,0,G,-\varepsilon_T^{-1}\left(\varepsilon_\sigma G-\varepsilon_q\frac{q}{\tau}\right),-\frac{q}{\tau}\right)^T,$$

where

$$\varepsilon_T\equiv\frac{\partial\varepsilon}{\partial T},\quad \varepsilon_e\equiv\frac{\partial\varepsilon}{\partial e},\quad \varepsilon_\sigma\equiv\frac{\partial\varepsilon}{\partial\sigma},\quad \varepsilon_q\equiv\frac{\partial\varepsilon}{\partial q}. \tag{6.46}$$

Easy calculations show that the eigenvalue problem for the matrix, **a**, of the coefficients of the spatial derivatives yields the following characteristic equation:

$$\mu\left(\mu^2-\frac{E}{\rho}\right)\left(\mu^2+\mu\varepsilon_T^{-1}\varepsilon_q\frac{K}{\tau}-\frac{1}{\rho_0}\varepsilon_T^{-1}\frac{K}{\tau}\right)=0. \tag{6.47}$$

The eigenvalues

$$\mu^{(1)}=0,\quad \mu^{(2),(3)}=\pm\sqrt{\frac{E}{\rho_0}},\quad \mu^{(4),(5)}=\frac{1}{2}\left(-\varepsilon_T^{-1}\varepsilon_q\frac{K}{\tau}\pm\sqrt{\Delta}\right),$$
$$\Delta\equiv\varepsilon_T^{-1}\frac{K}{\tau}\left(\varepsilon_q^2\varepsilon_T^{-1}\frac{K}{\tau}+\frac{4}{\rho_0}\right), \tag{6.48}$$

are all real provided the following sufficient conditions are fulfilled

$$E>0,\quad \varepsilon_T>0,\quad K>0,\quad \tau>0. \tag{6.49}$$

These conditions are connected with the thermodynamical admissibility of processes. Usually, they follow from the second law, and the stability of the thermodynamical equilibrium state.

Under the conditions (6.49), the eigenvalues (6.48) become the **characteristic speeds** of propagation of the disturbances along the characteristic curves. It is easy to show that the eigenvectors of the above eigenvalue problem are linearly independent, and consequently, the set of equations (6.43) is **hyperbolic**. The eigenvalues, $\mu^{(2),(3)}$, determine the speed of the **first sound**, and the eigenvalues, $\mu^{(4),(5)}$, the speed of the **second sound**.

Exercise 6.3. Find the right eigenvectors of the matrix **a**. Prove that they span the space of solutions $\mathbf{w}=(v,\sigma,T,e,q)^T$•

We proceed to investigate the second law of thermodynamics. For the purpose of this example, we assume a simplified form of the entropy inequality

$$\rho_0\frac{\partial\eta}{\partial t}+\frac{\partial}{\partial X}\left(\frac{q}{T}\right)\geq 0, \qquad \eta=\eta(\sigma,T,e,q), \tag{6.50}$$

which is supposed to be identically satisfied for all solutions of the set (6.43).

The assumption of the proportionality of the heat flux, and the entropy flux simplifies considerably the further considerations and does not limit the applicability of this model (see Chap. 8 for a more general case).

Bearing the Liu's Theorem in mind, we obtain

$$\begin{gathered}\exists\Lambda=\left(\Lambda^v,\Lambda^e,\Lambda^\sigma,\Lambda^\varepsilon,\Lambda^q\right)\in\Re^5\ \forall\mathbf{w}\in\Re^5,\boldsymbol{\mathcal{X}}\in\Re^{10}:\\ \rho_0\frac{\partial\eta}{\partial t}+\frac{\partial}{\partial X}\left(\frac{q}{T}\right)-\Lambda\cdot\left(\frac{\partial\mathbf{w}}{\partial t}-\mathbf{a}\frac{\partial\mathbf{w}}{\partial X}-\boldsymbol{f}\right)\geq 0,\end{gathered} \tag{6.51}$$

where

$$\boldsymbol{\mathcal{X}}\equiv\left(\frac{\partial\mathbf{w}}{\partial t},\frac{\partial\mathbf{w}}{\partial X}\right)^T. \tag{6.52}$$

The point iii/ of the Liu's Theorem leads to the following set of relations:

$$\begin{gathered}\Lambda^v=0,\\ \rho_0\frac{\partial\eta}{\partial e}-\Lambda^\varepsilon\rho_0\frac{\partial\varepsilon}{\partial e}-\Lambda^e+\Lambda^\sigma E=0,\qquad \rho_0\frac{\partial\eta}{\partial\sigma}-\Lambda^\varepsilon\rho_0\frac{\partial\varepsilon}{\partial\sigma}-\Lambda^\sigma=0,\\ \rho_0\frac{\partial\eta}{\partial T}-\Lambda^\varepsilon\rho_0\frac{\partial\varepsilon}{\partial T}=0,\qquad \rho_0\frac{\partial\eta}{\partial q}-\Lambda^\varepsilon\rho_0\frac{\partial\varepsilon}{\partial q}-\Lambda^q\tau=0,\end{gathered} \tag{6.53}$$

$$\Lambda^e-\Lambda^\varepsilon\sigma=0,\qquad -\frac{1}{T^2}q-\Lambda^q K=0,\qquad \frac{1}{T}-\Lambda^\varepsilon=0, \tag{6.54}$$

$$\Lambda^\sigma G-\Lambda^q q\geq 0. \tag{6.55}$$

It is obvious, that apart from the five equations for five multipliers, Λ, we obtain the four conditions for the **thermodynamical admissibility** of solutions of the set (6.43).

It is convenient to introduce the following function:

$$\psi\equiv\varepsilon-T\eta=\psi(\sigma,T,e,q). \tag{6.56}$$

As we will see, this analogue to the Helmholtz free energy function coincides with the usual thermodynamical potential only in particular cases.

According to this definition, the relations (6.53–6.55) yield

$$\Lambda^v = 0, \quad \Lambda^e = -\frac{\sigma}{T}, \quad \Lambda^\sigma = -\frac{\rho_0}{T}\frac{\partial\psi}{\partial\sigma}, \quad \Lambda^\varepsilon = \frac{1}{T}, \quad \Lambda^q = -\frac{q}{KT^2}, \tag{6.57}$$

and

$$\begin{gathered} \eta = -\frac{\partial\psi}{\partial T} \quad \Rightarrow \quad \varepsilon = \psi + T\eta = \psi - T\frac{\partial\psi}{\partial T}, \\ \frac{\partial\psi}{\partial e} + E\frac{\partial\psi}{\partial\sigma} = \frac{\sigma}{\rho_0}, \qquad \frac{\partial\psi}{\partial q} = \frac{\tau}{\rho_0 KT}q, \end{gathered} \tag{6.58}$$

$$-\frac{\partial\psi}{\partial\sigma}G + \frac{q^2}{\rho_0 KT} \geq 0. \tag{6.59}$$

It is clear that the function ψ in our example is the potential for the entropy, the internal energy, and the heat flux, but is not the potential for the stresses. This means that in this case the Gibbs equation cannot be constructed. This shows the weakness of thermodynamical models (e.g. of the plastic materials) constructed by the extensions of the Gibbs equation alone. This is frequently the case in the literature of the subject.

The equation $(6.58)_3$ can be solved immediately and we obtain

$$\psi = \psi_1\left(e - \tfrac{\sigma}{E}, T, q\right) + \tfrac{1}{2}\frac{\sigma^2}{\rho_0 E}. \tag{6.60}$$

Substitution of this result into $(6.58)_4$ and integration yield

$$\psi = \psi_2\left(e - \tfrac{\sigma}{E}, T\right) + \tfrac{1}{2}\frac{\sigma^2}{\rho_0 E} + \tfrac{1}{2}\frac{\tau}{\rho_0 KT}q^2. \tag{6.61}$$

Bearing in mind the relation $(6.58)_1$ for the entropy, we obtain the following relations:

$$\varepsilon = \psi_2 - T\frac{\partial\psi_2}{\partial T} + \tfrac{1}{2}\frac{\sigma^2}{\rho_0 E}, \qquad \eta = -\frac{\partial\psi_2}{\partial T} - \tfrac{1}{2}\frac{\tau}{\rho_0 KT^2}q^2. \tag{6.62}$$

Hence, along each characteristic $e-\sigma/E=$const., the internal energy has the same properties as those in the case of the thermoelastic material under isothermal conditions, the heat flux entering only the formula for the entropy. This means that the calorimetric measurements of the specific heat

$$c = \varepsilon_T = -T\frac{\partial^2\psi_2}{\partial T^2} = c\left(e - \tfrac{\sigma}{E}, T\right), \tag{6.63}$$

under some special initial conditions, namely for $e-\sigma/E$ and different temperatures, determine the internal energy. Then the knowledge of the elastic constant, E, the heat conductivity, K, and the relaxation time, τ, specifies the free energy, ψ.

Simultaneously, the differential of ψ yields

$$d\eta=\frac{1}{T}\left[\underline{d\varepsilon-\frac{\sigma}{\rho_0 E}d\sigma}-\frac{\tau q}{\rho_0 KT}dq-\frac{\partial\psi_2}{\partial e}d\left(e-\frac{\sigma}{E}\right)\right]. \tag{6.64}$$

The underlined terms correspond, directly, to the classical Gibbs equation of the thermoelastic materials in which σ=Ee. The two remaining terms describe the influence of the non-equilibrium thermal relaxation, and the stress-strain relaxation, respectively, which follow from the equations $(6.43)_5$ and $(6.43)_3$.

Furthermore, the entropy inequality delivers the residual inequality (6.59), which we proceed to investigate. Let us define

$$\Sigma\equiv-\frac{\partial\psi}{\partial\sigma}G+\frac{q^2}{\rho_0 KT}\geq 0,\qquad \Sigma=\Sigma(\sigma,T,e,q). \tag{6.65}$$

The function Σ is called the **dissipation** density. If we take both G=0 and q=0, then the function Σ attains its minimum value, zero. The first of these conditions means that the mechanical response is elastic, i.e. σ=Ee [see $(6.43)_3$]. The state in which σ=Ee, q=0 is called the **thermodynamical equilibrium**. Let us introduce the new variables (σ,T,e_n,q) instead of (σ,T,e,q), where $e_n=e-\sigma/E$ measures the „non-equilibrium deformation". Then the dissipation, Σ, in terms of new variables, satisfies the conditions

$$\begin{aligned}&\Sigma=\tilde{\Sigma}(\sigma,T,e_n,q)=\frac{1}{E}\left(\frac{\partial\psi_2}{\partial e_n}-\frac{\sigma}{\rho_0}\right)\tilde{G}+\frac{q^2}{\rho_0 KT}\geq 0,\qquad e_n=e-\frac{\sigma}{E},\\&\Sigma_E\equiv\tilde{\Sigma}(\sigma,T,0,0)=0,\qquad \tilde{G}=\tilde{G}(\sigma,T,e_n,q),\qquad \tilde{G}(\sigma,T,0,0)=0.\end{aligned} \tag{6.66}$$

Necessary conditions for the thermodynamical equilibrium then follow

$$\begin{aligned}&\left.\frac{\partial\tilde{\Sigma}}{\partial e_n}\right|_E=\frac{1}{E}\left(\left.\frac{\partial\psi_2}{\partial e_n}\right|_{e_n=0}-\frac{\sigma}{\rho_0}\right)\left.\frac{\partial\tilde{G}}{\partial e_n}\right|_E=0\;\Rightarrow\;\left.\frac{\partial\tilde{G}}{\partial e_n}\right|_E=0,\\&\left.\frac{\partial\tilde{\Sigma}}{\partial q}\right|_E=\frac{1}{E}\left(\left.\frac{\partial\psi_2}{\partial e_n}\right|_{e_n=0}-\frac{\sigma}{\rho_0}\right)\left.\frac{\partial\tilde{G}}{\partial q}\right|_E=0\;\Rightarrow\;\left.\frac{\partial\tilde{G}}{\partial q}\right|_E=0.\end{aligned} \tag{6.67}$$

These conditions show that, in a model describing processes near the thermodynamical equilibrium, $\tilde{G}$ must be non-linear in the variables e_n and q.

Further restrictions, following from the positive definiteness of the hessian of Σ, lead to conditions of higher derivatives of $\tilde{G}$, and to the positive definiteness of K ■

6.3 Hyperbolicity of Field Equations

Inspection of continuum models, appearing in various applications, shows that the set of field equations (5.59), like the set of partial differential equations, may be of different types. Let us name a few examples. The classical theories of heat conduction and diffusion are based on parabolic equations. The classical elasticity theory, and the theory of Eulerian fluids lead to hyperbolic equations. The equations of classical plasticity are of different types in different parts of the configuration space. This means that the requirement of hyperbolicity, which we are going to impose on the set (5.59), cannot be put on an equal footing with such principles as the conservation laws and the second law of thermodynamics, which are **universal** for macroscopical theories.

It cannot even be argued that non-hyperbolic models are of the lowest approximation of the „real" physical hyperbolic set of equations because the transition from the hyperbolic to non-hyperbolic models is connected with singular perturbations. For instance, the limit $\tau=0$ in the Cattaneo equation $(6.43)_5$ (the example 6.2.) yields the parabolic heat conduction equation whose solutions do not approximate to those of the hyperbolic system, in the sense of the regular perturbation series in powers of τ. Moreover, the classical solutions of the parabolic („approximate") equation exist globally, while the classical solutions of the hyperbolic quasilinear equation may cease to exist after a finite time (see: Appendix C).

The above arguments seem to show that hyperbolicity is much more a mathematical than physical requirement. It is certainly true that finite speeds of propagation of all disturbances which follow from the hyperbolicity correspond nicely with our intuition. However, many effects which propagate with infinite speeds within the non-relativistic models, such as, interaction forces (gravitational, Coulomb, etc.), do not disturb us so much as, say, the infinite speed of propagation of thermal disturbances. For this reason, this „physical" argument does not seem to hold water either.

However the mathematical requirement of well-posedness seems to justify the assumption of hyperbolicity, i.e. that at least locally

- the classical solution exists,
- it is unique,
- it depends continuously on the data,

in which case, the Cauchy initial value problem satisfies the above conditions, and is also well motivated by the physical arguments.

The extended thermodynamics is based on the assumption of hyperbolicity of the equations (5.59), and it will dominate our further considerations.

If we write the set of field equations (5.59) in the following explicit form of the set of quasilinear partial differential equations:

$$\mathbf{a}^0(\boldsymbol{w})\frac{\partial \boldsymbol{w}}{\partial t} + \mathbf{a}^k(\boldsymbol{w})\frac{\partial \boldsymbol{w}}{\partial x^k} = \boldsymbol{f}(\boldsymbol{w}), \tag{6.68}$$

with (N×N)-matrices $\mathbf{a}^0$, $\mathbf{a}^k$ defined by

$$\mathbf{a}^0(\boldsymbol{w}) \equiv \frac{\partial \boldsymbol{F}^0}{\partial \boldsymbol{w}}, \qquad \mathbf{a}^k(\boldsymbol{w}) \equiv \frac{\partial \boldsymbol{F}^k}{\partial \boldsymbol{w}}, \tag{6.69}$$

then we say that this set is **hyperbolic in the t-direction** if

(i) $\det \mathbf{a}^0 \neq 0$,

(ii) $\forall \mathbf{n} \in \mathcal{V}^3$, $|\mathbf{n}|=1$: the eigenvalue problem

$$\left(\mathbf{a}^k n_k - \mathbf{a}^0 \mu\right)\mathbf{r} = 0, \tag{6.70}$$

has only real eigenvalues $\mu=\mu(\boldsymbol{w},\mathbf{n})$ and N linearly independent eigenvectors $\mathbf{r}(\boldsymbol{w},\mathbf{n})$.

If the eigenvalues are all distinct, the system (6.68) is called **strictly hyperbolic**.

The system (6.68) is called **symmetric** if

$$\mathbf{a}^0 = \mathbf{a}^{0T}, \qquad \mathbf{a}^k = \mathbf{a}^{kT}, \qquad \mathbf{a}^0 - \text{positive definite.} \tag{6.71}$$

Such systems are automatically hyperbolic.

Hyperbolicity can be lost for three fundamental reasons:

(i) $\mathbf{a}^0$ is singular. Then some characteristic speeds of propagation are infinite, and the set of field equations is of the parabolic type.

(ii) Some eigenvalues are complex. In such cases, certain instabilities arise.

(iii) Eigenvectors do not span the space of solutions. In such cases either the initial discontinuities are constrained to some subspaces of $\Re^N$, or an immediate shock arises (lack of classical solutions).

Further discussion in this book concerning the models within extended thermodynamics is based on the **assumption** that the field equations of extended thermodynamics are **hyperbolic in the t-direction**.

It has been shown, by K. O. Friedrichs, that if, apart from the hyperbolic equations (6.68), some additional conservation laws are to be fulfilled by the solutions, then the system must be symmetric. The same assertion holds true in the case of the constraint in the form of the entropy inequality.

Namely, according to Liu's Theorem we have

$$\begin{aligned} &\exists \Lambda \in \Re^N, \quad \Lambda \neq 0, \quad \forall \boldsymbol{w} \in \Re^N: \\ &\frac{\partial \rho\eta}{\partial t} + \frac{\partial h^k}{\partial x^k} - \Lambda \cdot \left(\frac{\partial \boldsymbol{F}^0}{\partial t} + \frac{\partial \boldsymbol{F}^k}{\partial x^k} - \boldsymbol{f} \right) \geq 0. \end{aligned} \tag{6.72}$$

The chain rule of differentiation and the Liu's Theorem lead immediately to the relations

$$\frac{\partial \rho\eta}{\partial \boldsymbol{w}} = \left(\frac{\partial \boldsymbol{F}^0}{\partial \boldsymbol{w}} \right)^T \Lambda, \quad \frac{\partial h^k}{\partial \boldsymbol{w}} = \left(\frac{\partial \boldsymbol{F}^k}{\partial \boldsymbol{w}} \right)^T \Lambda. \tag{6.73}$$

Now we make the additional assumption that $\rho\eta$ is a **strictly concave** function of $\boldsymbol{w}$. This assumption extends the thermodynamical stability condition of the thermodynamical equilibrium on non-equilibrium processes (compare: Appendix A). Without loss of generality, we may choose $\boldsymbol{F}^0$ to coincide with $\boldsymbol{w}$. Then the relation $(6.73)_1$ implies

$$\Lambda = \frac{\partial \rho\eta}{\partial \boldsymbol{w}} \quad \Rightarrow \quad \frac{\partial \Lambda}{\partial \boldsymbol{w}} = \frac{\partial^2 \rho\eta}{\partial \boldsymbol{w} \partial \boldsymbol{w}} - \text{negative definite.} \tag{6.74}$$

This condition means that the map $\boldsymbol{w} \to \Lambda$ is globally univalent. Hence, we can change the variables from $\boldsymbol{w}$ to Λ and write the inequality (6.72) in the following form:

$$\frac{\partial H^0}{\partial t} + \boldsymbol{F}^0 \cdot \frac{\partial \Lambda}{\partial t} + \frac{\partial H^k}{\partial x^k} + \boldsymbol{F}^k \cdot \frac{\partial \Lambda}{\partial x^k} + \Lambda \cdot \boldsymbol{f} \geq 0, \tag{6.75}$$

where

$$\begin{aligned} H^0 &\equiv \rho\eta - \Lambda \cdot \boldsymbol{F}^0, \\ H^k &\equiv h^k - \Lambda \cdot \boldsymbol{F}^k. \end{aligned} \tag{6.76}$$

Using the chain rule of differentiation, we obtain

$$\begin{aligned} &\boldsymbol{F}^0 = -\frac{\partial H^0}{\partial \Lambda}, \quad \boldsymbol{F}^k = -\frac{\partial H^k}{\partial \Lambda}, \\ &\Lambda \cdot \boldsymbol{f}'(\Lambda) \geq 0, \quad \boldsymbol{f}'(\Lambda) \equiv \boldsymbol{f}(\boldsymbol{w}(\Lambda)), \end{aligned} \tag{6.77}$$

where H^0 and H^k are considered to be functions of Λ. Moreover, the relation $(6.76)_1$ shows that H^0 is the Legendre transformation of $\rho\eta$ and it is a strictly convex function of Λ. Namely,

$$\delta^2 \rho\eta = \delta\left(\frac{\partial \rho\eta}{\partial \boldsymbol{w}}\right) \cdot \delta \boldsymbol{w} = \delta\Lambda \cdot \delta \boldsymbol{w} = \delta\Lambda \cdot \delta\left(-\frac{\partial H^0}{\partial \Lambda}\right) = -\delta^2 H^0 < 0. \tag{6.78}$$

Hence, the matrix $\dfrac{\partial^2 H^0}{\partial \Lambda \partial \Lambda}$ is positive definite.

Thus, we arrive at the very important conclusion:

independently of the number of constitutive relations needed to construct the field equations (6.68) the second law of thermodynamics reduces the constitutive problem of extended thermodynamics to the constitutive relations for the **4-potential** (H^0, H^k).

Substitution of (6.77) into the field equations yields immediately

$$-\frac{\partial^2 H^0}{\partial \Lambda \partial \Lambda}\frac{\partial \Lambda}{\partial t} - \frac{\partial^2 H^k}{\partial \Lambda \partial \Lambda}\frac{\partial \Lambda}{\partial x^k} = \boldsymbol{f}'(\Lambda), \tag{6.78}$$

i.e. a symmetric system. In the theory of hyperbolic PDE, the multipliers Λ yielding the symmetric form of the field equations are called **main fields.**

The practical applications of the main fields in extended thermodynamics of specific materials are usually connected with serious technical difficulties. These arise owing to the necessity of inverting the variables back to the original fields, which, in contrast to the main fields, usually have a clear physical meaning and are measurable. An exact solution to this problem of the change of variables is known only in a few exceptional cases. For this reason, we must usually rely on some approximations as we will show in the following chapters.

6.4 Thermoelastic Materials with Mechanical Constraints

If we impose some additional constraints on the admissible fields, the procedure described above, i.e. the exploitation of the entropy inequality for the hyperbolic field equations, must be modified. Such constraints may include the incompressibility, the inextensibility in a chosen material direction or the non-conductivity of heat in a chosen direction. The problem of **material constraints** is still far from being solved, and we present in this section solely an example of the simplest and most common mechanical constraints.

The pioneering work by R. S. RIVLIN highlighted the importance of the general theory of constraints, and complete references to his work can be found in C. TRUESDELL, W. NOLL (1965). An important thermodynamic contribution to this subject was made by M. E. GURTIN and P. PODIO-GUIDUGLI (1973). In their work one can find the precise mathematical meaning of the notions discussed, in a rather lax manner, in this section. Formal foundations of the theory of constraints are contained in Appendix 3A of C. TRUESDELL's book (1984). This appendix has been written by G. CAPRIZ and P. PODIO-GUIDUGLI.

According to these considerations, the problem of material constraints has not only an engineering aspect connected, for instance, with some (almost) rigid reinforcements of soft matrices, but also influences the class of fundamental solutions applicable in experimental procedures for the constitutive relations. For instance, it is known that compressible elastic materials admit solely uniform universal solutions, i.e. such solutions whose form does not depend on particular material properties. On the other hand, the assumption of incompressibility extends this class of solutions to such important non-uniform solutions as the bending of bars and the torsion of circular shafts. This point shall be discussed again in remark 6.1.

In order to have a reference point for the considerations of mechanical constraints, first we present the structure of the simplest model of the non-linear solid without material constraints.

Example 6.3. *Non-linear Thermoelastic Materials*
We consider the class of processes which can be described by the fields of the velocity, $\mathbf{v}$, the deformation gradient, $\mathbf{F}$, the temperature, T, and the heat flux, $\mathbf{Q}$, defined on the reference configuration, $\mathcal{B}_0$, for all instants of time from the interval, $\mathcal{T}$. We use the Lagrangian description in which the conservation laws have the form [see (4.33), (4.39)]

$$\rho_0 \frac{\partial \mathbf{v}}{\partial t} = \operatorname{Div} \mathbf{P}, \quad \rho_0 \frac{\partial \varepsilon}{\partial t} + \operatorname{Div} \mathbf{Q} = \operatorname{tr} \mathbf{P} \left(\frac{\partial \mathbf{F}}{\partial t} \right)^T, \quad \frac{\partial \mathbf{F}}{\partial t} = \operatorname{Grad} \mathbf{v}, \tag{6.79}$$

and the constitutive relations are assumed to be as follows:

$$\mathbf{P} = \mathbf{F}\,\mathbf{T}(\mathbf{C},T), \quad \varepsilon = \varepsilon(\mathbf{C},T), \quad \mathbf{Q} = -K(\mathbf{C},T)\operatorname{Grad} T, \quad \mathbf{C} \equiv \mathbf{F}^T\mathbf{F}. \tag{6.80}$$

The constitutive relation for the heat flux $(6.80)_3$ shall be considered on the same footing as the other field equations in spite of its degenerate character. The most general form of these relations, for thermoelastic materials, is discussed extensively in the literature [e.g. see: I. MÜLLER (1985)] and we do not need to present it in this example. However, some aspects of the general problem shall be presented in Chap. 10.

The solutions of the above field equations are assumed to satisfy identically the entropy inequality

$$\rho_0 \frac{\partial \eta}{\partial t} + \operatorname{Div}\left(\frac{\mathbf{Q}}{T} \right) \geq 0, \quad \eta = \eta(\mathbf{C},T), \tag{6.81}$$

where it has been already anticipated that the entropy flux is proportional to the heat flux with the coldness, 1/T, as the coefficient.

In order to exploit this condition, we use the Lagrange multipliers. Then, according to Liu's Theorem, the inequality

$$\rho_0\frac{\partial\eta}{\partial t}+\mathrm{Div}\left(\frac{\mathbf{Q}}{T}\right)-\Lambda^{v}\cdot\left(\rho_0\frac{\partial\mathbf{v}}{\partial t}-\mathrm{Div}\,\mathbf{P}\right)-\Lambda^{\varepsilon}\left(\rho_0\frac{\partial\varepsilon}{\partial t}+\mathrm{Div}\,\mathbf{Q}-\mathrm{tr}\,\mathbf{P}\left(\frac{\partial\mathbf{F}}{\partial t}\right)^{T}\right)-$$
$$-\mathrm{tr}\,\Lambda^{F}\left(\frac{\partial\mathbf{F}}{\partial t}-\mathrm{Grad}\,\mathbf{v}\right)^{T}-\Lambda^{Q}\cdot\left(\mathbf{Q}+K\,\mathrm{Grad}\,T\right)\geq 0,\tag{6.82}$$

must hold for arbitrary fields.

We notice that the above inequality is linear with respect to the derivatives $\partial\mathbf{v}/\partial t$ and $\mathrm{Grad}\,\mathbf{v}$. Consequently,

$$\Lambda^{v}=0,\quad \Lambda^{F}=0.\tag{6.83}$$

The chain rule of differentiation now yields

$$\rho_0\left(\frac{\partial\eta}{\partial T}-\Lambda^{\varepsilon}\frac{\partial\varepsilon}{\partial T}\right)\frac{\partial T}{\partial t}+\mathrm{tr}\left(2\rho_0\mathbf{F}\frac{\partial\eta}{\partial\mathbf{C}}-\Lambda^{\varepsilon}2\rho_0\mathbf{F}\frac{\partial\varepsilon}{\partial\mathbf{C}}+\Lambda^{\varepsilon}\mathbf{P}\right)\left(\frac{\partial\mathbf{F}}{\partial t}\right)^{T}+$$
$$+\left(\frac{1}{T}-\Lambda^{\varepsilon}\right)\mathrm{Div}\,\mathbf{Q}-\left(\frac{\mathbf{Q}}{T^{2}}+K\Lambda^{Q}\right)\cdot\mathrm{Grad}\,T-\Lambda^{Q}\cdot\mathbf{Q}\geq 0.\tag{6.84}$$

Hence, bearing Liu's Theorem in mind, we obtain

$$\begin{gathered}\Lambda^{\varepsilon}=\frac{1}{T},\quad \Lambda^{Q}=-\frac{\mathbf{Q}}{T^{2}K},\\ \eta=-\frac{\partial\psi}{\partial T},\quad \varepsilon=\psi-T\frac{\partial\psi}{\partial T},\quad \mathbf{P}=2\rho_0\mathbf{F}\frac{\partial\psi}{\partial\mathbf{C}},\\ K\geq 0,\end{gathered}\tag{6.85}$$

where

$$\psi\equiv\varepsilon-T\eta=\psi(\mathbf{C},T),\tag{6.86}$$

is the **Helmholtz free energy** of the thermoelastic material.

In the particular case of isotropic materials, the Helmholtz free energy depends on the Cauchy-Green deformation tensor solely through its invariants. The relation $(6.85)_5$ for the Piola-Kirchhoff stress tensor can then be transformed to the following relation for the Cauchy stresses [see: (4.68)]:

$$\mathbf{T}=\Im_0\mathbf{1}+\Im_1\mathbf{B}+\Im_{-1}\mathbf{B}^{-1},\qquad \mathbf{B}=\mathbf{F}\mathbf{F}^{T},\tag{6.87}$$

where

$$\Im_0\equiv 2\rho\left(\frac{\partial\psi}{\partial \mathrm{II}^{C}}\mathrm{II}^{C}+\frac{\partial\psi}{\partial \mathrm{III}^{C}}\mathrm{III}^{C}\right),\quad \Im_1\equiv 2\rho\frac{\partial\psi}{\partial \mathrm{I}^{C}},\quad \Im_{-1}\equiv -2\rho\frac{\partial\psi}{\partial \mathrm{II}^{C}}\mathrm{III}^{C},\tag{6.88}$$

denote the **response coefficients** [see: C. TRUESDELL, W. NOLL (1965), Sect. 47] and $\rho \equiv \rho_0 J^{-1}$ is the current mass density. The invariants of the deformation tensor, **C**, are identical to the invariants of the deformation tensor, **B**, (see: the Table in Sect. 2.3.).

Exercise 6.4. Show that the eigenvectors of the Cauchy stress tensor, **T**, in the isotropic material coincide with the eigenvectors of the left Cauchy-Green deformation tensor, **B**, and the eigenvalues are given by the following relations:

$$\sigma^{(\alpha)} = \Im_0 + \Im_1 \lambda^{(\alpha)} + \Im_{-1} \frac{1}{\lambda^{(\alpha)}}, \quad \mathbf{T} = \sum_{\alpha=1}^{3} \sigma^{(\alpha)} \mathbf{k}^{(\alpha)} \otimes \mathbf{k}^{(\alpha)}, \tag{6.89}$$

where $\lambda^{(\alpha)}$ are the eigenvalues and $\mathbf{k}^{(\alpha)}$ are the eigenvectors of the deformation tensor, **B•**

We illustrate the constitutive relations (6.87) by two simple examples corresponding to the deformation of the prism in the homogeneous extension (example 2.1.) and in the simple shearing (example 2.2.).

In the first case the stress tensor, **T**, has the following form:

$$\mathbf{T} = \left(\Im_0 + \frac{1}{\lambda} \Im_1 + \lambda \Im_{-1} \right) (\mathbf{e}_1 \otimes \mathbf{e}_1 + \mathbf{e}_2 \otimes \mathbf{e}_2) + \left(\Im_0 + \lambda^2 \Im_1 + \frac{1}{\lambda^2} \Im_{-1} \right) \mathbf{e}_3 \otimes \mathbf{e}_3. \tag{6.90}$$

If we assume that the external loading is only applied on the upper and lower face of the prism, then we have

$$\begin{gathered} \mathbf{t}_n = t\mathbf{e}_3 \Rightarrow \Im_0 = -\left(\frac{1}{\lambda} \Im_1 + \lambda \Im_{-1} \right) \Rightarrow \mathbf{T} = \sigma^{33} \mathbf{e}_3 \otimes \mathbf{e}_3, \\ \sigma^{33} = \left(\lambda^2 - \frac{1}{\lambda} \right) \left(\Im_1 - \frac{1}{\lambda} \Im_{-1} \right). \end{gathered} \tag{6.91}$$

This relationship is plotted in Fig. 6.2. for **neo-Hookean material**, i.e. for $\Im_1$=const., $\Im_{-1}=0$. However, this model has a rather academic character because it deviates from the experimental curves too much to be of any use in practical applications. It shows, however, the qualitative behaviour of the tensile force for the present example.

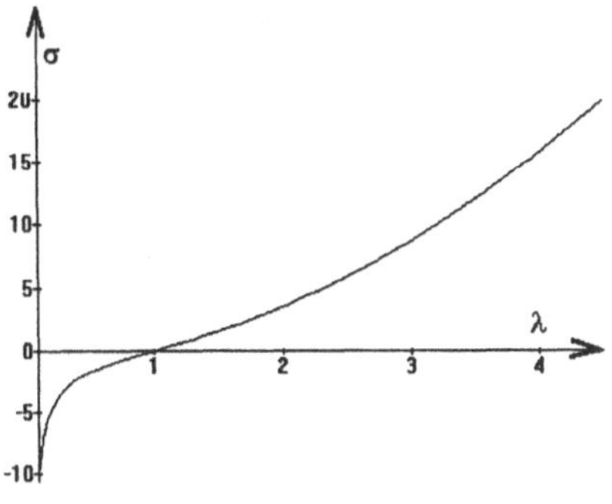

Fig. 6.2. *The normal stresses $\sigma \equiv \sigma^{33}/\Im_1$ as the function of stretch, λ, for the neohookean material*

In the case of the simple shearing, we have (compare: example 2.2.)

$$\begin{aligned}
&\mathbf{B} = \mathbf{e}_1 \otimes \mathbf{e}_1 + \left(1+\kappa^2\right)\mathbf{e}_2 \otimes \mathbf{e}_2 + \kappa\left(\mathbf{e}_3 \otimes \mathbf{e}_2 + \mathbf{e}_2 \otimes \mathbf{e}_3\right) + \mathbf{e}_3 \otimes \mathbf{e}_3, \\
&\mathbf{B}^{-1} = \mathbf{e}_1 \otimes \mathbf{e}_1 + \mathbf{e}_2 \otimes \mathbf{e}_2 - \kappa\left(\mathbf{e}_3 \otimes \mathbf{e}_2 + \mathbf{e}_2 \otimes \mathbf{e}_3\right) + \left(1+\kappa^2\right)\mathbf{e}_3 \otimes \mathbf{e}_3, \\
&\mathrm{I}^C = \mathrm{II}^C = 3+\kappa^2, \quad \mathrm{III}^C = 1 \quad \Rightarrow \quad \Im_\Gamma = \Im_\Gamma\left(\kappa^2\right), \ \Gamma = -1,0,1.
\end{aligned} \tag{6.92}$$

Substitution into the constitutive law (6.87) yields

$$\begin{aligned}
\mathbf{T} = &\left(\Im_0 + \Im_1 + \Im_{-1}\right)\mathbf{e}_1 \otimes \mathbf{e}_1 + \left(\Im_0 + \left(1+\kappa^2\right)\Im_1 + \Im_{-1}\right)\mathbf{e}_2 \otimes \mathbf{e}_2 + \\
&+\left(\Im_1 - \Im_{-1}\right)\kappa\left(\mathbf{e}_3 \otimes \mathbf{e}_2 + \mathbf{e}_2 \otimes \mathbf{e}_3\right) + \left(\Im_0 + \Im_1 + \left(1+\kappa^2\right)\Im_{-1}\right)\mathbf{e}_3 \otimes \mathbf{e}_3.
\end{aligned} \tag{6.93}$$

As expected, the shear stresses σ^{12} are the odd function of κ:

$$\sigma^{12} = \mu\left(\kappa^2\right)\kappa, \quad \mu\left(\kappa^2\right) \equiv \Im_1\left(\kappa^2\right) - \Im_{-1}\left(\kappa^2\right). \tag{6.94}$$

From the relation (6.93) we obtain the following pressure p:

$$p \equiv -\tfrac{1}{3}\operatorname{tr}\mathbf{T} = -\left(\Im_0 + \Im_1 + \Im_{-1}\right) - \tfrac{1}{3}\left(\Im_1 + \Im_{-1}\right)\kappa^2. \tag{6.95}$$

Consequently, the condition

$$\lim_{\kappa\to 0} p \equiv -\lim_{\kappa\to 0}\left(\Im_0 + \Im_1 + \Im_{-1}\right) = 0, \tag{6.96}$$

means that the undeformed state, $\kappa=0$, corresponds to the stress-free state. If the sum of response coefficients is differentiable in the neighbourhood of this state, then

$$\left|\lim_{\kappa\to 0}\frac{\Im_0 + \Im_1 + \Im_{-1}}{\kappa^2}\right| < \infty \tag{6.97}$$

and, according to the sign of the pressure, the sample under shear stresses must expand or contract. This is the so-called **Kelvin's effect**.

Simultaneously, the large shear deformation yields unequal normal stresses, σ^{22} and σ^{33}. According to (6.93), we have

$$\sigma^{22} = \sigma^{33} \quad \Rightarrow \quad \Im_1 - \Im_{-1} = 0 \quad \Rightarrow \quad \mu = 0, \tag{6.98}$$

and this would yield the absence of the shear stresses. This effect of unequal normal stresses is called a **Poynting effect**.

Certainly, both these effects appear only in the case of large deformations ■

After this brief excursus into non-linear elasticity of unconstrained materials, we are in a position to discuss the simplest material constraints. We assume, in contrast to the above examples of the thermoelastic materials, that the admissible field of the deformation gradient, **F**, is additionally constrained by the following scalar **constraint condition**:

$$\mathcal{g}(\mathbf{F}) = 0, \tag{6.99}$$

where the function $\mathcal{g}$ is assumed to be continuously differentiable. Geometrically, we can say that the changes of the deformation gradient, **F**, in the nine-dimensional space of its values are limited to the eight-dimensional hypersurface in this space defined by (6.99).

This additional condition means that the set of field equations is overdetermined. Consequently, we could expect difficulties with the existence of solutions if we were not able to introduce an additional scalar field to make the system determined again. Luckily, in the case of one-component systems we can introduce such a field. As we see, the constraint condition (6.99) requires the existence of an additional field of the **reaction force** sustaining the constraint of the motion of the body induced by (6.99).

The condition (6.99) is fulfilled if an arbitrary small change of the deformation gradient, $\delta\mathbf{F}$, compatible with the remaining field equations does not change the function $\mathcal{g}$, i.e.

$$\forall\,\delta\mathbf{F}\text{ - compatible with field equations:}\quad \frac{\partial \mathcal{g}}{\partial \mathbf{F}}\cdot(\delta\mathbf{F})=0,$$
$$\delta\mathbf{F}\equiv\frac{\partial \mathbf{F}}{\partial t}\delta t+(\operatorname{Grad}\mathbf{F})\delta\mathbf{X}. \tag{6.100}$$

It follows that the condition (6.99) for **F** can be equivalently replaced by the following local conditions:

$$\frac{\partial \mathbf{F}}{\partial t}\cdot\frac{\partial \mathcal{g}}{\partial \mathbf{F}}=0,\quad (\operatorname{Grad}\mathbf{F})^{\overset{13}{T}}\cdot\left(\frac{\partial \mathcal{g}}{\partial \mathbf{F}}\right)^T=0,\quad (\operatorname{Grad}\mathbf{F})^{\overset{13}{T}}\equiv\frac{\partial F^k{}_\alpha}{\partial X^\beta}\mathbf{e}_\beta\otimes\mathbf{e}_\alpha\otimes\mathbf{e}_k. \tag{6.101}$$

Again, the geometrical interpretation of these relations yields the result that arbitrary infinitesimal changes in the gradient, **F**, must be tangential to the hypersurface (6.99) because the derivative $\partial\mathcal{g}/\partial\mathbf{F}$ defines the nine-dimensional vector orthogonal to the hypersurface $\mathcal{g}=0$. Consequently, if we perform an arbitrary variation of **F**, which does not satisfy this condition, its component perpendicular to this hypersurface must be eliminated by the reaction force preventing this type of motion. Hence, the reaction force must be perpendicular to the surface of constraints (6.99).

Such a force does no work on the changes of deformation which are compatible with the constraint condition. Then we say that the constraints under consideration are **ideal**.

We proceed to exploit the entropy inequality. We shall do so for processes described in the example 6.3. above, i.e the field equations for (**v**,**F**,T,**Q**) follow from (6.79), and (6.80) with the additional relation (6.99). Consequently, we have to fulfil the inequality (6.84) with the additional constraints (6.101)

$$\rho_0\frac{\partial \eta}{\partial t}+\operatorname{Div}\left(\frac{\mathbf{Q}}{T}\right)-\Lambda^v\cdot\left(\rho_0\frac{\partial \mathbf{v}}{\partial t}-\operatorname{Div}\mathbf{P}\right)-\Lambda^\varepsilon\left(\rho_0\frac{\partial \varepsilon}{\partial t}+\operatorname{Div}\mathbf{Q}-\operatorname{tr}\mathbf{P}\left(\frac{\partial \mathbf{F}}{\partial t}\right)^T\right)-$$
$$-\operatorname{tr}\Lambda^F\left(\frac{\partial \mathbf{F}}{\partial t}-\operatorname{Grad}\mathbf{v}\right)^T-\Lambda^Q\cdot(\mathbf{Q}+K\operatorname{Grad}T)- \tag{6.102}$$
$$-\Lambda^{\mathcal{g}}\cdot\left((\operatorname{Grad}\mathbf{F})^{\overset{13}{T}}\cdot\left(\frac{\partial \mathcal{g}}{\partial \mathbf{F}}\right)^T\right)-\lambda^{\mathcal{g}}\left(\frac{\partial \mathbf{F}}{\partial t}\cdot\frac{\partial \mathcal{g}}{\partial \mathbf{F}}\right)\geq 0,$$

which must hold for **arbitrary** fields without any more constraints. Linearity of this inequality with respect to the derivatives $\partial \mathbf{v}/\partial t$ and $\operatorname{Grad}\mathbf{v}$, as well as with respect to $\operatorname{Grad}\mathbf{F}$, yields

$$\Lambda^{v}=0,\quad \Lambda^{F}=0,\quad \Lambda^{\mathcal{g}}=0. \tag{6.103}$$

Now the chain rule of differentiation yields

$$\rho_0\left(\frac{\partial \eta}{\partial T}-\Lambda^{\varepsilon}\frac{\partial \varepsilon}{\partial T}\right)\frac{\partial T}{\partial t}+\operatorname{tr}\left(2\rho_0\mathbf{F}\frac{\partial \eta}{\partial \mathbf{C}}-\Lambda^{\varepsilon}2\rho_0\mathbf{F}\frac{\partial \varepsilon}{\partial \mathbf{C}}+\Lambda^{\varepsilon}\mathbf{P}-\right.$$
$$\left.-\lambda^{\mathcal{g}}\frac{\partial \mathcal{g}}{\partial \mathbf{F}}\right)\left(\frac{\partial \mathbf{F}}{\partial t}\right)^{T}+\left(\frac{1}{T}-\Lambda^{\varepsilon}\right)\operatorname{Div}\mathbf{Q}-\left(\frac{\mathbf{Q}}{T^2}+K\Lambda^{Q}\right)\cdot\operatorname{Grad}T-\Lambda^{Q}\cdot\mathbf{Q}\geq 0. \tag{6.104}$$

Hence,

$$\Lambda^{\varepsilon}=\frac{1}{T},\quad \Lambda^{Q}=-\frac{\mathbf{Q}}{T^2K},$$
$$\eta=-\frac{\partial\psi}{\partial T},\quad \varepsilon=\psi-T\frac{\partial\psi}{\partial T},\quad \mathbf{P}=2\rho_0\mathbf{F}\frac{\partial\psi}{\partial\mathbf{C}}+T\lambda^{\mathcal{g}}\frac{\partial\mathcal{g}}{\partial\mathbf{F}}, \tag{6.105}$$
$$K\geq 0.$$

The Helmholtz free energy, ψ, is again defined by the relation (6.86).

These results are similar to those for the thermoelastic material considered in the example 6.3. [see: (6.85)], except for the additional term in the relation for the Piola-Kirchhoff stress. It contains additionally the multiplier $\lambda^{\mathcal{g}}$, whose presence follows from the constraint condition, and which remains thermodynamically undetermined. This means that the Helmholtz free energy, ψ, in contrast to the previously considered unconstrained model, is not the thermodynamical potential for stresses in the present case. If we introduce the following notion:

$$\boldsymbol{N}\equiv-\tfrac{1}{2}\mathrm{p}\left(\mathbf{F}\frac{\partial\mathcal{g}}{\partial\mathbf{F}^{T}}+\frac{\partial\mathcal{g}}{\partial\mathbf{F}}\mathbf{F}^{T}\right),\qquad \mathrm{p}\equiv-2\frac{T}{J}\lambda^{\mathcal{g}},\qquad \boldsymbol{N}^{T}=\boldsymbol{N}, \tag{6.106}$$

then the relation $(6.105)_5$ for the Piola-Kirchhoff stress yields the following formula for the Cauchy stresses $\mathbf{T}$:

$$\mathbf{T}-\boldsymbol{N}=2\rho\mathbf{F}\frac{\partial\psi}{\partial\mathbf{C}}\mathbf{F}^{T}. \tag{6.107}$$

The symmetric tensor, $\boldsymbol{N}$, is called the **reaction stress**, and its form follows neither from the constitutive laws, nor from the constraint condition. The unspecified part is due to the multiplier $\lambda^{\mathcal{g}}$ and has been denoted by p in the relation (6.106). This quantity must be found from the field equations. Thus, despite the additional equation furnished by the constraint condition, the set of field equations remains determined. The fields are now ($\mathbf{v}$, $\mathbf{F}$, T, $\mathbf{Q}$, p).

Let us notice that the condition $(6.101)_1$ yields

$$\operatorname{tr}\boldsymbol{N}\mathbf{L}^{T}=0. \tag{6.108}$$

This means that the reaction stress, $\boldsymbol{N}$, indeed does not work on the real deformations in the case of ideal constraints.

The above considerations can easily be extended to larger number of constraint conditions for the deformation gradient. If we replace the condition (6.99) by the following set:

$$\forall \mathbf{C}: \quad \mathscr{g}^{(\alpha)}(\mathbf{C}) = 0, \quad \alpha = 1,\ldots,a, \quad a \leq 6, \tag{6.109}$$

then the reaction stress follows in the form

$$\boldsymbol{N} = -\sum_{\alpha=1}^{a} p^{(\alpha)} \mathbf{F} \frac{\partial \mathscr{g}^{(\alpha)}}{\partial \mathbf{C}} \mathbf{F}^T, \tag{6.110}$$

where $p^{(\alpha)}$ are the constitutively undetermined coefficients, i.e. the fields of reactions on the constraints.

Example 6.4. *Incompressible Materials*
Let us assume the condition (6.99) to have the following form:

$$\mathscr{g}(\mathbf{F}) = (\det \mathbf{F})^2 - 1 \equiv \det \mathbf{C} - 1 = 0. \tag{6.111}$$

Then such a material, in contrast to the compressible materials in isochoric processes, cannot change the volume in *any real process of deformation*, therefore, it is called **incompressible**. We return to some aspects of this definition of incompressibility in Chap. 8. Now we have

$$\frac{\partial \mathscr{g}}{\partial \mathbf{C}} = \mathbf{C}^{-1} \det \mathbf{C} \quad \Rightarrow \quad \boldsymbol{N} = -p\mathbf{F}\left(\mathbf{C}^{-1} \det \mathbf{C}\right)\mathbf{F}^T = -p\mathbf{F}\left(\mathbf{F}^T\mathbf{F}\right)^{-1}\mathbf{F}^T = -p\mathbf{1}. \tag{6.112}$$

The above result of Poincaré shows that only the deviatoric part of the Cauchy stress tensor can be described by the constitutive law

$$\mathbf{T} + p\mathbf{1} = 2\rho\mathbf{F}\frac{\partial \psi}{\partial \mathbf{C}}\mathbf{F}^T, \qquad \psi = \psi(\mathbf{C}, T). \tag{6.113}$$

In the case of isotropic materials, the Helmholtz free energy, ψ, is solely the function of the invariants I^C, II^C, and T, because $III^C \equiv 1$. Consequently,

$$\begin{gathered} \frac{\partial \psi}{\partial \mathbf{C}} = \frac{\partial \psi}{\partial I^C}\mathbf{1} + \frac{\partial \psi}{\partial II^C}\left(I^C\mathbf{1} - \mathbf{C}\right) \quad \Rightarrow \\ \Rightarrow \quad \mathbf{T} = -p\mathbf{1} + \Im_1\mathbf{B} + \Im_{-1}\mathbf{B}^{-1}, \qquad p \equiv p - 2\rho\frac{\partial \psi}{\partial II^C}II^C, \end{gathered} \tag{6.114}$$

where the response coefficients $\Im_1$ and $\Im_{-1}$ are given by (6.88) with $III^C \equiv 1$ and p, owing to the reaction pressure p on its right-hand side, is constitutively unspecified. However, this is not the whole pressure because the traces of $\mathbf{B}$ and $\mathbf{B}^{-1}$ are not identically zero ■

Example 6.5. *Inextensible Materials*
The function $\mathcal{g}$ of the condition (6.80) is assumed to be

$$\mathcal{g}(\mathbf{F}) = (\mathbf{FK})\cdot(\mathbf{FK}) - 1 = \mathbf{K}\cdot\mathbf{CK} - 1, \tag{6.115}$$

where **K** is a chosen material vector field. It follows:

$$\frac{\partial \mathcal{g}}{\partial \mathbf{C}} = \mathbf{K}\otimes\mathbf{K} \quad \Rightarrow \quad N = -p\mathbf{F}(\mathbf{K}\otimes\mathbf{K})\mathbf{F}^{\mathrm{T}},$$

i.e.

$$N = N\left(\frac{\mathbf{FK}}{|\mathbf{FK}|}\right)\otimes\left(\frac{\mathbf{FK}}{|\mathbf{FK}|}\right), \qquad N \equiv -p|\mathbf{FK}|^2. \tag{6.116}$$

This result of Adkins and Rivlin shows that the traction in the direction **FK** cannot be described by the constitutive relations, and the result reflects the reaction force in the material on the **inextensibility** in this direction ■

Example 6.6. *Rigid Bodies*
If we assume that the material is inextensible in an arbitrary direction, the whole stress tensor, **T**, is not described by the constitutive law. The stress becomes the reaction on the constraints, which do not allow for any deformation of the body at all. Hence, we deal with the **rigid** body ■

Remark 6.1. *On the Universal Solutions – Part II*
The notion of incompressibility is connected with the universal solutions which have been mentioned in the second chapter (the remark 2.3). Its importance follows from the fact that the universal solutions for compressible elastic materials are, as we see further, solely homogeneous, while the incompressibility yields also heterogeneous universal solutions which are, in turn, important in the applications to the static experiments on non-linear materials.

The **universal solution** is the solution of the equation of equilibrium without body forces, and such that it is *independent* of the material properties. For instance, such solutions of the classical elasticity theory fulfil the following conditions:

$$\begin{aligned} \underset{\lambda,\mu}{\forall}\ \operatorname{div}\mathbf{T} = \operatorname{div}\big(\lambda(\mathbf{e}\cdot\mathbf{1})\mathbf{1} + 2\mu\mathbf{e}\big) = \mu\left(\frac{\lambda+\mu}{\mu}\operatorname{grad}\operatorname{div}\mathbf{u} + \operatorname{div}\operatorname{grad}\mathbf{u}\right) = 0 \ \Rightarrow \\ \Rightarrow \ \operatorname{grad}\operatorname{div}\mathbf{u} = 0, \quad \operatorname{div}\operatorname{grad}\mathbf{u} = 0, \end{aligned} \tag{6.117}$$

where **u** is the displacement vector, and λ, μ denote the Lamé constants of the linear elasticity. Consequently, in order for the displacement vector to be the solution for **all** linearly elastic materials, it must satisfy six partial differential equations of the second order. The universal solutions lend themselves to the determination of the material factor, $(\lambda+\mu)/\mu$. This is the reason why these solutions are the most valuable of all. That is, we must only find the tractions which produce these solutions. Then we can determine the material properties without solving complicated boundary-value problems. Obviously, any homogeneous deformation is universal.

In the case of non-linear elastic materials, the condition for the universal solutions follows from the equilibrium condition in the form

$$\mathrm{div}\left(\mathfrak{I}_0\mathbf{1}+\mathfrak{I}_1\mathbf{B}+\mathfrak{I}_{-1}\mathbf{B}^{-1}\right)=0, \tag{6.118}$$

which must be satisfied by **B** identically with respect to the response coefficients $\mathfrak{I}_\Gamma$. Consequently, the chain rule of differentiation renders the coefficients of $\mathfrak{I}_0$, $\partial\mathfrak{I}_0/\partial I^C$, $\partial\mathfrak{I}_0/\partial II^C,\ldots,\partial\mathfrak{I}_{-1}/\partial III^C$ equal to zero. Therefore, there are 12 conditions for six components of **B**. J. ERICKSEN*) showed that solely **B**=const. can satisfy these conditions.

The situation changes dramatically if we consider incompressible elastic materials. The research into this problem began in the late 1940s with the pioneering work of R. RIVLIN. In this case the relation (6.118) must be replaced by:

$$-\mathrm{grad}\,p+\mathrm{div}\left(\mathfrak{I}_1\mathbf{B}+\mathfrak{I}_{-1}\mathbf{B}^{-1}\right)=0 \quad\Rightarrow\quad \mathrm{curl}\,\mathrm{div}\left(\mathfrak{I}_1\mathbf{B}+\mathfrak{I}_{-1}\mathbf{B}^{-1}\right)=0. \tag{6.119}$$

In this case, again we obtain 12 conditions but the deformation tensor, **B**, possesses only five independent components. This yields the existence of many non-homogeneous universal solutions, which we have listed in remark 2.3 •

6.5 Solid Interfaces

We close this chapter with the presentation of the thermodynamical conditions for a certain class of singular surfaces appearing as interfaces during phase transformations, in the contact of two bodies, etc. For the interface to appear phases are assumed to be separated from each other. Owing to the microscopical incompatibilities, such as different symmetries of lattices constituting the phases, the transition region from one phase to another is characterized by the concentration of stresses and energy, which can be modelled by singular surfaces. These surfaces can move in the material and the motion is caused by the phase transformation.

An example of the interface in a solid can be seen, for instance, in the case of the so-called martensitic phase transformation, for which the interface separates the martensite (the low symmetry phase) from the austenite (the high symmetry phase), and it appears as a result of the incoherence of lattices of the martensite and austenite. In single crystals, the interfaces of this type are usually plain.

Another type of interface is created during the process of evaporation (condensation). The interface separates the liquid from the vapour, and its stability requires surface tension. Thus, such interfaces must be curved.

Here we will limit our attention mainly to the interfaces and contact surfaces of solids, and hence, use the Lagrangian description. This problem has recently been reconsidered by I-SHIH LIU (1990), and the following section contains some of his results.

Let us recall the Rankine-Hugoniot conditions (4.34) for the singular surface, $\mathcal{S}_0(t)$, which have the form

$$\begin{gathered} U[[\rho_0]]=0,\quad U[[\rho_0\mathbf{v}]]+[[\mathbf{P}]]\mathbf{N}=0,\\ U\left[\left[\rho_0\left(\tfrac{1}{2}v^2+\varepsilon\right)\right]\right]+\left[\left[\mathbf{P}^T\mathbf{v}-\mathbf{Q}\right]\right]\mathbf{N}=0,\quad -U[[\rho_0\eta]]+\left[\left[\frac{1}{T}\mathbf{Q}\right]\right]\mathbf{N}\geq 0, \end{gathered} \tag{6.120}$$

*) see: C. TRUESDELL, W. NOLL (1965), Sect. 91, where the account of the research on the incompressible materials can be found as well.

where, in the case of the Lagrangian description, the entropy inequality corresponds to the condition (6.14), and it is assumed that the entropy flux has the form **Q**/T.

We are interested in the interface for continuous mass density, ρ_0, and the continuous temperature,

$$[[\rho_0]] = 0, \quad [[T]] = 0. \tag{6.121}$$

Both conditions are well fulfilled for solid-solid transformations.

We also use the kinematical compatibility condition (4.42)

$$-U[[\mathbf{F}]] = [[\mathbf{v}]] \otimes \mathbf{N}, \tag{6.122}$$

which implies

$$[[\mathbf{v}]] = -U\boldsymbol{a}, \quad [[\mathbf{F}]] = \boldsymbol{a} \otimes \mathbf{N}, \quad \boldsymbol{a} \equiv [[\mathbf{F}]]\mathbf{N}. \tag{6.123}$$

As shown in Sect. 4.4., the energy balance on the singular surface, $\mathcal{S}_0(t)$, can be written in the form [compare (4.46)]

$$U\left(\rho_0[[\varepsilon]] - \langle \mathbf{PN} \rangle \cdot \boldsymbol{a}\right) = [[\mathbf{Q}]] \cdot \mathbf{N}, \tag{6.124}$$

where

$$\langle \mathbf{PN} \rangle \equiv \tfrac{1}{2}\left(\mathbf{P}^+ + \mathbf{P}^-\right)\mathbf{N}. \tag{6.125}$$

Under the assumption $(6.121)_2$, we can combine the relations $(6.120)_4$ and (6.124) to obtain

$$U\left(\rho_0[[\psi]] - \langle \mathbf{PN} \rangle \cdot \boldsymbol{a}\right) \equiv U f_N \geq 0. \tag{6.126}$$

The quantity f_N is called the **driving force** of the interface. The driving force can be written in a more useful alternative form by means of the following manipulations:

$$\begin{aligned}
&\langle \mathbf{PN} \rangle \cdot \boldsymbol{a} = [[(\mathbf{PN}) \cdot (\mathbf{FN})]] - [[\mathbf{PN}]] \cdot \langle \mathbf{FN} \rangle = \\
&= [[\mathbf{N} \cdot \mathbf{F}^T \mathbf{PN}]] - \rho_0 U^2 [[\mathbf{FN}]] \cdot \langle \mathbf{FN} \rangle = [[\mathbf{N} \cdot \mathbf{F}^T \mathbf{PN}]] - \tfrac{1}{2}\rho_0 U^2 [[(\mathbf{FN}) \cdot (\mathbf{FN})]] = \\
&= [[\mathbf{N} \cdot \mathbf{F}^T \mathbf{PN}]] - \tfrac{1}{2}\rho_0 U^2 \mathbf{N} \cdot [[\mathbf{C}]]\mathbf{N},
\end{aligned} \tag{6.127}$$

where $(6.120)_2$ and (6.122) have been used.

Substitution in (6.126) yields

$$f_N = \mathbf{N} \cdot \left[\left[\mathbf{M} + \frac{\rho_0}{2} U^2 \mathbf{C}\right]\right] \mathbf{N}, \quad U f_N \geq 0, \tag{6.128}$$

where

$$\mathbf{M} \equiv \rho_0 \psi \mathbf{1} - \mathbf{F}^T \mathbf{P}, \tag{6.129}$$

is called the **Eshelby's energy-momentum tensor**.

Most processes of the phase transformations in solids, which are considered by metallurgists, are assumed to be **quasistatic** (phase equilibrium processes), i.e.

(i) isothermal ([[T]]=0) and slow,
(ii) in mechanical equilibrium ([[**P**]]**N**=0),
(iii) without the surface entropy production.

Then

$$f_N = 0, \qquad f_N \approx \mathbf{N}\cdot[[\mathbf{M}]]\mathbf{N} = [[\rho_0\psi - \operatorname{tr}\mathbf{N}\otimes\langle\mathbf{PN}\rangle\mathbf{F}]] \tag{6.130}$$

This means that the continuity of the normal component of the Eshelby's energy-momentum tensor, **M**, characterizes interfaces in the conditions of the phase equilibrium.

Let us notice that for the ideal fluids

$$[[\mathbf{N}\cdot\mathbf{F}^{\mathrm{T}}\mathbf{PN}]] = [[J\mathbf{N}\cdot\mathbf{F}^{\mathrm{T}}\mathbf{T}\mathbf{F}^{-\mathrm{T}}\mathbf{N}]] = \left[\left[\frac{\rho_0}{\rho}\mathbf{N}\cdot\mathbf{F}^{\mathrm{T}}(-p\mathbf{1})\mathbf{F}^{-\mathrm{T}}\mathbf{N}\right]\right] = -\rho_0\left[\left[\frac{p}{\rho}\right]\right]. \tag{6.131}$$

Therefore,

$$\mathbf{N}\cdot[[\mathbf{M}]]\mathbf{N} = \rho_0\left[\left[\psi + \frac{p}{\rho}\right]\right] = \rho_0[[g]] = 0, \tag{6.132}$$

where g is the Gibbs free energy (i.e. the **chemical potential** for a single component system). The condition (6.132) constitutes the basis of the Gibbs consideration of phase transformations in fluids. For this reason, one may call **M** the **chemical potential tensor** when talking about quasistatic processes in solids. This name was most likely introduced by C. TRUESDELL in the first edition of his book on rational thermodynamics (1969).

7 Ideal Gases

7.1 Introduction

Bearing in mind the results of the previous chapters, we are in a position to construct extended thermodynamical models of various media, whose description requires the appearance of fields reflecting a substantial deviation from the thermodynamical equilibrium. We devote the present chapter to the extended thermodynamics of ideal dilute gases based on the field equations for 13 fields of mass, momentum and energy densities, the stress deviator, and the heat flux. We expect these field equations to have the same structure as those following from the kinetic theory indicated in the remark 5.1. The **transfer equations** of the kinetic theory, namely (5.110), (5.114), (5.120), and the equation for the third moment of the Boltzmann equation not quoted in this remark, were introduced by H. Grad (1958) (the so-called **13-moment approximation**), and it has been firmly established that these equations reflect the transport phenomena in dilute gases quite well.

However, in contrast to Grad's method, the model presented in this chapter is **macroscopical** as it requires a thermodynamical approach. We shall see that, in contrast to the models of the classical non-equilibrium thermodynamics, this model admits rapid processes with steep gradients. When there is strong deviation from the thermodynamical equilibrium it also offers an answer to the question of the usefulness of temperature as a thermodynamical variable. It shall be seen that in processes close to such an equilibrium the temperature is continuous across ideal walls (compare: Appendix A) and consequently, measurable. This is not the case in strongly non-equilibrium processes. Therefore temperature looses its usefulness in such processes.

Simultaneously, extended thermodynamics yields finite speeds of propagation of the disturbances. In the case of ideal gases, two out of three additional material coefficients, which render this property, are quite explicit, and this, in turn, is the consequence of the second law of thermodynamics.

Extended thermodynamics formulated within the framework of linear irreversible thermodynamics accounts for the finite speeds of propagation of the thermal disturbances, and shear waves in viscous fluids. This was initiated by the pioneering work of I. MÜLLER (1966). In the following presentation of the theory of ideal gases, we rely on the paper of I-SHIH LIU and I. MÜLLER (1983), which was the first work on this subject based on the systematic procedure of extended thermodynamics with the use of Lagrange multipliers.

To simplify the reference to the original paper, in this chapter we use the notation which was introduced by Liu and Müller for extended thermodynamics and which differs considerably from the standard notation of continuum mechanics. The same notation is used by I. MÜLLER in his book on thermodynamics (1985), where a brief presentation of extended thermodynamics can be found. It is also adopted in the book by I. MÜLLER and T. RUGGERI (1993). However, during the presentation we shall frequently refer to the notation used in the previous chapters.

For typographical reasons, it is convenient to work in a Cartesian (not necessarily inertial!) frame of reference, and place all indices as covariant. *We use parentheses to indicate general symmetrization, square brackets for antisymmetrization and angular brackets for the symmetrical and trace-free tensors.*

As mentioned above, the aim of the present model is to establish the field equations for 13 quantities:

$$\begin{aligned}&F \text{ - the mass density},\\&F_i \text{ - the momentum density},\\&F_{ij} \text{ - the momentum flux density}, F_{ij}=F_{ji},\\&\tfrac{1}{2}F_{ijj} \text{ - the energy flux density}.\end{aligned} \tag{7.1}$$

These fields are assumed to satisfy the **balance equations** which have the following tensorial hierarchical structure:

$$\begin{aligned}&\frac{\partial F}{\partial t}+\frac{\partial F_k}{\partial x_k}=0,\\&\frac{\partial F_i}{\partial t}+\frac{\partial F_{ik}}{\partial x_k}=F(b_i'+i_i)+2F_k\,\Omega_{ik},\\&\frac{\partial F_{ij}}{\partial t}+\frac{\partial F_{ijk}}{\partial x_k}=f_{\langle ij\rangle}+2F_{(i}\left(b_{j)}+i^0_{j)}\right)+4F_{k(i}\,\Omega_{j)k},\\&\frac{\partial F_{ijj}}{\partial t}+\frac{\partial F_{ijjk}}{\partial x_k}=f_{ijj}+3F_{(ij}\left(b_{j)}+i^0_{j)}\right)+6F_{kjj}\,\Omega_{ik}.\end{aligned} \tag{7.2}$$

The tensor, Ω, denotes the tensor of the relative angular velocity (3.60), and the vector $\mathbf{i}^0$ the inertial acceleration,

$$i^0_i\equiv\left(\dot{\Omega}_{ik}-\Omega_{ij}\Omega_{jk}\right)(x_k-c_k)-2\Omega_{ik}\dot{c}_k+\ddot{c}_i, \tag{7.3}$$

i.e. the contribution of the momentum to the inertial force, $\mathbf{i}^*$, has been separated from the contribution of the inertial acceleration [compare (4.107)].

The structure of the above set of equations is motivated by the structure of the kinetic transfer equations. For the same reason, the tensors, F_{ijk} and F_{ijkn}, are assumed to be **fully symmetric**. Inspection of the general set of balance equations (5.59) shows that the above set is a special case of (5.59) with a very peculiar symmetry. Namely, the flux in the nth tensorial equation acts as the density in the $(n+1)^{st}$ tensorial equation. We shall see that this symmetry limits the applicability of (7.2) solely to monatomic ideal gases.

On the other hand, it is seen from the lack of production terms, that the first two equations and the trace of the third equation are the **conservation laws** of mass, momentum and energy.

It is obvious that the set (7.2) does not form the field equations for the fields (7.1). In order to construct such equations we have to add **constitutive relations** for the following quantities:

$$\begin{aligned}&F_{\langle ij\rangle k}=F_{\langle ij\rangle k}\left(F,F_i,F_{ij},F_{ijj}\right),\\&F_{ijjk}=F_{ijjk}\left(F,F_i,F_{ij},F_{ijj}\right),\\&f_{\langle ij\rangle}=f_{\langle ij\rangle}\left(F,F_i,F_{ij},F_{ijj}\right),\\&f_{ijj}=f_{ijj}\left(F,F_i,F_{ij},F_{ijj}\right).\end{aligned} \tag{7.4}$$

If these 23 (9+6+5+3) functions were known. we could solve, in principle, the set of **field equations** (7.2), (7.4). These solutions are called **thermodynamical processes.**

According to the previous considerations in this book, we can limit the number of necessary constitutive relations by imposing additional conditions, characteristic for all macroscopic theories. These are

- the Galilean invariance of the field equations,
- the material frame indifference of the constitutive relations,
- the thermodynamical admissibility,
- the hyperbolicity.

The last condition, as we already know, is sufficient for finite speeds of propagation of pulses. We accept this condition for the present model, even though it does not possess the same status of generality as the other above listed requirements.

7.2 Galilean Invariance; Material Frame Indifference

The main result presented in Chap. 5 shows that the requirement of Galilean invariance of field equations yields a particular form of dependence of the fields on the velocity. Bearing in mind the formulae (5.85), (5.88), (5.92) and (5.100), we arrive in our present notation at the following relations:

$$\begin{aligned} F &= m, \\ F_i &= mv_i, \\ F_{ij} &= m_{ij} + mv_i v_j, \\ F_{ijk} &= m_{ijk} + 3m_{(ij}v_{k)} + mv_i v_j v_k, \\ F_{ijkn} &= m_{ijkn} + 4m_{(ijk}v_{n)} + 6m_{(ij}v_k v_{n)} + mv_i v_j v_k v_n, \end{aligned} \tag{7.5}$$

where the m-quantities are called the **non-convective densities**. It is obvious that m corresponds to the mass density, ρ, m_{ij} to the negative Cauchy stress tensor (the so-called **pressure tensor**), m_{ijj} to twice the heat flux. In addition, since the trace of the equation $(7.2)_3$ should be identical to the energy conservation law, we have the relation

$$m_{ii} = 3p = 2\rho\varepsilon \quad \Rightarrow \quad \rho\varepsilon = \tfrac{3}{2}p. \tag{7.6}$$

This relation holds true only for monatomic ideal gases, which proves the statement made in the Introduction about the limited character of the present model. During recent years G. M. Kremer and his coworkers have developed the extended thermodynamics of real gases, which is not constrained by the relation (7.6) [e.g. see: G. M. KREMER (1993)]. We shall not present this development in this book because of its enormous technical complexity.

Exercise 7.1. Find the trace of the equation $(7.2)_3$ and show that under the condition (7.6)

$$q_i = \tfrac{1}{2} m_{ijj} \bullet$$

To simplify the orientation in the notation, we collect the non-convective quantities and the corresponding classical fields into a table to which we refer later in this chapter.

Liu&Müller	m	m_{ij}	$½m_{ii}$	$½m_{iij}$	h	Φ_i
Classical	ρ	$-\sigma_{ij}$	$\rho\varepsilon=3/2p$	q_i	$\rho\eta$	h_i

(7.7)

The last two entries correspond to the entropy density, and the non-convective entropy flux which shall be considered in the next section.

Apart from the decomposition of fields, Ruggeri's Theorem also indicates that the following production quantities:

$$f_{\langle ij\rangle},\quad f_{iij}-2f_{\langle ij\rangle}v_j, \tag{7.8}$$

are non-convective.

It can be seen that the F-fields and m-non-convective quantities are in one-to-one correspondence. This means that the constitutive relations (7.4) can be reformulated as follows:

$$\begin{aligned} m_{\langle ij\rangle k} &= m_{\langle ij\rangle k}\left(m,m_{ij},m_{iij}\right),\\ m_{iijk} &= m_{iijk}\left(m,m_{ij},m_{iij}\right),\\ f_{\langle ij\rangle} &= f_{\langle ij\rangle}\left(m,m_{ij},m_{iij}\right),\\ f_{iij}-2f_{\langle ij\rangle}v_j &= \tilde{f}_{iij}\left(m,m_{ij},m_{iij}\right), \end{aligned} \tag{7.9}$$

where the dependence on F_i (i.e. on v_i) has been omitted because of the Galilean invariance considerations.

We proceed to discuss the condition of the material frame indifference. According to this requirement, the relations (7.9) must have the same form in the inertial and non-inertial frames. We already know, from the discussion in Chaps. 3 and 5 that this yields the isotropic structure of constitutive functions. In the explicit form [see: (3.55)]

$$\begin{aligned} O_{il}O_{jn}O_{km}m_{\langle ln\rangle m}\left(m,m_{ij},m_{iij}\right) &= m_{\langle ij\rangle k}\left(m,O_{ik}O_{jn}m_{kn},O_{in}m_{nkk}\right),\\ O_{jn}O_{km}m_{njjm}\left(m,m_{ij},m_{iij}\right) &= m_{jnnk}\left(m,O_{ik}O_{jn}m_{kn},O_{in}m_{nkk}\right),\\ O_{jn}O_{km}f_{\langle nm\rangle}\left(m,m_{ij},m_{iij}\right) &= f_{\langle jk\rangle}\left(m,O_{ik}O_{jn}m_{kn},O_{in}m_{nkk}\right),\\ O_{kn}\tilde{f}_{njj}\left(m,m_{ij},m_{iij}\right) &= \tilde{f}_{knn}\left(m,O_{ik}O_{jn}m_{kn},O_{in}m_{nkk}\right). \end{aligned} \tag{7.10}$$

The appropriate representations for these isotropic functions can be found in the tables quoted in Chap. 5. For instance, the productions $f_{<ij>}$ and $\tilde{f}_{inn}$ must be of the following form:

$$\begin{aligned} f_{\langle ij\rangle} &= Am_{\langle ij\rangle}+Bm_{n\langle i}m_{j\rangle n}+Cm_{kk\langle i}m_{j\rangle nn}+Dm_{knn}m_{k\langle i}m_{j\rangle pp}+\\ &\quad +Em_{knn}m_{kl}m_{l\langle i}m_{j\rangle pp},\\ f_{inn}-2f_{\langle in\rangle}v_n &= Rm_{inn}+Sm_{ij}m_{jnn}+Tm_{ij}m_{jn}m_{npp}, \end{aligned} \tag{7.11}$$

where the coefficients A to T may be arbitrary functions of invariants of the set of constitutive variables, i.e.

$$\left\{m,m_{ii},m_{ij}m_{ij},m_{ij}m_{jk}m_{ki},m_{ipp}m_{inn},m_{ipp}m_{ij}m_{jnn},m_{ipp}m_{ij}m_{jk}m_{knn}\right\}, \tag{7.12}$$

which is an irreducible set.

7.3 Entropy Inequality

The requirements for thermodynamical admissibility, which we listed among the general restrictions of constitutive relations, consist of two parts:

(i) all thermodynamical processes must fulfil identically the entropy inequality (6.11), (6.12) which for the purpose of this chapter we write in the following form:

$$\frac{\partial h}{\partial t}+\frac{\partial \varphi_k}{\partial x_k}\geq 0, \qquad \varphi_k \equiv h v_k + \Phi_k, \qquad h = h\left(m, m_{ij}, m_{ipp}\right), \quad \Phi_k = \Phi_k\left(m, m_{ij}, m_{ipp}\right), \tag{7.13}$$

(ii) the state of the thermodynamical equilibrium in which the inequality $(7.13)_1$ holds as the equality is stable, i.e. the entropy has a maximum in this state.

In this section, we concentrate on the first part of the above condition.

Let us first write the set of equations (7.2) in the following compact form:

$$\frac{\partial F_A}{\partial t}+\frac{\partial F_{Ak}}{\partial x_k} = f_A + b_A, \quad A = 1,\dots,13, \tag{7.14}$$

with

$$\begin{aligned}
(F_A) &\equiv \left(F, F_i, F_{ij}, F_{ijj}\right)^T,\\
(F_{Ak}) &\equiv \left(F_k, F_{ik}, F_{ijk}, F_{ijjk}\right)^T,\\
(f_A) &\equiv \left(\ ,\ , f_{\langle ij\rangle}, f_{ijj}\right)^T,\\
(\ \ ) &\equiv \left(\ , F\left(b_i + i_i\right) + F_k \quad , F_{(i}\left(b_{j)} + i_{j)}\right) + \right.\\
&\qquad \left. + F_{k(i} \quad)_k \; F_{(ij}\left(b_{j)} + i_{j)}\right) + F_{kjj} \quad \right)^T
\end{aligned}$$

$$\forall F_A : \sum_{A=1}^{13}\left(\frac{\partial h}{\partial F_A} - \Lambda_A\right)\frac{\partial F_A}{\partial t} + \sum_{B=1}^{13}\left(\frac{\partial \varphi_k}{\partial F_B} - \sum_{A=1}^{13}\Lambda_A \frac{\partial F_{Ak}}{\partial F_B}\right)\frac{\partial F_B}{\partial x_k} + \sum_{A=1}^{13}\Lambda_A\left(f_A + b_A\right) \geq 0, \tag{7.16}$$

where Λ_A, $A=1,\dots,13$, are the Lagrange multipliers, which are all functions of (F_A). Hence,

$$d h = \sum_{A=1}^{13}\Lambda_A d F_A, \quad d\varphi_k = \sum_{A=1}^{13}\sum_{B=1}^{13}\Lambda_A \frac{\partial F_{Ak}}{\partial F_B} d F_B. \tag{7.17}$$

Apart from the constitutive restrictions following from these relations, it can be readily seen that the Lagrange multipliers can be calculated from (7.17), and that they contain the convective (velocity-dependent), as well as non-convective parts.

There remains the residual inequality

$$\sum_{A=1}^{13} \Lambda_A (f_A + b_A) \geq 0. \tag{7.18}$$

We shall show in the following that the products with the body forces vanish from the inequality (7.18). This indicates the definition of the **thermodynamical equilibrium**

$$f_A\big|_E = 0 \;\Rightarrow\; f_{\langle ij\rangle}\big|_E = 0, \quad f_{inn}\big|_E = 0. \tag{7.19}$$

Inspection of the general representations (7.11) yields, consequently, the following necessary and sufficient conditions for the equilibrium

$$m_{\langle ij\rangle}\big|_E = 0, \quad m_{inn}\big|_E = 0, \tag{7.20}$$

i.e. in the equilibrium, the **deviatoric part of stresses**, and the **heat flux** must vanish.

The explicit evaluation of the conditions (7.17) and (7.18) for arbitrary constitutive relations (7.9), has never been carried through, and it is very doubtful, even if it could be done, whether the results would have any handy and clear form. For this reason, Liu and Müller exploited these thermodynamical admissibility conditions solely for processes near thermodynamical equilibrium. To this aim, the constitutive relations (7.9) have been assumed in this work to be approximated by **quadratic functions** in the non-equilibrium variables $m_{<ij>}$, m_{inn}. Namely, they are supposed to have the form

$$\begin{aligned}
m_{\langle ijk\rangle} &= \alpha\Big\{\underline{m_{ipp}m_{\langle jk\rangle} + m_{jpp}m_{\langle ik\rangle} + m_{kpp}m_{\langle ij\rangle}} - \\
&\quad - \tfrac{2}{5}\Big[\underline{m_{pnn}\big(m_{\langle pi\rangle}\delta_{jk} + m_{\langle pj\rangle}\delta_{ik} + m_{\langle pk\rangle}\delta_{ij}\big)}\Big]\Big\}, \\
m_{ippj} &= \Big[\beta_0 + \underline{\beta_1 m_{\langle pn\rangle}m_{\langle pn\rangle} + \beta_2 m_{knn}m_{kpp}}\Big]\delta_{ij} + \\
&\quad + \gamma m_{\langle ij\rangle} + \underline{\delta m_{ipp}m_{jnn} + \zeta m_{\langle ip\rangle}m_{\langle jp\rangle}}, \\
f_{\langle ij\rangle} &= \sigma_0 m_{\langle ij\rangle} + \underline{\sigma_1\big(m_{\langle ik\rangle}m_{\langle kj\rangle} - \tfrac{1}{3}m_{\langle pk\rangle}m_{\langle pk\rangle}\delta_{ij}\big)} + \\
&\quad + \underline{\sigma_2\big(m_{ipp}m_{jnn} - \tfrac{1}{3}m_{kpp}m_{knn}\delta_{ij}\big)}, \\
f_{inn} - 2f_{\langle in\rangle}v_n &= \tau_1 m_{inn} + \underline{\tau_2 m_{\langle ij\rangle}m_{jnn}},
\end{aligned} \tag{7.21}$$

where the coefficients α to τ_2 may be arbitrary functions of the equilibrium variables m and m_{jj}. All terms, that are second order in the non-equilibrium variables, have been underlined to expose the simplicity of the linear theory.

As we shall see later, the constitutive relations for the entropy and its flux (7.13) must contain third-order terms in order to deliver reliable information about the coefficients of the second-order approximation (7.21). Therefore,

$$
\begin{aligned}
h &= h_0 + h_1 m_{\langle kl\rangle} m_{\langle kl\rangle} + h_2 m_{knn} m_{kpp} + \\
&\quad \underline{+ h_3 m_{\langle kl\rangle} m_{\langle lj\rangle} m_{\langle jk\rangle} + h_4 m_{\langle kl\rangle} m_{knn} m_{lpp}}, \\
\Phi_k &= \varphi_1 m_{knn} + \varphi_2 m_{\langle kj\rangle} m_{jnn} + \underline{\varphi_3 m_{\langle lj\rangle} m_{\langle lj\rangle} m_{knn}} + \\
&\quad \underline{+ \varphi_4 m_{\langle kl\rangle} m_{\langle lj\rangle} m_{jpp} + \varphi_5 m_{lnn} m_{lpp} m_{kii}}.
\end{aligned}
\tag{7.22}
$$

Here the third-order terms have been underlined.

Even in this simple case, the calculations are extremely cumbersome. Here I shall quote only the final results. The technical details can be found in the paper by LIU and MÜLLER.

Let us first split the multipliers into convective and non-convective parts in the same manner as we did for the fields F_A [see: (7.5)]. Then for the set

$$
(\Lambda_A) = (\Lambda, \Lambda_i, \Lambda_{ij}, \Lambda_{ill}),
\tag{7.23}
$$

the form of this decomposition is as follows:

$$
\begin{aligned}
\Lambda_{mnp} &= \Lambda^I_{mnp}, \\
\Lambda_{mn} &= \Lambda^I_{mn} - \Lambda^I_{ill} 3v_{(i}\delta_{mn)}, \\
\Lambda_m &= \Lambda^I_m - 2\Lambda^I_{mn} v_n + \Lambda^I_{pnn} 3v_{(p} v_l \delta_{l)m}, \\
\Lambda &= \Lambda^I - \Lambda^I_m v_m + \Lambda^I_{mn} v_m v_n - \Lambda^I_{mll} v^2 v_m.
\end{aligned}
\tag{7.24}
$$

Substitution in (7.17) then yields

$$
\frac{\partial h}{\partial m} = \Lambda^I, \quad \frac{\partial h}{\partial m_{mn}} = \Lambda^I_{mn}, \quad \frac{\partial h}{\partial m_{mll}} = \Lambda^I_{mll},
\tag{7.25}
$$

and

$$
\begin{aligned}
\frac{\partial \Phi_k}{\partial m} &= \Lambda^I_{ipp} \frac{\partial m_{illk}}{\partial m} + \Lambda^I_{ij} \frac{\partial m_{\langle ij\rangle k}}{\partial m}, \\
\frac{\partial \Phi_k}{\partial m_{mn}} &= \Lambda^I_{ipp} \frac{\partial m_{illk}}{\partial m_{mn}} + \Lambda^I_{ij} \frac{\partial m_{\langle ij\rangle k}}{\partial m_{mn}} + \Lambda^I_m \delta_{nk}, \\
\frac{\partial \Phi_k}{\partial m_{mll}} &= \Lambda^I_{ipp} \frac{\partial m_{illk}}{\partial m_{mnn}} + \Lambda^I_{ij} \frac{\partial m_{\langle ij\rangle k}}{\partial m_{mnn}} + \Lambda^I_{ij} \tfrac{3}{5} \delta_{m(i}\delta_{jk)},
\end{aligned}
\tag{7.26}
$$

as well as

$$
\begin{aligned}
& m\Lambda^I_m + 3m_{(ik}\delta_{k)m}\Lambda^I_{ill} = 0, \\
& h = m\Lambda^I + \tfrac{5}{3}\Lambda^I_{ij} m_{ij} + 2\Lambda^I_{ill} m_{ipp}, \\
& m_{\langle ij\rangle k}\Lambda^I_{llj} = -\Lambda^I_{j\langle i} m_{k\rangle j} - \tfrac{9}{10}\Lambda^I_{ll\langle i} m_{k\rangle nn}, \\
& 2\Lambda^I_{i[k} m_{p]i} + m_{ii[p}\Lambda^I_{k]ll} = 0.
\end{aligned}
\tag{7.27}
$$

The conditions (7.25-7.27) are equivalent to (7.17). On the other hand, for the residual inequality we obtain

$$\Lambda^{I}_{\langle ij\rangle} f_{\langle ij\rangle} + \Lambda^{I}_{pll}\left(f_{pll} - 2v_l f_{\langle lp\rangle}\right) + 2\left(2\Lambda^{I}_{mp} m_{km} + m_{nnk}\Lambda^{I}_{pll}\right)\Omega_{kp} \geq 0. \tag{7.28}$$

Owing to the antisymmetry of Ω_{km}, we see that the condition $(7.24)_4$ eliminates the second part of the inequality (7.28). This means that the body forces indeed do not contribute to the residual inequality, as we anticipated in the definition of the thermodynamical equilibrium. Hence,

$$\Lambda^{I}_{\langle ij\rangle} f_{\langle ij\rangle} + \Lambda^{I}_{pnn}\left(f_{pkk} - 2v_k f_{\langle kp\rangle}\right) \geq 0. \tag{7.29}$$

Simultaneously, the relations (7.25-7.26) corroborate the earlier statement on the necessity of third-order terms in the approximations for entropy and entropy flux. Indeed, the differentiation on the left-hand side lowers the order of these approximations while on the right-hand side we have, of course, terms of second order.

It can be seen from $(7.22)_1$ that h_0 is the equilibrium value of the entropy:

$$h_0 = h_0(m, m_{ii}). \tag{7.30}$$

At the same time substitution of the approximations (7.21) and (7.22) into (7.26-7.28) delivers a number of relationships between the coefficients of expansions which we list as necessary. Among these identities one obtains

$$m^2 \frac{\partial\left(\frac{h_0}{m}\right)}{\partial m} + \tfrac{5}{3} m_{pp} \frac{\partial h_0}{\partial m_{kk}} = 0. \tag{7.31}$$

Hence, the differential of (7.30) can be written in the form

$$d\left(\frac{h_0}{m}\right) = \frac{\partial\left(\frac{h_0}{m}\right)}{\partial m_{kk}}\left(d\, m_{pp} - \tfrac{5}{3}\frac{m_{pp}}{m} d\, m\right). \tag{7.32}$$

If we return to the classical notation (7.7) we obtain

$$d\eta\big|_E = \frac{\partial h_0}{\partial\left(\frac{1}{2} m_{kk}\right)}\left(d\varepsilon - \frac{p}{\rho^2} d\rho\right). \tag{7.33}$$

This is, certainly, the **Gibbs equation** for the ideal gas in equilibrium (see: Appendix A). Hence, we can identify

$$\frac{1}{T} = \frac{\partial h_0}{\partial\left(\frac{1}{2} m_{kk}\right)}. \tag{7.34}$$

The temperature is, of course, a function of m and m_{kk}. We can invert this relation to replace m_{kk} by T, which is the more common and suggestive variable. If we do so the relations following from (7.25-7.28) can be written in the convenient form

$$h_1 = -\frac{1}{4pT}, \quad h_2 = -\frac{1}{10pT}\frac{1}{\gamma - \frac{5p}{\rho}},$$
$$\underline{h_3 = \frac{1}{6p^2 T}}, \quad \underline{h_4 = \frac{9}{50}\frac{1}{p^2 T}\frac{1}{\gamma - \frac{5p}{\rho}}}, \tag{7.35}$$

$$\varphi_1 = \frac{1}{2T}, \quad \varphi_2 = -\frac{1}{5}\frac{1}{pT},$$
$$\underline{\varphi_3 = -\frac{47}{120}\frac{1}{p^2 T}\frac{\gamma - \frac{175}{47}\frac{p}{\rho}}{\gamma - \frac{5p}{\rho}} + \frac{1}{12pT^2}\frac{1}{\frac{\partial p}{\partial T}}, \quad \varphi_4 = \frac{8}{175}\frac{1}{p^2 T}\frac{\gamma - \frac{35}{4}\frac{p}{\rho}}{\gamma - \frac{5p}{\rho}}}, \tag{7.36}$$
$$\underline{\varphi_5 = \frac{-\delta - \frac{3}{25p} + \frac{3}{4T}\frac{1}{\frac{\partial p}{\partial T}} - \frac{1}{2\rho}\frac{1}{\gamma - \frac{5p}{\rho}}}{5pT\left(\gamma - \frac{5p}{\rho}\right)}},$$

$$\frac{\partial \beta_0}{\partial T} = \frac{5}{7}\frac{\partial (p\gamma)}{\partial T}, \quad \frac{\partial \beta_0}{\partial \rho} = \frac{5}{7}\frac{\partial (p\gamma)}{\partial \rho}, \tag{7.37}$$

$$\frac{\partial \gamma}{\partial \rho} = -\frac{1}{p}\frac{\partial p}{\partial \rho}\left(\gamma - 7\frac{p}{\rho}\right), \tag{7.38}$$

$$\underline{\zeta = \frac{2}{7}\frac{\gamma}{p}}, \tag{7.39}$$

$$\underline{\alpha = \frac{9}{35p}\frac{\gamma - \frac{35}{9}\frac{p}{\rho}}{\gamma - \frac{5p}{\rho}}}, \tag{7.40}$$

$$\underline{\frac{\partial \gamma}{\partial T} = -\left(\frac{1}{p}\frac{\partial p}{\partial T} - \frac{7}{2T}\right)\gamma - \frac{21}{2T}\frac{\partial p}{\partial \rho}}, \tag{7.41}$$

$$\underline{\beta_2 - \frac{1}{2}\delta = \frac{2}{15p} - \frac{3}{4T}\frac{1}{\frac{\partial p}{\partial T}} + \frac{1}{6p}\frac{\gamma - \frac{2p}{\rho}}{\gamma - \frac{5p}{\rho}}},$$
$$\underline{\beta_1 = \frac{5}{21p}\left(\gamma - \frac{7}{2}\frac{p}{\rho}\right) - \frac{5}{12T}\frac{1}{\frac{\partial p}{\partial T}}\left(\gamma - \frac{5p}{\rho}\right)}, \tag{7.42}$$

$$-\frac{5}{2} + \frac{3}{2}\frac{\rho}{p}\frac{\partial p}{\partial \rho} + \frac{T}{p}\frac{\partial p}{\partial T} = 0. \tag{7.43}$$

Again we have underlined those coefficients and relations which are connected with the non-linear part of approximations.

The last equation can be integrated readily, and doing so yields

$$p = T^{\frac{5}{2}} F\left(\frac{\rho}{T^{\frac{3}{2}}}\right). \tag{7.44}$$

This shows that, as in the classical thermodynamics of ideal gases, the present model implies the **thermal state equation** to within a single function F of the single variable $\rho/T^{\frac{3}{2}}$. To find the form of this function, one must refer either to experiments or to statistical mechanics.

For brevity, let us denote the single variable of F by z. Then we can integrate (7.38), (7.41) and, consequently, (7.37). We obtain

$$\gamma = 7\frac{T}{F}\left(\int \frac{FF'}{z} dz + C_\gamma\right), \tag{7.45}$$

$$\beta_0 = 5T^{\frac{7}{2}}\left(\int \frac{FF'}{z} dz + C_\gamma\right) + C_\beta, \tag{7.46}$$

and according to the Gibbs equation (7.33)

$$\eta|_E = \frac{3}{2}\int\left(\frac{F'}{z} - \frac{5}{3}\frac{F}{z^2}\right) dz + C_\eta. \tag{7.47}$$

It follows that a given function F determines $\eta|_E$, γ, β_0 to within **three constants of integration**. The other conditions (7.35-7.42) deliver α, ζ, β_1, $\beta_2 - {}^1/_2\delta$, h_1 to h_4 and φ_1 to φ_4. The coefficient φ_5 contains the unknown function $\delta(\rho,T)$.

This completes the exploitation of the relations (7.17). We proceed to the residual inequality (7.29) which, up to the second-order terms, yields

$$2h_1\sigma_0 m_{\langle ij\rangle} m_{\langle ij\rangle} + 2h_2\tau_1 m_{ipp} m_{ikk} \geq 0. \tag{7.48}$$

Hence, if we account for $(7.35)_1$, (7.44) and (7.45),

$$h_1\sigma_0 = -\frac{\sigma_0}{4pT} \geq 0, \quad h_2\tau_1 = -\frac{\tau_1}{10pT}\frac{1}{\gamma - \frac{5p}{\rho}} \geq 0. \tag{7.49}$$

Thus, it follows

$$\sigma_0 \leq 0, \quad \tau_1 \leq 0, \tag{7.50}$$

provided we assume hyperbolicity which is the subject of the next section.

Let us list the results obtained above for the **first-order theory:**

$$p = T^{\frac{5}{2}} F\left(\frac{\rho}{T^{\frac{3}{2}}}\right), \qquad m_{\langle ij\rangle k} = 0,$$

$$m_{ikkj} = \left\{5T^{\frac{7}{2}}\left(\int \frac{FF'}{z} dz + C_\gamma\right) + C_\beta\right\}\delta_{ij} + 7\frac{T}{F}\left(\int \frac{FF'}{z} dz + C_\gamma\right) m_{\langle ij\rangle},$$

$$f_{\langle ij\rangle} = \sigma_0 m_{\langle ij\rangle}, \qquad f_{ikk} - 2f_{\langle ik\rangle} v_k = 2\tau_1 q_i,$$

$$h = \rho\left(\frac{3}{2}\int\left(\frac{F'}{z} - \frac{5}{3}\frac{F}{z^2}\right) dz + C_\eta\right) - \frac{m_{\langle ij\rangle} m_{\langle ij\rangle}}{4T^{\frac{7}{2}} F(z)} - \tag{7.52}$$

$$- \frac{2q_i q_i}{5T^{\frac{7}{2}} F\left\{7\frac{T}{F}\left(\int \frac{FF'}{z} dz + C_\gamma\right) - 5\frac{T^{\frac{5}{2}}}{\rho} F\right\}},$$

$$\Phi_i = \frac{q_i}{T} - \frac{2}{5}\frac{m_{\langle ij\rangle}}{T^{\frac{7}{2}}}\frac{q_j}{F}.$$

Let us notice that in this set of constitutive relations, only the functions $\sigma_0(\rho,T)$, $\tau_1(\rho,T)$ and F do not follow from the entropy principle. As we shall see, σ_0 is related to the viscosity with the negative sign, and τ_1 is connected with the heat conductivity. This connection will be made explicit in the next section.

Among the above results, the last relation for the non-convective entropy flux is particularly important because of its practical implications.

The usefulness of the temperature in classical thermodynamics follows from the fact that it is easily measurable, i.e. by bringing two bodies into thermal contact - one whose temperature is to be measured, and the other one which serves as a thermometer. If the surface of contact does not contribute to the result of the measurement, it is claimed that the temperature of both bodies on this surface is the same. This result follows from the continuity of the normal components of heat and entropy fluxes on such **ideal walls** between the bodies, as well as from the classical relation between both fluxes

$$n_i[[q_i]] = 0, \quad n_i[[\Phi_i]] = 0, \quad \Phi_i = \frac{1}{T} q_i \quad \Rightarrow \quad [[T]] = 0. \tag{7.53}$$

It is easy to see from the relation $(7.52)_7$, that the above conclusion is false for the present model. However, the first two relations (7.53) also hold in this case and we obtain

$$\left[\left[\frac{1}{T}\right]\right] = \frac{2}{5}\left[\left[\frac{m_{\langle ij\rangle} q_j}{T^{\frac{7}{2}} F}\right]\right]\frac{n_i}{q_p n_p}. \tag{7.54}$$

This is the implicit relation between the values T^+ and T^- of the temperature on both sides of the contact surface. These values are not equal unless the stress deviator vanishes.

This property demonstrates that temperature is of little use as a thermodynamical variable in processes which strongly deviate from the thermodynamical equilibrium.

7.4 Hyperbolicity

In problems involving arbitrary systems of partial differential equations, the requirement of hyperbolicity usually leads to rather complicated technical considerations. Extended thermodynamics simplifies the form of this condition, considerably, owing to the **additional requirement** of thermodynamical admissibility. This requirement reduces the hyperbolic system of field equations to a **symmetric hyperbolic system** for main fields, and the condition assumes a simple form, (6.74). In our present notation this condition can be written as follows:

$$\sum_{A,B=1}^{13} \frac{\partial^2 h}{\partial F_A \partial F_B} \delta F_A \delta F_B < 0, \tag{7.55}$$

where δF_A denotes an arbitrary variation of fields.

The second law of thermodynamics and the approximations of the previous section have made the condition (7.55) quite explicit. In the case of the first-order theory, substitution of $(7.22)_1$ and (7.35) yields, after easy calculations, the following conditions:

$$h_1 = -\frac{1}{4pT} < 0, \quad h_2 = -\frac{1}{10pT}\frac{1}{\gamma - \frac{5p}{\rho}} < 0,$$
$$\frac{\partial p}{\partial \rho} > 0, \quad \frac{\partial \varepsilon}{\partial T} > 0. \tag{7.56}$$

The inequalities $(7.56)_{3,4}$ represent the usual **thermodynamical stability** conditions. This means that the hyperbolicity requirement automatically takes care of the second part of the thermodynamical admissibility condition, which we mentioned at the beginning of Sect. 7.3.

The first condition is, certainly, fulfilled if the pressure is positive. However, there remains the inequality

$$\gamma - \frac{5p}{\rho} > 0, \tag{7.57}$$

which we already anticipated in the previous section, when deriving the condition (7.51).

We conclude this chapter with a brief remark on speeds of pulses as predicted by the extended thermodynamics model. The governing set of equations for the fields ρ, v_i, T, $\sigma_{<ij>}$, q_i in the linearized theory follows from the above considerations after easy manipulations in the following form:

$$
\begin{aligned}
&\frac{\partial \rho}{\partial t}+\rho\frac{\partial v_k}{\partial x_k}=0,\\
&\rho\frac{\partial v_i}{\partial t}+\frac{\partial}{\partial x_k}\left(p\delta_{ik}-\sigma_{\langle ik\rangle}\right)=0,\\
&\rho\frac{\partial \varepsilon}{\partial t}+\frac{\partial q_k}{\partial x_k}+\left(p\delta_{kl}-\sigma_{\langle kl\rangle}\right)\frac{\partial v_k}{\partial x_l}=0, \qquad (7.58)\\
&\frac{\partial \sigma_{\langle ij\rangle}}{\partial t}-\tfrac{4}{5}\frac{\partial q_{\langle i}}{\partial x_{j\rangle}}-2p\frac{\partial v_{\langle i}}{\partial x_{j\rangle}}=\sigma_0\sigma_{\langle ij\rangle},\\
&\frac{\partial q_i}{\partial t}+\tfrac{5}{4}\frac{p}{T}\left(\gamma-\frac{5p}{\rho}\right)\frac{\partial T}{\partial x_i}-\tfrac{1}{2}\left(\gamma-\frac{5p}{\rho}\right)\frac{\partial \sigma_{\langle ik\rangle}}{\partial x_k}=\tau_1 q_i,
\end{aligned}
$$

where, for simplicity, the body forces have been neglected. Now we can easily identify the **viscosity**, μ, and the **heat conductivity**, K, in the present model:

$$\mu=-\frac{p}{\sigma_0}\geq 0,\qquad K=-\frac{5}{4}\frac{1}{\tau_1}\frac{p}{T}\left(\gamma-\frac{5p}{\rho}\right)\geq 0, \qquad (7.59)$$

the inequalities following from (7.50).

The hyperbolicity requirement yields the finite characteristic speeds for the system (7.58), which means that both temperature disturbances and shear pulses propagate with finite speeds. This result was already obtained by I. MÜLLER in his earliest version of extended thermodynamics presented in his PhD Thesis, quoted at the beginning of this chapter.

Exercise 7.2. Consider the following plane harmonic wave solution of (7.58):

$$\rho=\tilde{\rho}+\overline{\rho}e^{i(\omega t-kx)},\quad \mathbf{v}=(\overline{v},0,0)^T e^{i(\omega t-kx)},\quad T=\tilde{T}+\overline{T}e^{i(\omega t-kx)},$$

$$\mathbf{T}^D=\left(\sigma_{\langle ij\rangle}\right)=\begin{pmatrix}\overline{\sigma} & 0 & 0\\ 0 & -\frac{\overline{\sigma}}{2} & 0\\ 0 & 0 & -\frac{\overline{\sigma}}{2}\end{pmatrix}e^{i(\omega t-kx)},\quad \mathbf{q}=\left(q_i\right)=(\overline{q},0,0)^T e^{i(\omega t-kx)}, \qquad (7.60)$$

where $x\equiv x^1$, $\tilde{\rho},\tilde{T}$ are constants describing the initial equilibrium state of the gas and

$$p=\frac{k\rho}{M}T. \qquad (7.61)$$

Find the dispertion relation for the set (7.58) and prove that the limits of phase velocities $V_{ph}=\frac{\omega}{k}$ when $\omega\rightarrow\infty$ remain finite•

8 Maxwellian Fluids; Viscoelastic Solids

Maxwellian fluids are viscous fluids possessing certain elastic properties which yield so-called normal stress effects (Weissenberg effects). The latter are very reminiscent of the properties of solids. On the other hand, the viscoelastic solids are elastic solids which, in addition, possess certain viscous properties. Both classes of materials have no sharp dividing line and, from the point of view of the construction of the thermodynamical model, should be considered jointly. However, this is not done for both historical and practical reasons. We shall also not do so in this chapter. Its first part is devoted to the construction of a model of the Maxwellian fluid. It is done „almost" by means of extended thermodynamics, as explained earlier in this book. There is, however, a deviation from the method in the construction of the constitutive relations for the sources. On the other hand, the extended thermodynamics of viscoelastic solids is an illustration of the method without any deviations, but it shall only be presented very briefly in the second part of this chapter.

PART 1: MAXWELLIAN FLUIDS

8.1 Background of the Models of Maxwellian Fluids

During the Second World War the discovery of napalm (na(phthene)+palm(itate)) started the development of fluid models capable of describing the motion of such substances. It was observed that during the extrusion from pipes some fluids, for example napalm, increase their cross-section in contrast to normal viscous Newtonian fluids. Fortunately, this property is characteristic not only for napalm but also for numerous other less harmful fluids – various polymer solutions, suspensions, etc., which have important industrial applications. These **viscoelastic fluids** are called also **non-Newtonian**, and their most distinguishable feature is the so-called **Weissenberg effect**. Namely, stirring of such a fluid in a round container by a rotating shaft causes this fluid to climb along the shaft rather than to be driven in a radial direction to the walls of the container as a usual Newtonian fluid would be. Responsible for this effect, as well as for the above mentioned effect of swelling, are the normal stresses whose influence on the motion is accounted for in the constitutive relations by at least two material coefficients, in addition to the viscosity coefficients for the Newtonian fluids*).

*) An extensive presentation of phenomena and of the early rheological models of viscoelastic fluids can be found in Chap. E of C. TRUESDELL, W. NOLL (1965).

In the simplest models, the so-called **fluids of the Nth grade**, the constitutive relations for the Cauchy stresses for, for instance, N=2 and 3, are assumed to have the following form:

- 2nd grade:

$$\mathbf{T} = -p\mathbf{1} + \mu\mathbf{A}_1 + \alpha_1\mathbf{A}_2 + \alpha_2\mathbf{A}_1^2,$$

- 3rd grade: (8.1)

$$\mathbf{T} = -p\mathbf{1} + \mu\mathbf{A}_1 + \alpha_1\mathbf{A}_2 + \alpha_2\mathbf{A}_1^2 + \\ + \beta_1\mathbf{A}_3 + \beta_2(\mathbf{A}_1\mathbf{A}_2 + \mathbf{A}_2\mathbf{A}_1) + \beta_3(\mathrm{tr}\mathbf{A}_1^2)\mathbf{A}_2,$$

$$\mathbf{A}_1 \equiv \mathbf{L} + \mathbf{L}^T, \quad \mathbf{A}_n \equiv \dot{\mathbf{A}}_{n-1} + \mathbf{A}_{n-1}\mathbf{L} + \mathbf{L}^T\mathbf{A}_{n-1},$$

where, as before, **T** denotes the Cauchy stress tensor, p the pressure, **L** the velocity gradient, $\mathbf{A}_n$ the Rivlin-Ericksen tensors [see: formula (3.37)]; the dot is the material time derivative, μ the viscosity, and α_1, α_2, β_1, β_2, β_3 are the material coefficients, the first two of which are the so-called **normal stress coefficients**.

These material parameters are measured in rheological experiments, e.g. in rheometers in which the thin film (~1 mm) of the fluid is moved, for instance, between two discs or a disc and a plate (torsional flows) rotating relative to each other with a *constant* angular velocity. It has been firmly established that μ and α_2 are positive and α_1 is negative. As an example, in the table 8.1. we quote data for the typical polymer solution, κ being the so-called **rate of shearing**, the sole relevant component of the velocity gradient, **L**.

Table 8.1. *Viscosity and normal stress coefficients in 6.5 wt.% of polyisobutylene in decalin*

κ [1/s]	μ [Ns/m²]	α_1 [Ns²/m²]	α_2 [Ns²/m²]
5	6.481	-0.416	0.720
10	5.654	-0.244	0.382
20	4.585	-0.147	0.256

The relations for stresses (8.1) have been found to be an approximation of a functional dependence of stresses on the history of deformation for the so-called **retarded** motions [see: B. D. COLEMAN, W. NOLL (1960)]. This functional dependence is typical for **materials with memory**. Multilinear functionals (multiple time integrals) for stresses in various materials with memory were introduced by A. E. GREEN and R. S. RIVLIN (1957) and since then this approach has been developed for many rheological materials. However, we shall not discuss this subject in these notes.

There is another way to account for the memory effects in non-Newtonian fluids, and in fact, it has already been developed by Maxwell and Boltzmann. It is based on the assumption that the stresses satisfy a certain **evolution equation**, which may have, for instance, the following form:

$$\tau\frac{D\mathbf{T}}{Dt} + \mathbf{T} = 2\mu\,\mathbf{D}, \quad \frac{D\mathbf{T}}{Dt} \equiv \dot{\mathbf{T}} - (\mathbf{L}\mathbf{T} + \mathbf{T}\mathbf{L}^T) + 2\xi(\mathbf{D}\mathbf{T} + \mathbf{T}\mathbf{D}), \\ 0 \le \xi \le 1, \tag{8.2}$$

where the time derivative, $^{D}/_{Dt}$, coincides with the Oldroyd derivative (3.83) for $\xi=0$ (the **lower Maxwellian model**), with the Jaumann-Zaremba derivative (3.71) – for $\xi=\frac{1}{2}$. The **upper Maxwellian model** follows for $\xi=1$. The material parameters τ, μ, and ξ denote the **relaxation time**, the **viscosity**, and the **shielding coefficient**, respectively*).

In general, the fluids described by such evolution equations are called **Maxwellian**.

Again, it can be shown that relation (8.1) approximates the solutions of equation (8.2) for the class of slow motions.

Despite their formal connection with materials with memory, and with Maxwellian fluids, there are serious problems with such models as defined by the relations (8.1). It has been shown by E. DANN and R. L. FOSDICK (1974) that the thermodynamical stability requirement for fluids of the 2nd grade yields: $\alpha_1>0$ and $\alpha_1+\alpha_2=0$. Neither of these conditions is confirmed by experiments, as the inspection of the above table shows. Even worse, all such approximations have unstable rest states, lead to the infinite speeds of shear pulses and possess many other non-physical features.

Example 8.1. *One-dimensional N^{th} Grade Models*
We illustrate the above remarks with an instructive one-dimensional model [see: K. WILMANSKI (1986)]. In the case of the simplest shearing flow, defined in the Cartesian frame by the following conditions for components of the velocity field **v**

$$v^1=0, \quad v^2=v^2\left(x^1\right), \quad v^3=0, \tag{8.3}$$

the Maxwellian fluid is defined by the equation for shear stresses:

$$\tau\frac{\partial\sigma}{\partial t}+\sigma=\mu\kappa, \quad \sigma\equiv\sigma^{12}, \quad \kappa\equiv\frac{\partial v^2}{\partial x^1}, \tag{8.4}$$

where μ is the constant shear viscosity and τ the constant relaxation time. If we assume that the shear stress is zero at $t\to-\infty$, this equation has the following Boltzmann solution:

$$\sigma(t)=\frac{\mu}{\tau}\int_0^\infty\kappa(t-s)\exp\left(-\frac{s}{\tau}\right)ds. \tag{8.5}$$

Hence, the present value of the stress is dependent on the whole **history** of shearing, κ, mollified by the exponential weighting factor. The Maxwellian fluid is, therefore, an example of **material with memory**.

Let us mention in passing that many rheologists claim the above model to be too simplified. It seems to follow from experiments that fluids with memory are characterized by more than one relaxation time. This would indicate the non-linear functional dependence of stresses on the history of shearing. If this is the case, the evolution of stresses cannot be described by a first order differential equation of the form (8.4).

Such solutions of equations for stresses like (8.5) cannot be constructed for general rheological models in such a straightforward manner. On the contrary, for the overwhelming majority of these models they cannot be constructed at all in any closed form. For this reason, Rivlin and Ericksen proposed the approximation mentioned before, which in our example corresponds to the expansion of $\kappa(t\text{-}s)$ using the power series in

*) For details see, e.g. H. GIESEKUS (1984).

terms of the variable, s, and truncating the series at the Nth term. For the relation (8.5), we easily obtain

$$\sigma \approx \mu \sum_{k=0}^{N-1} (-\tau)^k \kappa^{(k)}, \quad \kappa^{(k)} \equiv \frac{d^k \kappa}{d t^k}(t). \tag{8.6}$$

In order to obtain the above result we must assume **uniform convergence** of the series. There are certain problems with uniformity, particularly in the case of the infinite support of the memory integral in (8.5), which we shall not discuss in this book [see: R. S. RIVLIN, K. WILMANSKI (1987)].

The models of fluids based on constitutive relations similar to (8.6) describe the so-called **N^{th}-order fluids**. The fluids described by (8.6) without the assumption that the constitutive law (8.6) approximates the Maxwellian fluid are called the **N^{th}-grade fluids**. In such cases, τ need not be considered as a relaxation time, and consequently, does not have to be positive.

In our example we consider incompressible N^{th} order fluids, occasionally referring to the Nth-grade fluids. The momentum balance equation, in the one-dimensional case under consideration, is of the form

$$\rho \frac{\partial v}{\partial t} = \frac{\partial \sigma}{\partial x}, \quad v \equiv v^2, \quad x \equiv x^1, \tag{8.7}$$

where ρ is constant. The equations (8.4) and (8.6) now yield:

- the Maxwellian fluid (M),

$$\tau \rho \frac{\partial^2 v}{\partial t^2} + \rho \frac{\partial v}{\partial t} - \mu \frac{\partial^2 v}{\partial x^2} = 0, \tag{8.8}$$

- the Nth-order fluid (N),

$$\rho \frac{\partial v}{\partial t} - \mu \frac{\partial^2}{\partial x^2} \sum_{k=1}^{N-1} (-\tau)^k \frac{\partial^k v}{\partial t^k} = 0. \tag{8.9}$$

It is easy to notice that (8.8) yields the following **speeds of shear pulses**:

$$V_{ph}^{\infty} = \sqrt{\frac{\mu}{\tau \rho}} = \sqrt{-\frac{\mu^2}{\alpha_1 \rho}}, \quad \alpha_1 \equiv -\mu \tau. \tag{8.10}$$

Let us introduce the following dimensionless variables:

$$\xi = \sqrt{\frac{\rho}{\tau \mu}} x, \quad \eta = \frac{t}{\tau}. \tag{8.11}$$

Then we obtain the following form of the governing equations:

$$\frac{\partial^2 v}{\partial \eta^2}+\frac{\partial v}{\partial \eta}-\frac{\partial^2 v}{\partial \xi^2}=0, \qquad \text{(M)}$$
$$\frac{\partial v}{\partial \eta}-\frac{\partial^2}{\partial \xi^2}\sum_{k=1}^{N-1}(-1)^k\frac{\partial^k v}{\partial \eta^k}-\frac{\partial^2 v}{\partial \xi^2}=0. \qquad \text{(N)} \tag{8.12}$$

We seek the solution of these equations of the form

$$v(\xi,\eta)=\bar{v}\,e^{i(k\xi-\omega\eta)}, \tag{8.13}$$

where $\bar{v}$ denotes the amplitude of the velocity, k the wave number, ω the frequency, all of them are in general, complex.

Substitution of (8.13) into (8.12) leads to the following **dispersion relations**:

$$i\omega(1-i\omega)=k^2, \qquad \text{(M)}$$
$$i\omega(1-i\omega)=k^2\left[1-(i\omega)^N\right] \qquad \text{(N)} \tag{8.14}$$

Bearing the convergence requirement for the Nth order fluids in mind, we obtain

$$|i\omega|<1 \;\Rightarrow\; (Re\,\omega)^2+(Im\,\omega)^2<1. \tag{8.15}$$

This means that N^{th} order fluids admit monochromatic waves (8.13) if the frequency is sufficiently small.

Now, we investigate the properties of the dispersion relations (8.14) for two types of boundary conditions shown in Fig. 8.1.

Fig. 8.1. *Shearing flows discussed in the Example 8.1.*

In the case of channel flow shown on the left-hand side of this figure, the flow at first as a certain (e.g. sinusoidal) velocity profile develops spontaneously in time. In the case shown on the right-hand side of Fig. (8.1), the flow is induced in the half-space by the plate at ξ=0 vibrating with a given frequency *Re*ω=ν.

Let us first consider the flow in a channel of the thickness h. Owing to the geometry of the problem, we have

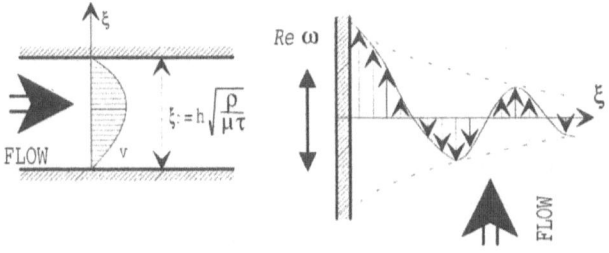

$$k = \frac{\pi}{h}\sqrt{\frac{\mu\tau}{\rho}}, \tag{8.16}$$

and this wave number is real. Let us split the frequency, ω, into the real part, ν, (the vibration frequency), and the imaginary part, β, (the damping):

$$\omega = \nu + i\beta. \tag{8.17}$$

Then the dispersion relation for a Maxwellian fluid yields

$$\begin{aligned} &\nu(1+2\beta) = 0, \\ &k^2 = \nu^2 - \beta(1+\beta). \end{aligned} \quad \text{(M)} \tag{8.18}$$

There are two classes of solutions of these equations (c.f. Fig. 8.2.), namely

$$\begin{aligned} &a)\ \nu = 0 \Rightarrow k^2 = -\beta(1+\beta) \quad i.e.\quad k \le \tfrac{1}{2} \Rightarrow h \ge 2\pi\sqrt{\frac{\mu\tau}{\rho}}, \\ &b)\ \nu \ne 0 \Rightarrow \beta = -\tfrac{1}{2}, k^2 = \nu^2 + \tfrac{1}{4} \quad i.e.\quad k > \tfrac{1}{2} \Rightarrow h < 2\pi\sqrt{\frac{\mu\tau}{\rho}}. \end{aligned} \tag{8.19}$$

Obviously, the second class can appear only for narrow channels.
The velocity functions corresponding to (8.19), follow easily

$$\begin{aligned} &a)\ \nu = 0: \quad v = \left(A_1 e^{\beta_1 \eta} + A_2 e^{\beta_2 \eta}\right)\cos\frac{\pi}{\xi_0}\xi, \qquad \xi_0 \equiv \frac{\pi}{k} = h\sqrt{\frac{\rho}{\mu\tau}}, \\ &b)\ \nu \ne 0: \quad v = e^{-\frac{\eta}{2}}\left(A_1 \sin \nu\eta + A_2 \cos \nu\eta\right)\cos\frac{\pi}{\xi_0}\xi, \end{aligned} \tag{8.20}$$

where A_1, A_2 are constants.

We proceed to the Nth order fluids. The dispersion relation $(8.14)_2$ can be written in the form

$$(1 - i\omega)\left[1 + i\omega\left(1 - \frac{1}{k^2}\right) + (i\omega)^2 + \cdots + (i\omega)^{N-1}\right] = 0. \tag{8.21}$$

Hence, the first possible solution independent of the order of the fluid is as follows:

$$\nu = 0, \quad \beta = -1. \tag{8.22}$$

This is the limit of convergence of the series in powers of relaxation time if (8.6) is considered to be an approximation of the Maxwellian model. If we ignore this problem, solution (8.13) must be of the form

$$v(\xi,\eta)=\overline{v}e^{-\eta}e^{ik\xi} \quad \Rightarrow \quad v(x,t)=\overline{v}e^{-\frac{t}{\tau}}\exp\left(ik\sqrt{\frac{\rho}{\tau\mu}}x\right). \tag{8.23}$$

Hence, we would obtain solely damped solutions without any harmonic vibrations and consequently, without any waves, in contrast to the solution (8.20) for the Maxwellian fluid.

The second possible solution appears when $1-i\omega\neq 0$. For such cases the dispersion relations for N=2,...,6 are listed in the Table 8.2. below.

Table 8.2.

N=2	$v\left[\left(1-\frac{1}{k^2}\right)\right]=0,$ $\beta\left(1-\frac{1}{k^2}\right)-1=0,$
N=3	$v\left[-2\beta+\left(1-\frac{1}{k^2}\right)\right]=0,$ $-\beta^2+\beta\left(1-\frac{1}{k^2}\right)-1+v^2\{1\}=0,$
N=4	$v\left[-v^2+3\beta^2-2\beta+\left(1-\frac{1}{k^2}\right)\right]=0,$ $\beta^3-\beta^2+\beta\left(1-\frac{1}{k^2}\right)-1+v^2\{-3\beta+1\}=0,$
N=5	$v\left[-v^2(-4\beta+1)-4\beta^3+3\beta^2-2\beta+\left(1-\frac{1}{k^2}\right)\right]=0,$ $-\beta^4+\beta^3-\beta^2+\beta\left(1-\frac{1}{k^2}\right)-1+v^2\{-v^2+6\beta^2-3\beta+1\}=0,$
N=6	$v\left[-v^2\left(-v^2+10\beta^2-4\beta+1\right)+5\beta^4-4\beta^3+3\beta^2-2\beta+\left(1-\frac{1}{k^2}\right)\right]=0,$ $\beta^5-\beta^4+\beta^3-\beta^2+\beta\left(1-\frac{1}{k^2}\right)-1+v^2\{5v^2\beta-v^2-10\beta^3+6\beta^2-3\beta+1\}=0.$

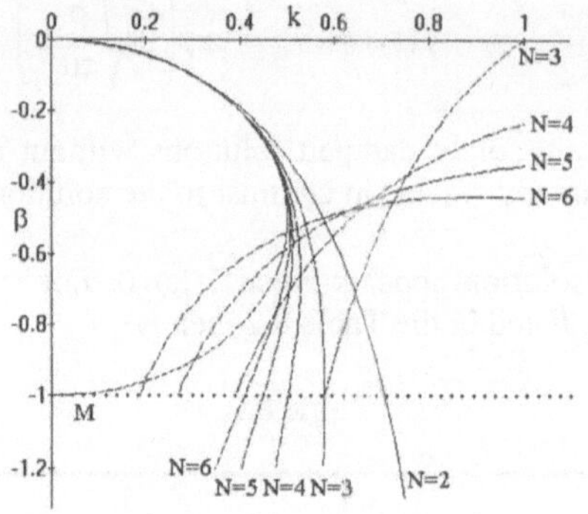

Fig. 8.2. *The negative solutions for damping, β≡Imω(<0), as the function of wave number* k *for the first boundary conditions.*

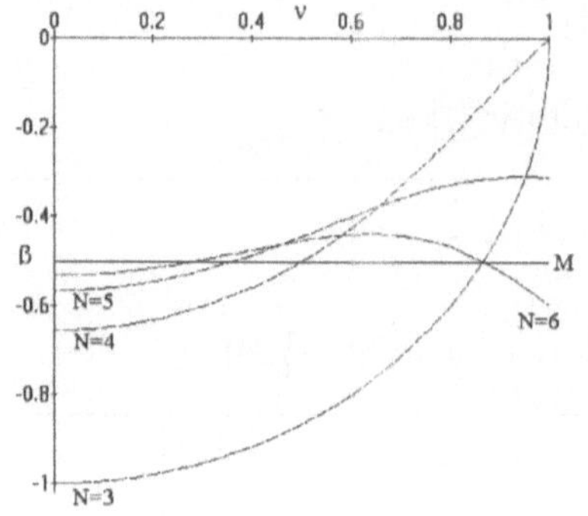

Fig. 8.3. *The negative solutions for the damping, β≡Imω(<0), as the function of vibration frequency, ν≡Reω, for the first boundary conditions.*

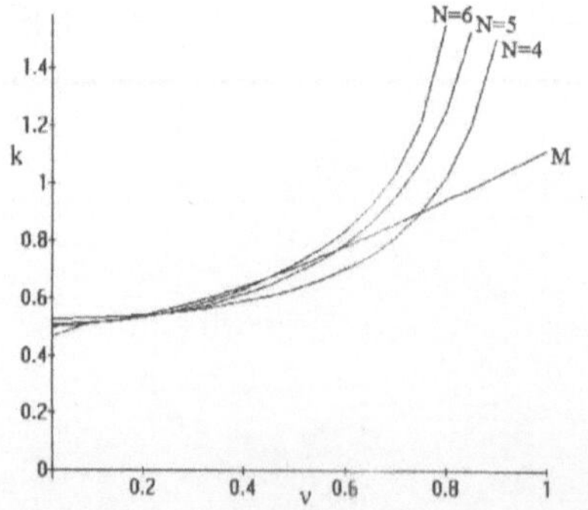

Fig. 8.4. *The wave number, k, as the function of vibration frequency, ν≡Reω, for the first boundary conditions.*

In Figs. 8.2.–8.4 we show the solutions of these equations in the range of negative values of β (damping!). If the Maxwellian model is considered to be the reference, it can be seen that higher order fluid models do not improve results, even within the range of convergence. Even worse, for all N≥2, the imaginary part of ω, i.e. the „damping", β, has also a positive branch for

$$\begin{aligned} &N=2: \quad \nu=0, k>1 \quad \Rightarrow \quad h<\pi\sqrt{\frac{\mu\tau}{\rho}}, \\ &N=3: \quad \nu\in(0,1), k>1 \quad \Rightarrow \quad h<\pi\sqrt{\frac{\mu\tau}{\rho}}, \\ &N=4: \quad \nu=0, k-\text{arbitrary}, \quad \text{etc.}, \end{aligned} \tag{8.24}$$

which is not shown in these figures. Since β can be positive this yields unbounded growth of the amplitude of velocity which is physically unacceptable. We return to this point after the discussion of the second type of the flow shown in Fig. 8.1.

Let us note that the 2nd order fluid does not admit any waves at all, since $\nu=0$. For the other models monochromatic waves could, in principle, exist. However, the case N=3 lies on the limit of convergence. Their velocity, $V_{ph}=\nu/k$, is shown in Fig. 8.5. Here again, the behaviour for different N is rather unpredictable, and also higher values of N do not lead to much of the improvement.

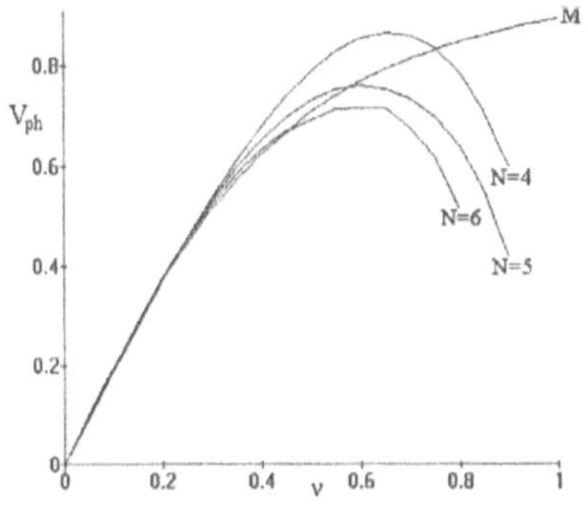

Fig. 8.5. *Phase velocities as the functions of vibration frequency, $\nu=Re\omega$, for the first boundary conditions*

Let us now discuss the second example of the boundary conditions shown in Fig. 8.1. We expect the solution (8.13) to have the form

$$\omega=\nu, \quad k=\chi+i\gamma, \tag{8.25}$$

where ν, χ, γ are real. Then the corresponding dispersion relations have the form

$$\begin{aligned} i\nu(1-i\nu) &= k^2, && (M) \\ i\nu(1-i\nu) &= k^2\left[1-(i\nu)^N\right] && (N) \end{aligned} \tag{8.26}$$

whose solutions are collected in the table below.

Table 8.3.

	γ	χ
M:	$\sqrt{\frac{\nu}{2}\left(\sqrt{1+\nu^2}-\nu\right)}$	$\sqrt{\frac{\nu}{2}\left(\sqrt{1+\nu^2}+\nu\right)}$
N – odd:	$\sqrt{\frac{\nu}{2}\left(\sqrt{\frac{1+\nu^2}{1-\left(-\nu^2\right)^N}}-\nu\frac{1-\left(-\nu^2\right)^{\frac{N-1}{2}}}{1-\left(-\nu^2\right)^N}\right)}$	$\sqrt{\frac{\nu}{2}\left(\sqrt{\frac{1+\nu^2}{1-\left(-\nu^2\right)^N}}+\nu\frac{1-\left(-\nu^2\right)^{\frac{N-1}{2}}}{1-\left(-\nu^2\right)^N}\right)}$
N – even:	$\sqrt{\frac{\nu}{2}\frac{\sqrt{1+\nu^2}-\nu}{1-\left(-\nu^2\right)^{\frac{N}{2}}}}$	$\sqrt{\frac{\nu}{2}\frac{\sqrt{1+\nu^2}+\nu}{1-\left(-\nu^2\right)^{\frac{N}{2}}}}$

The dependence of the wave vector, χ, and the attenuation, γ, on the frequency, ν, is shown in Figs. 8.6. and 8.7. Again, due to the convergence requirement, the condition $\nu<1$ must be satisfied. It is seen that the behaviour of the Nth order fluids near the value $\nu=1$ becomes completely wild.

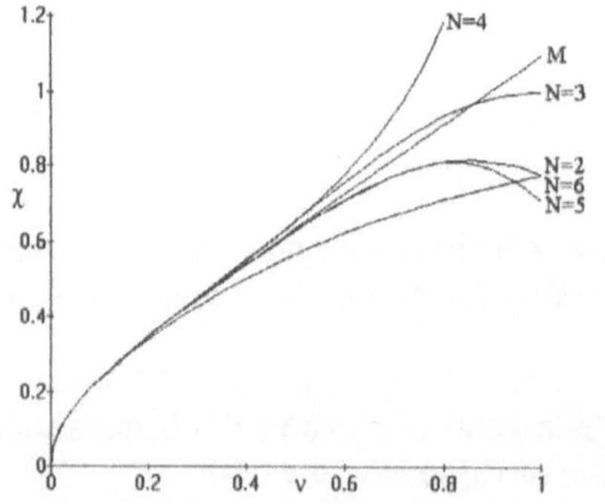

Fig. 8.6. *The wave vector, χ, as the function of frequency, ν, for the second boundary conditions*

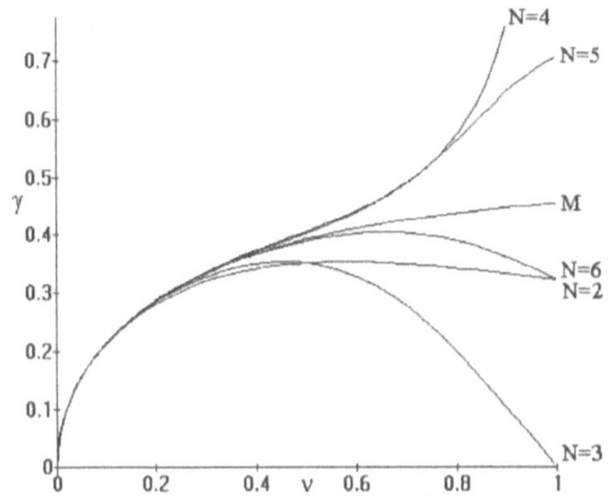

Fig. 8.7. *The attenuation, γ, as the function of frequency ν, for the second boundary conditions*

We complete this example with the calculation of the phase velocities. We obtain easily

$$V_{ph}^{M} = \sqrt{\frac{2}{\sqrt{1+\frac{1}{\nu^2}}+1}}, \qquad \lim_{\nu\to\infty} V_{ph}^{M} = 1,$$

$$V_{ph}^{N} = V_{ph}^{M}\sqrt{1-\left(-\nu^2\right)^{\frac{N}{2}}}, \quad N-\text{even}, \tag{8.27}$$

$$V_{ph}^{N} = \sqrt{2}\left[\sqrt{1+\frac{1}{\nu^2}}+\frac{1-\left(-\nu^2\right)^{\frac{N-1}{2}}}{\sqrt{1-\left(-\nu^2\right)^{N}}}\right]^{-\frac{1}{2}}\left(1-\left(-\nu^2\right)^{N}\right)^{\frac{1}{4}}, \quad N-\text{odd}.$$

These are shown in Fig. 8.8.

Inspection of the above results allows us to formulate the following conclusions:

1. There is comparatively good qualitative agreement between the Maxwell fluid model and the Nth order fluid models for very small frequencies, e.g. $\nu<0.2$, the upper limit corresponds to ~0.127Hz. Otherwise, the properties of the Nth order fluids depend very strongly on N, and for $\nu>1$ they behave very differently and do not converge to the Maxwellian fluid any more. Therefore, these values of ν are admissible only in the Nth grade fluids;

2. The shear pulses propagate in the Maxwellian fluid with finite speed, V=1 ($\sim 1^{m}/_{s}$), whereas, due to the cutoff in the spectrum of frequencies ($\nu=1$) the Nth order fluids do not admit such pulses at all;

3. The Nth grade fluids (i.e. for $\nu>1$) predict numerous instabilities which do not appear in Maxwellian fluids and do not seem to have any physical relevance.

4. The 2nd and 3rd order fluids predict the critical damping (ν=0) for flows in channels.■

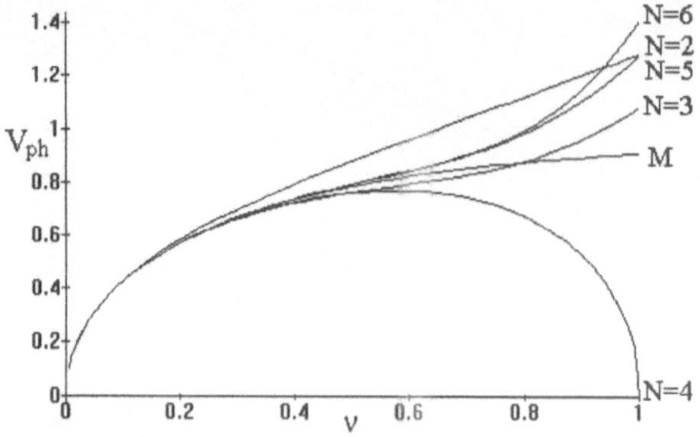

Fig. 8.8. *The phase velocities as the function of frequency,* ν, *in the case of the second boundary conditions*

To complete this picture of disaster, let us mention that D. JOSEPH [1981] proved the instability of the rest state of the fluids for an arbitrary grade N, in a completely general case. It is amazing that despite these devastating properties, the Nth grade models were used, and are still used, to describe „successfully" experimental results for non-Newtonian fluids. This is due to the fact, that accidentally, the steady flows of the Maxwellian and the Nth grade fluids are described by the same equations for stresses, and simultaneously, almost all rheological experiments are done in such steady conditions.

The above remarks show the necessity for a proper thermodynamical model of non-Newtonian fluids in which stresses are described by some balance equations, and hence, the material has a memory, the speeds of shear pulses are finite, no unexpected instabilities appear, etc.

Such a model has been constructed within the frame of extended thermodynamics for non-conducting fluids [see: I. MÜLLER, K. WILMANSKI (1986)] and then extended to a certain class of the heat conducting fluids [see: K. WILMANSKI (1988)]. Further in this chapter, we shall restrict our considerations to non-conducting fluids and follow in the presentation the first of the above mentioned papers.

8.2 Constitutive Relations for a Non-Conducting Non-Newtonian Fluid; Field Equations

We proceed to construct a thermodynamical model of a non-conducting non-Newtonian fluid in the manner characteristic of extended thermodynamics and demonstrated for ideal gases in the previous chapter.

The objective of thermodynamics, within this frame, is the determination of 10 fields:

- the mass density, ρ,
- the velocity field, $\mathbf{v}$,
- the Cauchy stress tensor, $\mathbf{T}$.

If we ignore the heat flux $\mathbf{q}$, and consequently, the second equation (5.104), the set of field equations follows from (5.103) and (5.104). In an arbitrary non-inertial reference frame we have that

$$\begin{aligned}
&\dot{\rho} + \rho \operatorname{div} \mathbf{v} = 0, \\
&\rho \dot{\mathbf{v}} - \operatorname{div} \mathbf{T} = \rho\left(\mathbf{b} + \mathbf{i}^0\right) + 2\rho \Omega \mathbf{v}, \\
&\rho \dot{\varepsilon} - \operatorname{tr} \mathbf{T}\mathbf{L}^T = 0, \\
&\rho\left\langle \overset{\bullet}{\overline{\frac{1}{\rho}\hat{\mathbf{F}}}} \right\rangle - \Omega\left\langle \hat{\mathbf{F}} \right\rangle + \left\langle \hat{\mathbf{F}} \right\rangle \mathbf{W} - \langle \mathbf{T} \rangle (\Omega - \mathbf{W}) + (\Omega - \mathbf{W}) \langle \mathbf{T} \rangle - \\
&\quad - \langle \mathbf{T} \rangle \mathbf{D} - \mathbf{D} \langle \mathbf{T} \rangle + \tfrac{2}{3} \operatorname{tr}(\langle \mathbf{T} \rangle \mathbf{D}) \mathbf{1} + 2p \langle \mathbf{D} \rangle + \left\langle \operatorname{div} \hat{\mathbf{G}} \right\rangle = \left\langle \hat{\mathbf{f}} \right\rangle,
\end{aligned} \tag{8.28}$$

where the dot denotes the material time derivative $\left(\frac{\partial}{\partial t} + \mathbf{v} \cdot \operatorname{grad}\right)$, $\langle \cdot \rangle$ denotes the traceless parts of tensors

$$\begin{aligned}
&\left\langle \hat{\mathbf{F}} \right\rangle \equiv \hat{\mathbf{F}} - \tfrac{1}{3} \operatorname{tr} \hat{\mathbf{F}} \mathbf{1} = \hat{F}_{\langle kl \rangle} \mathbf{e}^k \otimes \mathbf{e}^l, \\
&\langle \mathbf{T} \rangle \equiv \mathbf{T} - \tfrac{1}{3} \operatorname{tr} \mathbf{T} \mathbf{1} = \sigma_{\langle kl \rangle} \mathbf{e}^k \otimes \mathbf{e}^l, \\
&\left\langle \operatorname{div} \hat{\mathbf{G}} \right\rangle \equiv \operatorname{div} \hat{\mathbf{G}} - \tfrac{1}{3} \operatorname{tr}\left(\operatorname{div} \hat{\mathbf{G}}\right) \mathbf{1} = \frac{\partial}{\partial x^m} G_{\langle kl \rangle}{}^m \mathbf{e}^k \otimes \mathbf{e}^l, \\
&\left\langle \hat{\mathbf{f}} \right\rangle \equiv \hat{\mathbf{f}} - \tfrac{1}{3} \operatorname{tr} \hat{\mathbf{f}} \mathbf{1} = \hat{f}_{\langle kl \rangle} \mathbf{e}^k \otimes \mathbf{e}^l.
\end{aligned} \tag{8.29}$$

The relative angular velocity tensor of the non-inertial frame is denoted Ω (comp. (3.60)), and $\mathbf{i}^0$ denotes the velocity-independent part of the inertial acceleration [see:(7.3)]. $\mathbf{W}$ and $\mathbf{D}$ denote the spin and stretching tensors, respectively [see: (3.28)].

Exercise 8.1. Derive the equations (8.28) from (5.103-5.104) changing the frame to the non-inertial one•

We assume that the thermal state equation in the thermodynamical equilibrium of a fluid

$$p = p(\rho, T), \tag{8.30}$$

is known. Then, we can change the fields to be determined by replacing ρ with T:

$$\{T, \mathbf{v}, p, \langle \mathbf{p} \rangle\}, \qquad \langle \mathbf{p} \rangle \equiv -\langle \mathbf{T} \rangle. \tag{8.31}$$

The tensor $\langle \mathbf{p} \rangle$ is called the **pressure deviator**. In rheology, it is customary to replace the stress deviator by the pressure deviator because of its kinetic foundations, where the pressure tensor, $\mathbf{p}$, rather than the stress tensor, $\mathbf{T}$, is used. The replacement of the mass density, ρ, by the temperature as the field is connected with our further discussion of the definition of incompressibility of the non-Newtonian fluids within the frame of extended thermodynamics.

In order to close system (8.28) for the fields (8.31) we need constitutive relations for the following quantities:

$$\left\{\rho, \varepsilon_{kl}, \hat{G}_{\langle kl\rangle m}, \hat{f}_{\langle kl\rangle}\right\}, \qquad \varepsilon_{kl} \equiv \tfrac{1}{3}\varepsilon\delta_{kl} + \frac{1}{2\rho}\hat{F}_{\langle kl\rangle}. \tag{8.32}$$

We shall refer to ε as the **internal energy tensor** and to its deviatoric part as the **internal energy deviator**.

According to the procedure of extended thermodynamics, we assume the first three quantities in $(8.32)_1$ to depend only on the fields

$$\begin{aligned} \rho &= \rho(T, \mathbf{v}, \mathbf{p}), \\ \varepsilon_{kl} &= \varepsilon_{kl}(T, \mathbf{v}, \mathbf{p}), \\ \hat{G}_{\langle kl\rangle m} &= \hat{G}_{\langle kl\rangle m}(T, \mathbf{v}, \mathbf{p}). \end{aligned} \tag{8.33}$$

Consistently, the source term, $\hat{f}_{\langle kl\rangle}$, should satisfy a similar equation. This would be proper for the non-conducting viscous fluids without the normal stress effects (elastic reactions).

However, we have already indicated in Sect. 8.1. that non-Newtonian fluids possess a memory. If the kinetic theory of such fluids were known, the memory effects would have to arise from the collision operator and not from the kinetic part of the equation (compare the remark 5.1.). This means that the source, $\hat{f}_{\langle kl\rangle}$, must depend on at least a time derivative of the pressure deviator $p_{<kl>}$. This is the break in the equipresence of the variables between (8.33) and

$$\hat{f}_{\langle kl\rangle} = \hat{f}_{\langle kl\rangle}\left(T, \mathbf{v}, \mathbf{p}, \langle \mathbf{p}^{\Delta}\rangle\right), \tag{8.34}$$

where

$$\langle \mathbf{p}^{\Delta}\rangle \equiv \left\langle \frac{D_J \mathbf{p}}{Dt}\right\rangle \equiv \langle \dot{\mathbf{p}}\rangle - \mathbf{W}\langle \mathbf{p}\rangle + \langle \mathbf{p}\rangle \mathbf{W} \tag{8.35}$$

is the corotational (Jaumann-Zaremba) time derivative [see (3.71)]. The choice of the corotational derivative, rather than any other objective time derivative, is immaterial within the frame of the present model.

In this model the break in equipresence is tolerated for the sake of simplicity. It was argued in the paper by I. MÜLLER and K. WILMANSKI (1986) that the normal stress effects require this type of dependence. Further on in the second part of this chapter we shall see that the presence of this time derivative is connected with the proper identification of the parameter α_1 measured in rheological experiments.

The constitutive relations (8.33-34) are to be restricted by the following requirements:

- the material frame indifference,
- the entropy principle,
- the thermodynamical stability.

The last two conditions shall be discussed in the next section. We proceed to exploit the first condition.

We have already seen that the material frame indifference for fluids yields the isotropic structure of constitutive relations, and moreover, forbids the dependence on the velocity field.

We restrict our considerations to processes close to thermodynamical equilibrium, defined by the conditions

$$p_{\langle kl\rangle}\Big|_E = 0, \quad \overset{\Delta}{p}_{\langle kl\rangle}\Big|_E = 0. \tag{8.36}$$

Furthermore, we assume that the time changes of the pressure deviator are small near the equilibrium. Hence, we only account for terms quadratic in $p_{<kl>}$ and linear in $\overset{\Delta}{p}_{\langle kl\rangle}$. Under these assumptions, the representation theorems of Chap. 5 yield

$$\begin{aligned}
&\rho = \rho_0 + \rho_1 P, \qquad P \equiv \mathrm{tr}\langle \mathbf{p}\rangle^2,\\
&\varepsilon = \varepsilon_0 + \varepsilon_1 P,\\
&\langle \varepsilon\rangle = e_2\langle \mathbf{p}\rangle + e_3\left(\langle \mathbf{p}\rangle^2 - \tfrac{1}{3}P\mathbf{1}\right),\\
&\langle \hat{\mathbf{f}}\rangle = \beta_1\langle \mathbf{p}\rangle + \beta_2\langle \mathbf{p}^{\Delta}\rangle + \beta_3\left(\langle \mathbf{p}\rangle^2 - \tfrac{1}{3}P\mathbf{1}\right),
\end{aligned} \tag{8.37}$$

and

$$\hat{G}_{\langle kl\rangle m} = 0. \tag{8.38}$$

All coefficients $\rho_0,\ldots,\beta_3$ may be functions of p and T [*caution*: the coefficients β are, in the present model, not connected with the coefficients β of the 3rd grade fluid model described by the relation $(8.1)_2$!].

The last very restrictive result of the vanishing flux, $\hat{\mathbf{G}}$, will not hold in the case of heat conducting fluids.

Bearing (8.38) in mind, let us rewrite the equations (8.28) in our present notation

$$\begin{aligned}
&\dot{\rho} + \rho \operatorname{div} \mathbf{v} = 0,\\
&\rho\,\dot{\mathbf{v}} + \operatorname{div}\mathbf{p} = \rho\left(\mathbf{b} + \mathbf{i}^0\right) + 2\rho\boldsymbol{\Omega}\,\mathbf{v},\\
&2\rho\left\{\dot{\varepsilon} + \langle\varepsilon\rangle\boldsymbol{\Omega} - \boldsymbol{\Omega}\,\langle\varepsilon\rangle\right\} + \langle\mathbf{p}\rangle\left(\mathbf{L}^T + \boldsymbol{\Omega}\right) + (\mathbf{L} - \boldsymbol{\Omega})\langle\mathbf{p}\rangle + 2p\mathbf{D} = \langle\hat{\mathbf{f}}\rangle.
\end{aligned} \tag{8.39}$$

Insertion of (8.37) into (8.39) gives a set of field equations for T, **v**, **p**, whose solutions are called the **thermodynamical processes**.

8.3 Entropy Principle

In general, the entropy inequality is of the form (6.11–6.12), i.e. for our fields we have

$$\eta=\eta(T,\mathbf{v},\mathbf{p}),\quad \mathbf{h}=\mathbf{h}(T,\mathbf{v},\mathbf{p}). \tag{8.40}$$

The material frame indifference yields immediately, if we again restrict ourselves to small deviations from the thermodynamical equilibrium

$$\eta=\eta_0+\eta_1 P,\quad \mathbf{h}=0, \tag{8.41}$$

where η_0, η_1 may depend on p and T.

The entropy inequality must hold for all thermodynamical processes, i.e. solutions of the field equations. To get rid of this qualification, we apply Liu's Theorem and obtain

$$\begin{aligned}\forall(T,\mathbf{v},\mathbf{p}):\ &\rho\dot{\eta}-\lambda(\dot{\rho}+\rho\,\mathrm{div}\,\mathbf{v})-\boldsymbol{\lambda}\cdot\left(\rho\dot{\mathbf{v}}+\mathrm{div}\,\mathbf{p}-\rho\left(\mathbf{b}+\mathbf{i}^0\right)-2\rho\boldsymbol{\Omega}\,\mathbf{v}\right)-\\ &-\mathrm{tr}\,\boldsymbol{\Lambda}\left[2\rho(\dot{\boldsymbol{\varepsilon}}+\langle\boldsymbol{\varepsilon}\rangle\boldsymbol{\Omega}-\boldsymbol{\Omega}\langle\boldsymbol{\varepsilon}\rangle)+\langle\mathbf{p}\rangle\left(\mathbf{L}^T+\boldsymbol{\Omega}\right)+\right.\\ &\left.+(\mathbf{L}-\boldsymbol{\Omega})\langle\mathbf{p}\rangle+2p\mathbf{D}-\langle\hat{\mathbf{f}}\rangle\right]\ge 0,\qquad \boldsymbol{\Lambda}=\boldsymbol{\Lambda}^T.\end{aligned} \tag{8.42}$$

It is clear that the substitution of the constitutive relations (8.37) in (8.42) yields the linear structure of this inequality in the derivatives

$$\left\{\dot{T},\dot{\mathbf{p}},\dot{\mathbf{v}},\mathbf{L}\equiv\mathrm{grad}\,\mathbf{v}\right\}. \tag{8.43}$$

Therefore, the coefficients of these derivatives must vanish identically, and we obtain

$$\begin{aligned}&\lambda=0,\\ &\frac{\partial\eta}{\partial T}-\frac{\lambda}{\rho}\frac{\partial\rho}{\partial T}-2\Lambda^{kl}\frac{\partial\varepsilon_{kl}}{\partial T}=0,\\ &\left(\frac{\partial\eta}{\partial p_{kl}}-\frac{\lambda}{\rho}\frac{\partial\rho}{\partial p_{kl}}-2\Lambda^{ij}\frac{\partial\varepsilon_{ij}}{\partial p_{kl}}\right)\delta_{kl}=0,\end{aligned} \tag{8.44$_1$}$$

$$\begin{aligned}&\left(\frac{\partial\eta}{\partial p_{kl}}-\frac{\lambda}{\rho}\frac{\partial\rho}{\partial p_{kl}}-2\Lambda^{ij}\frac{\partial\varepsilon_{ij}}{\partial p_{kl}}\right)\left(\delta_k^n\delta_l^m-\tfrac{1}{3}\delta_{kl}\delta^{mn}\right)-\frac{\beta_2}{\rho}\Lambda^{\langle nm\rangle}=0,\\ &\rho\lambda\delta_{nm}+2\Lambda^k_{\ n}p_{\langle km\rangle}+2p\Lambda_{nm}+\beta_2\Lambda^k_{\ n}p_{\langle km\rangle}-\beta_2\Lambda^k_{\ m}p_{\langle kn\rangle}=0.\end{aligned} \tag{8.44$_2$}$$

There remains the following residual inequality

$$\mathrm{tr}\,\boldsymbol{\Lambda}\left\{-4\rho\langle\boldsymbol{\varepsilon}\rangle\boldsymbol{\Omega}-2\langle\mathbf{p}\rangle\boldsymbol{\Omega}+\beta_1\langle\mathbf{p}\rangle+\beta_3\left(\langle\mathbf{p}\rangle^2-\tfrac{1}{3}P\mathbf{1}\right)\right\}\ge 0. \tag{8.45}$$

It is seen that the multipliers λ and Λ can be functions of T and **p**, but not of $\langle\mathbf{p}^{\Delta}\rangle$ and **v**. Simultaneously, it does not make much sense to seek general solutions for these multipliers as the constitutive relations (8.37) and (8.41) are already approximations. For this reason we write these multipliers in the form

$$\frac{1}{\rho}\lambda = l_0 + l_1 P,$$
$$\Lambda = \tfrac{1}{3}(\lambda_0 + \lambda_1 P)\mathbf{1} + \Lambda_2\langle \mathbf{p}\rangle + \Lambda_3\left(\langle \mathbf{p}\rangle^2 - \tfrac{1}{3}P\mathbf{1}\right), \tag{8.46}$$

where l_0, l_1, λ_0, λ_1, Λ_2, Λ_3 can be functions of p, T. Insertion of (8.46) in (8.44) yields up to the second order terms in $\langle \mathbf{p}\rangle$

$$\frac{\partial\eta_0}{\partial T} + \frac{\partial\eta_1}{\partial T}P - l_0\frac{\partial\rho_0}{\partial T} - l_1 P\frac{\partial\rho_0}{\partial T} - l_0\frac{\partial\rho_1}{\partial T}P -$$
$$-\tfrac{2}{3}\lambda_0\frac{\partial\varepsilon_0}{\partial T} - \tfrac{2}{3}\lambda_0\frac{\partial\varepsilon_1}{\partial T}P - \tfrac{2}{3}\lambda_1 P\frac{\partial\varepsilon_0}{\partial T} - 2\Lambda_2 P\frac{\partial e_2}{\partial T} = 0,$$
$$\frac{\partial\eta_0}{\partial p} + \frac{\partial\eta_1}{\partial p}P - l_0\frac{\partial\rho_0}{\partial p} - l_1 P\frac{\partial\rho_0}{\partial p} - l_0\frac{\partial\rho_1}{\partial p}P -$$
$$-\tfrac{2}{3}\lambda_0\frac{\partial\varepsilon_0}{\partial p} - \tfrac{2}{3}\lambda_0\frac{\partial\varepsilon_1}{\partial p}P - \tfrac{2}{3}\lambda_1 P\frac{\partial\varepsilon_0}{\partial p} - 2\Lambda_2 P\frac{\partial e_2}{\partial p} = 0, \tag{8.47}$$
$$\left(\eta_1 - l_0\rho_1 - \Lambda_2 e_2 - \tfrac{2}{3}\varepsilon_1\lambda_0 + \frac{\beta_2}{2\rho_0}\Lambda_2\right)\langle \mathbf{p}\rangle = 0,$$

$$\left[\rho^2(l_0 + l_1 P) + \tfrac{2}{3}\Lambda_2 P + \tfrac{2}{3}p(\lambda_0 + \lambda_1 P)\right]\mathbf{1} +$$
$$+\left[\tfrac{2}{3}\lambda_0 + 2p\Lambda_2\right]\langle \mathbf{p}\rangle + \left[2\Lambda_2 + 2p\Lambda_3\right]\left(\langle \mathbf{p}\rangle^2 - \tfrac{1}{3}P\mathbf{1}\right) = 0. \tag{8.47$_{\rm cont.}$}$$

Collecting terms of the same order of magnitude relative to $\langle \mathbf{p}\rangle$ together, we obtain the following conditions:

$$\frac{\partial\eta_0}{\partial T} - l_0\frac{\partial\rho_0}{\partial T} - \tfrac{2}{3}\lambda_0\frac{\partial\varepsilon_0}{\partial T} = 0,$$
$$\frac{\partial\eta_0}{\partial p} - l_0\frac{\partial\rho_0}{\partial p} - \tfrac{2}{3}\lambda_0\frac{\partial\varepsilon_0}{\partial p} = 0, \tag{8.48}$$

$$\frac{\partial\eta_1}{\partial T} - l_1\frac{\partial\rho_0}{\partial T} - l_0\frac{\partial\rho_1}{\partial T} - \tfrac{2}{3}\lambda_0\frac{\partial\varepsilon_1}{\partial T} - \tfrac{2}{3}\lambda_1\frac{\partial\varepsilon_0}{\partial T} - 2\Lambda_2\frac{\partial e_2}{\partial T} = 0,$$
$$\frac{\partial\eta_1}{\partial p} - l_1\frac{\partial\rho_0}{\partial p} - l_0\frac{\partial\rho_1}{\partial p} - \tfrac{2}{3}\lambda_0\frac{\partial\varepsilon_1}{\partial p} - \tfrac{2}{3}\lambda_1\frac{\partial\varepsilon_0}{\partial p} - 2\Lambda_2\frac{\partial e_2}{\partial p} = 0, \tag{8.49}$$

$$\eta_1 - l_0\rho_1 - \Lambda_2 e_2 - \tfrac{2}{3}\varepsilon_1\lambda_0 + \frac{\beta_2}{2\rho_0}\Lambda_2 = 0, \tag{8.50}$$

$$\rho_0^2 l_0 + \tfrac{2}{3}p\lambda_0 = 0, \tag{8.51}$$

$$\Lambda_2 + p\lambda_1 + \tfrac{2}{3}\rho_0^2 l_1 + 3\rho_0\rho_1 l_0 = 0, \tag{8.52}$$

$$\tfrac{2}{3}\lambda_0 + 2p\Lambda_2 = 0, \tag{8.53}$$

$$\Lambda_2 + p\Lambda_3 = 0. \tag{8.54}$$

Simultaneously, the residual inequality (8.45) exploited up to quadratic terms in $\langle \mathbf{p} \rangle$ yields

$$\beta_1 \Lambda_2 P \geq 0. \tag{8.55}$$

We combine (8.48) and (8.51) to obtain

$$d\eta_0 = \tfrac{2}{3}\lambda_0\left(d\varepsilon_0 - \frac{p}{\rho_0^2} d\rho_0\right). \tag{8.56}$$

This is, certainly, the Gibbs equation for a non-Newtonian fluid in thermodynamical equilibrium. Hence,

$$\lambda_0 = \tfrac{3}{2}\frac{1}{T}, \tag{8.57}$$

where T is the absolute temperature (see: Appendix A). As the integrability condition for (8.56), we obtain

$$\frac{\partial \rho_0}{\partial T} = \frac{\rho_0^2}{T}\frac{\partial \varepsilon_0}{\partial p} - \frac{p}{T}\frac{\partial \rho_0}{\partial p}. \tag{8.58}$$

By means of (8.51–53) we can now identify l_0, Λ_2, Λ_3

$$l_0 = -\frac{p}{\rho_0^2 T}, \quad \Lambda_2 = -\frac{1}{2pT}, \quad \Lambda_3 = \frac{1}{2p^2 T}, \tag{8.59}$$

and according to (8.50) we have

$$\varepsilon_1 - T\eta_1 - \frac{p}{\rho_0^2}\rho_1 = \frac{1}{2p}\left(e_2 - \frac{\beta_2}{2\rho_0}\right). \tag{8.60}$$

Yet the relations (8.49) and (8.52) remain unexploited. Bearing (8.57-8.60) in mind, after easy manipulations we obtain

$$\begin{aligned} \tfrac{2}{3}\lambda_1 T c_p &= -\frac{\partial}{\partial T}\left[\frac{1}{2p}\left(e_2 - \frac{\beta_2}{2\rho_0}\right)\right] - \eta_1 - \frac{1}{3p\rho_0^2}\frac{\partial \rho_0}{\partial T} + \frac{1}{p}\frac{\partial e_2}{\partial T}, \\ \tfrac{2}{3}\lambda_1 T\left(\frac{T}{\rho_0^2}\frac{\partial \rho_0}{\partial T}\right) &= -\frac{\partial}{\partial p}\left[\frac{1}{2p}\left(e_2 - \frac{\beta_2}{2\rho_0}\right)\right] - \frac{\rho_1}{\rho_0^2} - \frac{1}{3p\rho_0^2}\frac{\partial \rho_0}{\partial p} + \frac{1}{p}\frac{\partial e_2}{\partial p}, \end{aligned} \tag{8.61}$$

$$l_1 T = -\tfrac{2}{3}\lambda_1 T\frac{p}{\rho_0^2} + 2\frac{p}{\rho_0^3}\rho_1 + \frac{1}{3p\rho_0^2}, \tag{8.62}$$

with

$$c_p \equiv \frac{\partial \varepsilon_0}{\partial T} - \frac{p}{\rho_0^2}\frac{\partial \rho_0}{\partial T}, \tag{8.63}$$

being the specific heat at constant pressure.

The set (8.61), (8.62) allows us to calculate λ_l and l_1. If we divide $(8.61)_1$ by $(8.61)_2$, there remains a complicated restriction upon the constitutive relations.

Finally, the substitution of $(8.59)_2$ in (8.55) yields

$$\beta_1 \leq 0. \tag{8.64}$$

Let us summarize the above results.

First of all, it is easy to see that useful results are rather scarce. We have identified the Lagrange multipliers which is not much of the result because the multipliers are auxiliary quantities anyway. Apart from that, we have the integrability (8.58) and the result (8.60) which has an important bearing on thermodynamical stability, to be discussed in the following section. We shall also use (8.61) in the discussion of incompressibility.

8.4. Thermodynamical Stability; Shear Pulses

By means of a quadratic approximation of the constitutive relations we can construct an explicit constitutive expression for the **Gibbs free energy function (free enthalpy)**

$$\begin{aligned} g &= \varepsilon - T\eta + \frac{p}{\rho} = \varepsilon_0 + \varepsilon_1 - T(\eta_0 + \eta_1 P) + \frac{p}{\rho_0 + \rho_1 P} \approx \\ &\approx \varepsilon_0 - T\eta_0 + \frac{p}{\rho_0} + \left(\varepsilon_1 - T\eta_1 - \frac{p}{\rho_0^2}\rho_1\right)P = g_0 + \left(\varepsilon_1 - T\eta_1 - \frac{p}{\rho_0^2}\rho_1\right)P, \end{aligned} \tag{8.65}$$

which yields the proper thermodynamical potential at equilibrium for the thermodynamical variables T and p.

The requirement of **thermodynamical stability of the equilibrium** under these variables means that the free enthalpy must attain its minimum at $\langle \mathbf{p} \rangle = 0$ (i.e. P=0). Hence,

$$\varepsilon_1 - T\eta_1 - \frac{p}{\rho_0^2}\rho_1 > 0, \tag{8.66}$$

or according to relation (8.60),

$$e_2 - \frac{\beta_2}{2\rho_0} > 0. \tag{8.67}$$

Apart from this condition, we obtain the usual inequalities of thermostatics which follow from the requirement that g_0 must increase under virtual changes of T and p (compare Appendix A). In our case, these conditions read

$$\frac{\partial \eta_0}{\partial T} > 0, \quad \frac{\partial \rho_0}{\partial p} > 0, \quad \left(\frac{\partial \rho_0}{\partial T}\right)^2 < \rho_0^2 \frac{\partial \eta_0}{\partial T} \frac{\partial \rho_0}{\partial p}, \tag{8.68}$$

and express the positivity of the specific heat and compressibility, as well as an upper bound for the thermal expansion.

The physical meaning of inequality (8.67) can be seen from the field equation $(8.39)_3$. Let us write this equation in an inertial frame, omit the non-linear terms, then its traceless part has the following form:

$$2\rho_0 e_2 \langle \dot{\mathbf{p}} \rangle = \beta_1 \langle \mathbf{p} \rangle + \beta_2 \langle \dot{\mathbf{p}} \rangle,$$

i.e.

$$\left[\frac{2\rho_0}{\beta_1}\left(e_2 - \frac{\beta_2}{2\rho_0}\right)\right] \langle \dot{\mathbf{p}} \rangle = \langle \mathbf{p} \rangle. \tag{8.69}$$

According to the inequalities (8.64) and (8.67), the expression in the square brackets is negative. This assures the **relaxation** of the pressure deviator. Consequently, the thermodynamical stability requirement yields the physically desired result.

The condition (8.67) also plays an important role in the wave analysis of the model. We showed in Sect. 8.1. that, in contrast to the Maxwellian fluid, N^{th} grade fluids do not admit propagation of shear pulses. Now we show that our model yields the finite speed of propagation of such pulses. This can be done directly by the analysis of the eigenvalue problem for the set (8.39), which is a hyperbolic set of equations. We proceed, in a different way, by investigating a **small disturbance** of the rest state described by the following relations:

$$\mathbf{v} = (0, \bar{v}, 0)^T e^{i(\omega t - kx)}, \quad T = \tilde{T} + \bar{T} e^{i(\omega t - kx)}, \quad x \equiv x^1,$$
$$p = \tilde{p} + \bar{p} e^{i(\omega t - kx)}, \quad \left(p_{\langle kl \rangle}\right) = \begin{pmatrix} 0 & \bar{\pi} & 0 \\ \bar{\pi} & 0 & 0 \\ 0 & 0 & 0 \end{pmatrix} e^{i(\omega t - kx)}. \tag{8.70}$$

Here $\tilde{T}, \tilde{p}$ are constant fields, describing the rest state, and $\bar{v}, \bar{T}, \bar{p}, \bar{\pi}$ are constant amplitudes of the disturbance with magnitude small enough to make all their products negligible. Obviously, relations (8.70) describe the **shear waves** because the amplitude of the velocity field is perpendicular to the direction of propagation.

The linearized form of (8.39) around the rest state in the inertial frame is:

$$\frac{\partial \rho}{\partial t} + \tilde{\rho} \operatorname{div} \mathbf{v} = 0,$$
$$\tilde{\rho} \frac{\partial \mathbf{v}}{\partial t} + \operatorname{grad} p + \operatorname{div} \langle \mathbf{p} \rangle = 0,$$
$$\tilde{\rho} \frac{\partial \varepsilon}{\partial t} + \tilde{p} \operatorname{div} \mathbf{v} = 0, \tag{8.71}$$
$$2\tilde{\rho} \tilde{e}_2 \frac{\partial}{\partial t} \langle \mathbf{p} \rangle + 2\tilde{p} \langle \mathbf{D} \rangle = \beta_1 \langle \mathbf{p} \rangle + \beta_2 \frac{\partial}{\partial t} \langle \mathbf{p} \rangle.$$

Substitution of (8.70) into (8.71) yields the relation

$$\begin{pmatrix} \frac{\partial\tilde{\rho}}{\partial T}i\omega & \frac{\partial\tilde{\rho}}{\partial p}i\omega & 0 & 0 \\ 0 & 0 & \tilde{\rho}i\omega & ik \\ \tilde{\rho}\frac{\partial\tilde{\varepsilon}}{\partial T}i\omega & \tilde{\rho}\frac{\partial\tilde{\varepsilon}}{\partial p}i\omega & 0 & 0 \\ 0 & 0 & -2\tilde{p}ik & \left(2\tilde{\rho}\tilde{e}_2-\tilde{\beta}_2\right)i\omega-\tilde{\beta}_1 \end{pmatrix}\begin{pmatrix}\overline{T}\\ \overline{p}\\ \overline{v}\\ \overline{\pi}\end{pmatrix}=0, \tag{8.72}$$

and for the first component of the momentum balance equation

$$-ik\overline{p}=0. \tag{8.73}$$

It follows

$$\overline{p}=0,\quad \overline{T}=0, \tag{8.74}$$

and the remaining non-trivial equations (8.72) lead to the dispersion relation

$$\frac{k^2}{\omega^2}=\frac{\tilde{\rho}^2}{\tilde{p}}\left(\tilde{e}_2-\frac{\tilde{\beta}_2}{2\tilde{\rho}}\right)+i\frac{\tilde{\rho}\tilde{\beta}_1}{2\tilde{p}}\frac{1}{\omega}. \tag{8.75}$$

The phase speed, V_{ph}, and the attenuation, γ, are given by

$$V_{ph}=Re\left(\frac{\omega}{k}\right),\quad \gamma=Im\,k. \tag{8.76}$$

The speed of propagation of the shear pulses follows immediately

$$V_{ph}^{\infty}=\lim_{\omega\to\infty}V_{ph}=\sqrt{\frac{p}{\rho^2}\frac{1}{e_2-\frac{\beta_2}{2\rho}}}, \tag{8.77}$$

where the „wriggles" have been dropped.

Hence, the inequality (8.67) assures **finite speed** of propagation. Its value depends on the contribution of memory effects reflected by the coefficient β_2. However, even in the absence of these effects, i.e. for Newtonian fluids the speed (8.77) would be not equal to zero. This is, certainly, connected with the structure of the field equations of extended thermodynamics which remain hyperbolic also for $\beta_2=0$. In the simplest case of a monatomic ideal gas

$$e_2=\frac{1}{2\rho}\quad\Rightarrow\quad V_{ph}^{\infty}=\sqrt{2\frac{p}{\rho}}, \tag{8.78}$$

which is only slightly greater than the speed of the compression waves in the gas $\sqrt{\frac{5}{3}\frac{p}{\rho}}$.

However, a caution is required if the relation (8.77) is to be compared quantitatively with the experimental data. Our approximation of the memory effects by the time derivative $\langle\mathbf{p}^{\Delta}\rangle$ may be appropriate for small frequencies, but it is bound to fail for high frequencies.

8.5. Incompressibility; 2nd Grade Fluids.

We mentioned in Chap. 6 that not only do many fluids and solids behave in the range of large deformations as though they were incompressible, but also that such an assumption yields, in non-linear elasticity, many important universal heterogeneous solutions. In this discussion we defined the incompressibility as a constraint which allows the body to undergo solely isochoric motions, i.e. div$\mathbf{v}$=0.

Simultaneously, the transition from compressible to incompressible materials in their thermodynamical theory is singular if the mass density is chosen as the constitutive variable because the results obtained by the variation of ρ cannot be transfered to incompressible materials. This was the reason for choosing pressure as the variable in our constitutive theory presented in this chapter. We also showed in chapter 6 that the pressure becomes the reaction on the incompressibility constraint. This means, that apart from the condition on the isochoric motion, in incompressible materials the superposition of a time-dependent pressure, $\pi(t)$, upon an existing pressure field, $p(\mathbf{x},t)$, should not change the fields $T(\mathbf{x},t)$, $\mathbf{v}(\mathbf{x},t)$, and $\langle \mathbf{p}(\mathbf{x},t) \rangle$.

This means that neither the pressure nor its time derivative may occur in the field equations

$$\begin{aligned}
&\dot{\rho} + \rho \operatorname{div} \mathbf{v} = 0, \\
&\rho \dot{\mathbf{v}} + \operatorname{grad} p + \operatorname{div}\langle \mathbf{p} \rangle = \rho\left(\mathbf{b} + \mathbf{i}^0\right) + 2\rho\,\boldsymbol{\Omega}\,\mathbf{v}. \\
&\rho \dot{\varepsilon} + \rho \operatorname{div} \mathbf{v} + \operatorname{tr}\left(\langle \mathbf{p} \rangle \mathbf{D}\right) = 0, \qquad (8.79) \\
&2\rho\left\{\left\langle \dot{\varepsilon} \right\rangle + \langle \varepsilon \rangle \boldsymbol{\Omega} - \boldsymbol{\Omega}\langle \varepsilon \rangle\right\} + \langle \mathbf{p} \rangle\left(\mathbf{L}^{\mathrm{T}} + \boldsymbol{\Omega}\right) + \left(\mathbf{L} - \boldsymbol{\Omega}\right)\langle \mathbf{p} \rangle - \tfrac{2}{3}\operatorname{tr}\left(\langle \mathbf{p} \rangle \mathbf{D}\right)\mathbf{1} + \\
&\qquad \underline{\underline{+ 2p\langle \mathbf{D} \rangle = \beta_1 \langle \mathbf{p} \rangle + \beta_2\left\langle \mathbf{p}^{\Delta} \right\rangle + \beta_3\left(\langle \mathbf{p} \rangle^2 - \tfrac{1}{3}P\mathbf{1}\right)}}
\end{aligned}$$

This condition cannot be fulfilled by the above equations. It is seen that in the last equation there are some terms with the explicit presence of pressure and some other, quite explicit terms, which do not contain pressure.

However, we will argue, in due course, that the last equation can be approximated by neglecting all terms which are not underlined. In such an equation the leading terms can be made pressure-independent by assuming that

$$\beta_1, \beta_2, \beta_3 \textit{ are proportional to } p. \qquad (8.80)$$

The other three equations satisfy the condition if we assume that

$$\rho_0, \rho_1, \varepsilon_0, \varepsilon_1 \textit{ are independent of } p. \qquad (8.81)$$

According to this condition, (8.58) yields the result that ρ_0 is independent of T. Hence, the mass balance equation can be written in the form

$$\frac{d\rho_1}{dT}\dot{T} + \rho_1 \dot{P} = 0, \qquad (8.82)$$

which yields

$$\rho_1 = 0, \quad \frac{d\rho_1}{dT} = 0. \tag{8.83}$$

Inspection of $(8.61)_2$ leads to the equation

$$\frac{\partial e_2}{\partial p} = -\frac{e_2}{p}, \tag{8.84}$$

i.e. for incompressible materials

$$e_2 = \frac{C(T)}{p}, \tag{8.85}$$

which is the definite restriction on the constitutive relation for $\langle \varepsilon \rangle$ following from the entropy inequality.

Now we return to the field equations (8.79) and identify the material coefficients. We saw in the previous chapter, and also from our experience with kinetic theories, that the constitutive relation for $\langle \mathbf{p} \rangle$ can be derived from $(8.79)_4$ by replacing $\langle \mathbf{p} \rangle$ on the left-hand side by its equilibrium value, namely zero. In this way we obtain the approximating equation for $\langle \mathbf{p} \rangle$

$$2p\langle \mathbf{D} \rangle = \beta_1 \langle \mathbf{p} \rangle + \beta_2 \langle \mathbf{p}^{\Delta} \rangle + \beta_3 \left(\langle \mathbf{p} \rangle^2 - \tfrac{1}{3} P\mathbf{1} \right). \tag{8.86}$$

which is a type of **evolution equation for Maxwellian fluids.**

We derive the constitutive law for $\langle \mathbf{p} \rangle$ by applying the so-called **Maxwell iteration procedure** to (8.86). The leading term on the right-hand side is the first one. Hence, in the first step of iteration we have

$$\langle \mathbf{p} \rangle = -2\mu \langle \mathbf{D} \rangle, \quad \mu \equiv -\frac{p}{\beta_1}, \tag{8.87}$$

i.e. the **Navier-Stokes** equation with the viscosity, μ.

In the second step of iteration we insert (8.87) into the second and third terms of (8.86) and obtain

$$\langle \mathbf{p} \rangle = -2\mu \langle \mathbf{D} \rangle - 2\alpha_1 \langle \mathbf{D}^{\Delta} \rangle - 4\alpha_2 \left(\langle \mathbf{D} \rangle^2 - \tfrac{1}{3} \operatorname{tr} \langle \mathbf{D} \rangle^2 \mathbf{1} \right), \tag{8.88}$$

where

$$\alpha_1 \equiv \frac{\beta_2}{p} \mu^2, \quad \alpha_2 \equiv -\frac{\beta_3}{p} \mu^3. \tag{8.89}$$

Now it is obvious that the assumption (8.80) yields the coefficients μ, α_1, α_2 to be pressure-independent which is reasonably well confirmed by experiments. In the above presented approximation they are also independent of the shearing which is less satisfactory, as inspection of the table at the beginning of this chapter shows. However, the conditions (8.64) and (8.67) yield

$$\mu \geq 0, \quad \alpha_1 < \frac{\mu^2}{p} 2\rho_0 e_2, \tag{8.90}$$

and so, in contrast to the results for standard thermodynamics of 2nd grade fluids, α_1 may be negative and, in fact, is strictly negative if we accept the equation (8.86).

Now we are in the position to estimate the order of magnitude of the terms appearing in our model. To this aim we consider the one-dimensional steady flow

$$\mathbf{v} = \left(0, v^2\left(x^1\right), 0\right)^T. \tag{8.91}$$

Insertion into (8.88) yields

$$\begin{aligned} p_{\langle 11\rangle} &= \left(-\tfrac{4}{3}\alpha_1 - \tfrac{1}{3}\alpha_2\right)\kappa^2, \qquad \kappa \equiv \frac{\partial v^2}{\partial x^1}, \\ p_{\langle 22\rangle} &= \left(\tfrac{2}{3}\alpha_1 - \tfrac{1}{3}\alpha_2\right)\kappa^2, \\ p_{\langle 33\rangle} &= \left(\tfrac{2}{3}\alpha_1 + \tfrac{2}{3}\alpha_2\right)\kappa^2. \end{aligned} \tag{8.92}$$

In viscometric experiments, it is common to measure the following **normal stress differences**

$$p_{\langle 33\rangle} - p_{\langle 11\rangle} = \left(2\alpha_1 + \alpha_2\right)\kappa^2, \qquad p_{\langle 33\rangle} - p_{\langle 22\rangle} = \alpha_2\kappa^2. \tag{8.93}$$

Hence, the coefficients of our model can be written in the form

$$\begin{aligned} \frac{\beta_1}{\kappa} &= -\frac{p}{\mu\kappa}, \quad \beta_2 = p\frac{\alpha_1}{\mu^2} = p\frac{p_{\langle 22\rangle} - p_{\langle 11\rangle}}{2\kappa^2\mu^2}, \\ \beta_3\mu &= -p\frac{\alpha_2}{\mu^2} = -p\frac{p_{\langle 33\rangle} - p_{\langle 22\rangle}}{\kappa^2\mu^2}. \end{aligned} \tag{8.94}$$

Let us notice, that in order to justify the transition from equation $(8.79)_4$ to (8.86), and consequently, to (8.88), we have to compare β_1/κ, β_2 and $\beta_3\mu$ to 1. Let us select the value κ=10 $^1/_s$ from the experimental data (see: Table 8.1.). The corresponding values of the normal stress differences are

$$\begin{aligned} p_{\langle 33\rangle} - p_{\langle 11\rangle} &= -10.549 \text{ Pa}, \\ p_{\langle 33\rangle} - p_{\langle 22\rangle} &= 47 \text{ Pa}, \qquad \mu = 5.654 \text{ Pa s}. \end{aligned} \tag{8.95}$$

Then for the atmospheric pressure

$$p = 10^5 \text{Pa} \quad \Rightarrow \quad \frac{\beta_1}{\kappa} = -1768, \quad \beta_2 = -778, \quad \beta_3\mu = -1218. \tag{8.96}$$

Therefore, it is normally justified to ignore the terms on the left-hand side of $(8.79)_4$ which involve $\langle \mathbf{p} \rangle$. These terms would be comparable to those left in (8.86) if the pressure was dropped to the value ~10^3 Pa. The same effect of magnifying the influence of those terms could be achieved by rotating the turntable with the speed ~100Hz which is due to the frame dependence of this part of the equation $(8.79)_4$.

Finally let us mention that the speed of shear pulses, given by (8.77) for the data quoted above would be

$$V_{ph}^{\infty} \approx \sqrt{\frac{2p}{\rho}\frac{1}{\beta_2}} \approx \sqrt{2\frac{p}{\rho}10^{-3}} = 0.45\,\tfrac{m}{s}. \tag{8.97}$$

Despite the qualification made in Sect. 8.4., this result seems to agree well with the observations for polymer solutions.

PART 2: VISCOELASTIC SOLIDS

8.6. Field Equations in the Lagrangian Description

In this part, we focus attention on the extended thermodynamics of isotropic solids whose mechanical properties are viscoelastic. Such a model was investigated for the first time by I-SHIH LIU (1989) who applied the Lagrangian description in extended thermodynamics. This is made possible by assuming that the deformation gradient, **F**, is an unknown field, in addition to the usual velocity field.

We shall also anticipate the existence of certain **viscous deviatoric stresses**, **S**, which are a part of the Cauchy stress tensor, **T**, relaxing to zero at thermodynamical equilibrium, i.e.

$$\mathbf{S} = \mathbf{T} - \mathbf{T}|_E, \tag{8.98}$$

where $\mathbf{T}|_E$ is the Cauchy stress tensor in the equilibrium state to be defined in the following.

Consequently, the fields which are the objective in the present model, are as follows:

$$\{\mathbf{v}, \mathbf{F}, \mathbf{S}, T, \mathbf{Q}\}, \tag{8.99}$$

where T is the absolute temperature, and **Q** is the heat flux in the Lagrangian description.

Instead of (8.29) in the Eulerian description now we have the following balance equations:

$$\begin{aligned}
&\frac{\partial F_{i\alpha}}{\partial t} - \frac{\partial v_i}{\partial X^\alpha} = 0,\\
&\frac{\partial G_i}{\partial t} - \frac{\partial P_{i\alpha}}{\partial X^\alpha} = \rho_0\left(b_i + i_i^0\right) + 2\Omega_{ij}G_j, \qquad G_i \equiv \rho_0 v_i,\\
&\frac{\partial G_{ij}}{\partial t} + \frac{\partial H_{ij\alpha}}{\partial X^\alpha} = f_{\langle ij\rangle} + 2G_{(i}\left(b_{j)} + i_{j)}^0\right) - 2\Omega_{k(i}\left(G_{j)k} - P_{j)\alpha}F_{k\alpha}\right),\\
&\frac{\partial G_{ijj}}{\partial t} + \frac{\partial H_{ijj\alpha}}{\partial X^\alpha} = f_{ijj} + 3G_{(ij}\left(b_{j)} + i_{j)}^0\right) - 3\Omega_{k(i}\left(G_{jj)}v_k + G_{jj)k} + H_{jj)\alpha}F_{k\alpha}\right).
\end{aligned} \tag{8.100}$$

The first one of these equations is the integrability condition (4.39). The Piola - Kirchhoff stress tensor, **P,** appearing in $(8.100)_3$, and the Cauchy stress tensor, **T**, are connected by the relation $(4.68)_2$, i.e.

$$J^{-1}P_{k\alpha}F_{l\alpha}=T_{kl},\qquad J\equiv\frac{\rho_0}{\rho}. \tag{8.101}$$

In contrast to the equations (7.2) for ideal gases, the fields $F_{i\alpha}$, G_i, G_{ij}, and G_{ijj} are not identical with the fluxes appearing in subsequent equations in the tensorial hierarchy. Here the fluxes, $v_j\delta_{ij}$, $P_{i\alpha}$, $H_{ij\alpha}$, $H_{ijj\alpha}$, are different. This situation is similar to that for real gases which we mentioned in Sect. 7.2. The trace of $(8.100)_3$ is, certainly, the energy conservation law. Therefore, the production term, $f_{<ij>}$, consists solely of the deviatoric part. The last equation is the vectorial part of the third-rank tensorial equation in the tensorial hierarchy. Both the deviatoric part of $(8.100)_3$ and $(8.100)_4$ are supposed to describe the behaviour of the additional fields of the model: the viscous stresses, **S**, and the heat flux, **Q**. These balance equations are, certainly, not conservation laws because we expect these fields to relax to zero in the thermodynamical equilibrium and this relaxation is driven by the production terms.

In order to specify the velocity dependence, we shall now use Ruggeri's results on Galilean invariance (see: Sect. 5.3.). We obtain

$$\begin{aligned}
&G_{ij}=\rho_0 v_i v_j+m_{ij}, \qquad G_{ijj}=\rho_0 v_i v^2+3m_{(ij}v_{j)}+m_{ijj},\\
&H_{ij\alpha}=-2v_{(i}P_{j)\alpha}+M_{ij\alpha},\ H_{ijj\alpha}=-3v_{(i}v_jP_{j)\alpha}+3v_{(i}M_{jj)\alpha}+M_{ijj\alpha},\\
&f_{\langle ij\rangle}=\ell_{\langle ij\rangle},\qquad f_{ijj}=3\ell_{(ij}v_{j)}+\ell_{ijj}.
\end{aligned} \tag{8.102}$$

The dependence on the velocity **v** is now solely explicit. In the classical notation we have, as before, that

$$\varepsilon=\frac{1}{2\rho_0}m_{ii},\quad Q_\alpha=\tfrac{1}{2}M_{ii\alpha}. \tag{8.103}$$

Certainly, ε denotes the specific internal energy, and Q_α the components of the material heat flux connected with the usual heat flux by the relation $(4.84)_3$

$$q_k=J^{-1}F_{k\alpha}Q_\alpha. \tag{8.104}$$

In order to construct the field equations we need the constitutive relations for the following quantities:

$$P_{i\alpha},G_{ij},H_{ij\alpha},f_{\langle ij\rangle},G_{ijj},H_{ijj\alpha},f_{ijj}. \tag{8.105}$$

In the spirit of extended thermodynamics, we assume that the above quantities are sufficiently smooth functions of the fields (8.99). We do not assume the dependence on any gradients or time derivatives. This is different from the model of the Maxwellian fluid in which we made the assumption (8.34). Let us comment on this difference. The lack of dependence on the time derivative of the deviatoric pressure $\langle \mathbf{p}^\Delta\rangle$ would mean that the material parameter of the Maxwellian model , β_2, had to vanish [compare $(8.37)_5$]. This means the α_1-parameter, measured in the normal stress experiments, is determined solely by the e_2-parameter of the model [see: $(8.37)_4$]. On the other hand, this would mean that the normal stress effects would be solely controlled by the kinetic part of the balance equation. Such effects seem to be too small in fluids to be responsible for the normal stress effects (elastic response). In the case of solids the situation is different because there is an equilibrium elastic response which is, certainly, much stronger than the similar effect caused by the non-equilibrium flow. Consequently, we may expect to obtain the proper description of viscoelastic solids without accounting for the elasticity in the production terms. This is exactly what happens in the presented model of I-Shih Liu.

We require that the constitutive relations obey the second law of thermodynamics, the principle of the material frame indifference and the assumption of the hyperbolicity of the field equations.

The second law requires that all solutions of the field equations should satisfy the following entropy inequality

$$\rho_0 \frac{\partial \eta}{\partial t} + \frac{\partial H_\alpha}{\partial X^\alpha} \geq 0, \tag{8.106}$$

where H_α are the components of the entropy flux **H** in the Lagrangian description.

We exploit this requirement using Lagrange multipliers. Simple calculations yield the following results.

There exists the 4-potential (h', H'_α) such that

$$\begin{aligned} d h' &= m_{ij}\, d\Lambda^I_{ij} + m_{ijj}\, d\Lambda_{ill} - \lambda_{i\alpha}\, dF_{i\alpha}, \\ d H'_\alpha &= -P_{i\alpha}\, d\Lambda^I_i + M_{ij\alpha}\, d\Lambda^I_{ij} + M_{ijj\alpha}\, d\Lambda_{ill}, \end{aligned} \tag{8.107}$$

where

$$\begin{aligned} & h' \equiv m_{ij}\Lambda^I_{ij} + m_{ijj}\Lambda_{ill} - \rho_0 \eta, \\ & H'_\alpha \equiv -P_{i\alpha}\Lambda^I_i + M_{ij\alpha}\Lambda^I_{ij} + M_{ijj\alpha}\Lambda_{ill} - H_\alpha, \\ & \Lambda^I_{ij} \equiv \Lambda_{ij} + 3\Lambda_{kll} v_{(k}\delta_{ij)}, \quad \Lambda^I_i \equiv \Lambda_i + 2\Lambda_{ik} v_k + 3\Lambda_{kll} v_{(k} v_m \delta_{m)j} \quad \Rightarrow \\ & \Rightarrow \quad \Lambda^I_i = -\frac{3}{\rho_0} m_{(ij}\Lambda_{j)ll}, \quad \lambda_{i\alpha} = -2\Lambda^I_{ij} P_{j\alpha} + 3\Lambda_{mll}\delta_{m(j} M_{ij)\alpha}. \end{aligned} \tag{8.108}$$

The multipliers $\lambda_{i\alpha}$, Λ_i, Λ_{ij}, Λ_{ill} eliminate the constraint of the inequality (8.106) to the solutions of field equations. They correspond to the balance equations (8.100) and are constitutive quantities. On the other hand, the intrinsic multipliers, defined by the relations $(8.108)_{3,4}$ are independent of the velocity, similarly to all other constitutive quantities in the relations (8.107). This follows from the requirement of the frame indifference.

Let us notice that the 4-potential, defined in (8.108), yields the simple form of the differentials (8.107) due to the change of independent variables. Instead of the original constitutive variables Lagrange multipliers are used here to describe the constitutive relations. This is the problem of the main fields mentioned in Chap. 6. Such a transformation of variables requires certain additional assumptions, for example the hyperbolicity of the field equations being sufficient. We skip the discussion of this problem for the present model.

The simplicity of the thermodynamical relations is, however, connected with the lack of simple physical interpretations. Consequently, the final results must be transformed back to the original constitutive variables, which is usually the most difficult task of extended thermodynamics.

Apart from the existence of the 4-potential, the second law yields the following identity:

$$\varepsilon_{mik}\left[2\Lambda^{I}_{ij}\left(m_{jk} - P_{j\alpha}F_{k\alpha}\right) + \Lambda_{ill}\left(m_{jjk} + M_{jj\alpha}F_{k\alpha}\right) + 2\Lambda_{jll}M_{ij\alpha}F_{k\alpha}\right] = 0. \qquad (8.109)$$

There remains the residual inequality which determines the dissipation, $\mathcal{D}$,

$$\mathcal{D} \equiv \Lambda^{I}_{ij}\ell_{\langle ij\rangle} + \Lambda_{ill}\ell_{ijj} \geq 0. \qquad (8.110)$$

These are the complete results following from the entropy inequality. The residual inequality can be, certainly, exploited further because it is the statement that the dissipation reaches its minimum, zero, in the state

$$\ell_{\langle ij\rangle}\Big|_E = 0, \qquad \ell_{ijj}\Big|_E = 0. \qquad (8.111)$$

This is, of course, the state of **thermodynamical equilibrium**. Hence, according to the definition of $\mathcal{D}$, we obtain the necessary conditions for this state to appear

$$\left.\frac{\partial \mathcal{D}}{\partial \ell_{\langle ij\rangle}}\right|_E = \Lambda^{I}_{\langle ij\rangle}\Big|_E = 0, \qquad \left.\frac{\partial \mathcal{D}}{\partial \ell_{ijj}}\right|_E = \Lambda^{I}_{ijj}\Big|_E = 0. \qquad (8.112)$$

Bearing the relations $(8.107)_1$ and $(8.108)_{5,6}$ in mind, we get the following **Gibbs equation** for the equilibrium

$$d\eta\big|_E = \tfrac{2}{3}\Lambda^{I}_{ii}\left(d\varepsilon\big|_E - \frac{1}{\rho_0}P_{i\alpha}\big|_E\, dF_{i\alpha}\right) \quad \Rightarrow \quad \tfrac{2}{3}\Lambda^{I}_{ii} = \frac{1}{T}, \qquad (8.113)$$

where T is the absolute temperature.

We can now introduce the equilibrium Helmholtz free energy function, in the standard way,

$$\psi\big|_E \equiv \varepsilon\big|_E - T\eta\big|_E \quad \Rightarrow$$
$$\Rightarrow \quad P_{i\alpha}\big|_E = \rho_0\frac{\partial \psi|_E}{\partial F_{i\alpha}}, \quad \varepsilon\big|_E = \psi\big|_E - T\frac{\partial \psi|_E}{\partial T}, \quad h'\big|_E = \rho_0\frac{\psi|_E}{T}. \qquad (8.114)$$

Moreover, differentiation of identity (8.109) with respect to $\Lambda^{I}_{\langle ij\rangle}$ and evaluation in the equilibrium yields

$$m_{\langle ij\rangle}\Big|_E = \frac{\rho_0}{\rho}T_{\langle ij\rangle}\Big|_E, \qquad (8.115)$$

where the relation (8.101) has been used.

Further, we shall restrict the considerations to isotropic constitutive functions and to the processes which are assumed to deviate not too far from the thermodynamical equilibrium.

8.7. The Second-Order Model of Isotropic Materials

In the case of second-order models, we retain the terms in the constitutive relations which contain at most second order terms in the multipliers $\Lambda^I_{\langle ij\rangle}$ and $\Lambda^I{}_{ijj}$ measuring the deviation from the thermodynamical equilibrium. For isotropic materials we have the result that

$$\begin{aligned} \frac{1}{\rho_0}h' &= \frac{\psi|_E}{T} + \frac{1}{\rho}T_{\langle ij\rangle}\Big|_E \Lambda^I_{\langle ij\rangle} + \tfrac{1}{2}h_{ijkl}\Lambda^I_{\langle ij\rangle}\Lambda^I_{\langle kl\rangle} + \tfrac{1}{2}k_{ij}\Lambda_{imm}\Lambda_{jnn}, \\ H_l' &= \left(\phi_{kl} + \phi'_{ijkl}\Lambda^I_{\langle ij\rangle}\right)\Lambda_{kmm}, \end{aligned} \tag{8.116}$$

where $\psi|_E$, $T_{<ij>}|_E$ as well as the coefficients h_{ijkl}, k_{ij}, ϕ_{kl}, and ϕ'_{ijkl} are isotropic functions of the deformation tensor, **B**, and the temperature, T. In addition, the coefficients should satisfy certain symmetry relations which we shall not quote for this presentation.

Now it is easy to show that the Helmholtz free energy is given by the following constitutive relation

$$\psi \equiv \varepsilon - T\eta = \psi|_E - \frac{T}{2}h_{ijkl}\Lambda^I_{\langle ij\rangle}\Lambda^I_{\langle kl\rangle} - \frac{T}{2}k_{ij}\Lambda_{imm}\Lambda_{jnn}. \tag{8.117}$$

Simultaneously,

$$\lambda_{k\alpha} = -\frac{\rho_0}{T}\frac{\partial \psi|_E}{\partial F_{k\alpha}} - \frac{\partial\left(P_{i\beta}|_E F_{j\beta}\right)}{\partial F_{k\alpha}}\Lambda^I_{\langle ij\rangle} - \frac{\rho}{2}\left(\frac{\partial h_{ijln}}{\partial F_{k\alpha}}\Lambda^I_{\langle ij\rangle}\Lambda^I_{\langle ln\rangle} + \frac{\partial k_{ij}}{\partial F_{k\alpha}}\Lambda_{ill}\Lambda_{jnn}\right). \tag{8.118}$$

Bearing $(8.108)_6$ in mind, we finally obtain

$$\begin{aligned} S_{ij} \equiv T_{ij} - T_{ij}\Big|_E &= T C_{ijkl}\Lambda^I_{\langle kl\rangle}, \qquad C_{ijkl} \equiv A_{ijkl} - T_{jl}\Big|_E \delta_{ik}, \\ A_{ijkl} &\equiv \rho \frac{\partial^2 \psi|_E}{\partial F_{i\alpha}\partial F_{k\beta}} F_{j\alpha}F_{l\beta}. \end{aligned} \tag{8.119}$$

The fourth-rank tensors, **A**, and **C**, are called the **spatial elasticity tensors** or the **first** and **second** elasticity tensors [see: J. E. MARSDEN, T. J. R. HUGHES (1983)]. The above result specifies the **viscous stress tensor**, **S**, whose existence had been anticipated at the beginning of these considerations.

It can be shown, by means of (8.109), that the fourth-rank coefficient h_{ijkl} in (8.117) must have the following simple form:

$$h_{ijkl} = \frac{T}{\rho}\left(C_{\langle ij\rangle\langle kl\rangle} + \gamma\delta_{\langle ij\rangle\langle kl\rangle}\right), \qquad \delta_{\langle ij\rangle\langle kl\rangle} \equiv \tfrac{1}{2}\left(\delta_{ik}\delta_{jl} + \delta_{jk}\delta_{il} - \tfrac{2}{3}\delta_{ij}\delta_{kl}\right), \tag{8.120}$$

where γ is the isotropic scalar function of $\mathbf{B}$ and T.

We can also calculate the constitutive relations for fluxes. It follows

$$\begin{gathered} q_l = \left(\chi_{kl} + \chi'_{ijkl}\Lambda^I_{\langle ij\rangle}\right)\Lambda_{kmm}, \qquad M_{\langle ij\rangle l} = \alpha_{ijkl}\Lambda_{kmm}, \\ M_{kjjl} = \beta_{kl} + \beta'_{mnkl}\Lambda^I_{\langle mn\rangle}, \end{gathered} \tag{8.121}$$

where

$$\begin{gathered} \chi_{ij} \equiv -T^2\left(\frac{\partial r_{im}}{\partial T}T_{mj}\Big|_E + \frac{\partial \phi_{ij}}{\partial T}\right), \qquad r_{kl} \equiv -\frac{10}{3}\varepsilon|_E\delta_{kl} - \frac{2}{\rho}T_{\langle kl\rangle}\Big|_E, \\ \chi'_{ijkl} \equiv -T^2\left(\frac{\partial r'_{ijkm}}{\partial T}T_{ml}\Big|_E + \frac{\partial \phi'_{ijkl}}{\partial T} + TC_{ml\langle ij\rangle}\frac{\partial r_{km}}{\partial T}\right), \\ r'_{ijkl} \equiv \frac{10}{3}\frac{T^2}{\rho}\frac{\partial T_{\langle ij\rangle}\Big|_E}{\partial T}\delta_{kl} - 2h_{ijkl}, \\ \alpha_{ijkl} \equiv r'_{ijkm}T_{ml}\Big|_E + \phi'_{ijkl}, \\ \beta_{ij} \equiv r_{im}T_{mj}\Big|_E + \phi_{ij}, \qquad \beta'_{ijkl} \equiv \alpha_{ijkl} + Tr_{km}C_{ml\langle ij\rangle}. \end{gathered} \tag{8.122}$$

If we limit the considerations to the second-order terms then not all of these objects are needed to determine the specific energy, ε, the Helmholtz free energy, ψ, the specific entropy, η, the stress tensor, $\mathbf{T}$, the heat flux, $\mathbf{q}$, and the entropy flux, $\mathbf{h}$ as well as the other flux terms. As a matter of fact, we need two scalars $\psi|_E$ and γ, one tensor of the second rank, β_{ij}, and one tensor of the fourth rank, K_{ijkl}, defined by the relation

$$K_{ijkl} \equiv \delta_{ij}\left(\rho k_{kl} + \tfrac{10}{3}\chi_{kl}\right) + 2\alpha_{ijkl}. \tag{8.123}$$

In addition, we have to specify the constitutive relations for production terms for which thermodynamical considerations yield solely the inequalities. In the second-order approximation they can be written in the form

$$\ell_{\langle ij\rangle} = \frac{\rho_0}{T}\sigma_{ijkl}C^{-1}_{\langle kl\rangle\langle mn\rangle}S_{\langle mn\rangle}, \qquad \ell_{ijj} = \rho_0\tau_{ij}\chi^{-1}_{jm}q_m, \tag{8.124}$$

where the tensors, σ_{ijkl} and τ_{ij}, are isotropic functions of $\mathbf{B}$ and T.

Even though it is good to know that we can construct a model of such generality for viscoelastic materials, the structure of Liu's model described above is, certainly, too complex to expose clearly the basic properties of the field equations. For this reason, we consider, in more detail, the fully linear model which is, most likely, the only one of the practical bearing.

8.8. Linear Viscoelastic Solids

The isotropic tensor functions, appearing in the constitutive relations above, have the following structure

$$\begin{aligned} \beta_{ij} &= \beta\delta_{ij}, \qquad K_{ijkl} = \kappa_1\delta_{i(k}\delta_{l)j} + \kappa_2\delta_{ij}\delta_{kl}, \\ \tau_{ij} &= \tau\delta_{ij}, \qquad \sigma_{ijkl} = \sigma\,\delta_{i(k}\delta_{l)j}, \end{aligned} \tag{8.125}$$

where β, κ_1, κ_2, τ, and σ are, in general, functions of **B** and T.

If we assume, in addition, that the deformations are small (linear model), we obtain the following constitutive relations:

$$\begin{aligned} & T_{ij} = \lambda\varepsilon_{kk}\delta_{ij} + 2\mu\varepsilon_{ij} + S_{\langle ij\rangle}, \qquad \varepsilon = e(T), \qquad \varepsilon_{ij} \approx \tfrac{1}{2}\left(B_{ij} - \delta_{ij}\right), \\ & m_{\langle ij\rangle} = 2\mu\varepsilon_{\langle ij\rangle} + \left(1 + \frac{\gamma}{2\mu}\right)S_{\langle ij\rangle}, \qquad m_{ill} = -\left(\frac{10}{3} + \frac{\kappa_2}{T^2}\left(\frac{\partial\beta}{\partial T}\right)^{-1}\right)q_i, \\ & M_{\langle ij\rangle l} = \frac{1}{4}\frac{\kappa_1}{T^2}\left(\frac{\partial\beta}{\partial T}\right)^{-1}\left(q_i\delta_{jl} + q_j\delta_{il} - \tfrac{2}{3}q_l\delta_{ij}\right), \\ & M_{kjjl} = \beta\delta_{kl} - \tfrac{10}{3}e\left(\lambda\varepsilon_{jj}\delta_{kl} + 2\mu\varepsilon_{kl}\right) + \left(\frac{\kappa_1}{4\mu T} - \tfrac{10}{3}e\right)S_{\langle kl\rangle}, \\ & \ell_{\langle ij\rangle} = \frac{\rho}{2\mu T}\sigma S_{\langle ij\rangle}, \qquad \ell_{ijj} = -\frac{\rho}{T^2}\left(\frac{\partial\beta}{\partial T}\right)^{-1}\tau q_i. \end{aligned} \tag{8.126}$$

In these relations λ, μ denote the Lamé constants. Otherwise, the material parameters must satisfy inequalities which follow from the second law and from the requirement of hyperbolicity

$$\begin{aligned} & \frac{de}{dT} > 0, \quad \lambda + \tfrac{2}{3}\mu > 0, \quad \mu > 0, \quad 2\mu + \gamma < 0, \\ & \kappa_2 + \tfrac{10}{3}T^2\frac{d\beta}{dT} < 0, \quad \sigma \geq 0, \quad \tau \geq 0. \end{aligned} \tag{8.127}$$

The field equations for the viscous stresses, **S**, and for the heat flux, **q**, now have the form

$$\begin{aligned} & S_{\langle ij\rangle} - 2\nu\dot{\varepsilon}_{\langle ij\rangle} + \tau_S\dot{S}_{\langle ij\rangle} = -2\nu\frac{\kappa_1}{4\mu T^2}\left(\frac{d\beta}{dT}\right)^{-1}\frac{\partial q_{\langle i}}{\partial x_{j\rangle}}, \\ & q_i + \kappa\frac{\partial T}{\partial x_i} + \tau_q\dot{q}_i = -\kappa T\frac{\kappa_1}{4\mu T^2}\left(\frac{d\beta}{dT}\right)^{-1}\frac{\partial S_{\langle ij\rangle}}{\partial x_j}, \end{aligned} \tag{8.128}$$

where

$$\nu \equiv \frac{2\mu^2 T}{\rho\sigma}, \quad \kappa \equiv \frac{T^2}{\rho\tau}\left(\frac{d\beta}{dT}\right)^2,$$
$$\tau_S \equiv -\frac{T}{\rho\sigma}(2\mu+\gamma), \quad \tau_q \equiv -\frac{1}{\rho\tau}\left(\kappa_2 + \tfrac{10}{3}T^2\frac{d\beta}{dT}\right). \tag{8.129}$$

Consequently, ν is the **viscosity coefficient**, κ, the **thermal conductivity**, τ_S, the **stress relaxation time**, τ_q, the **thermal relaxation time** and due to (8.127) all of them are non-negative.

The equations in (8.128) describe a generalization of the classical models. The first equation, without the coupling with the heat flux, defines the **standard linear solid** within viscoelasticity. The second one, without the coupling with viscous stresses, is the Cattaneo equation, which we have already mentioned in this book. The thermal waves which are described by this equation shall be discussed in the ninth chapter.

9 Second Sound

9.1 Preliminary Remarks

We already indicated in the introduction to this book that the processes on the atomic level of description are connected with the motion of molecules, which is usually limited to a small part of the configuration space. For instance, in the case of crystalline solids, the molecules constituting the lattice vibrate around their equilibrium positions and their maximum displacement from these positions, being of the order of magnitude of 10^{-11}m (0.1 Å), is small – even in comparison with the lattice constants. In some cases, such as the free electrons in metals or the particles of rarified gases, these displacements may be larger but they are still much smaller than the characteristic dimensions limiting the applicability of macroscopic continuous models (long wave approximation!).

Simultaneously, all transport and relaxation processes described by the continuum thermodynamics are due to these microscopic motions. For instance, the propagation of elastic waves in a rod results from the harmonic approximation of the lattice vibrations as shown in section 1.2. In this example the wave was a superposition of monochromatic waves characterized by the frequency ω and the wave vector (the wave number in the one-dimensional case) $\mathbf{k}$. We also mentioned that due to the linearity of the model (the harmonic approximation) there is no interaction between these monochromatic waves and the Hamiltonian of the system is the sum of Hamiltonians corresponding to each monochromatic wave (the harmonic oscillator). Due to this property the energy, which can be supplied to a chosen oscillator, would remain accumulated in this oscillator forever without being transferred to other levels of frequency (i.e. to other oscillators). This means that such an ideal system cannot relax spontaneously into any sort of an equilibrium state.

This is, certainly, not the case in real systems. Due to interactions the energy is transmitted to all possible forms of vibrations of molecules until some equilibrium distribution of frequencies is reached. This means that in order to account for the thermodynamical relaxation, some **anharmonic** effects must be taken into account.

Unfortunately, the harmonic oscillators are rather exceptional for macroscopical models in that their vibration properties are the same in classical and quantum mechanics. This is not the case if we include anharmonic effects which require a quantum mechanical description. However, the most important macroscopical properties of thermodynamical systems can be described within the frame of the classical formulation with very few additional assumptions whose justification would require a quantum mechanical approach. This is the type of a model which we aim to present in this chapter.

Motivated by quantum mechanics, we shall construct a thermodynamical model of the **phonon gas**. In the simple case considered in the introduction, the phonon (the normal vibration) was characterized by the frequency ω and the wave vector $\mathbf{k}$, and this description is characteristic for particles within quantum mechanics. However, in contrast to a normal particle, the phonon cannot exist in a vacuum as it is immanently connected with some material background such as a crystalline lattice. Therefore, it is sometimes called a **quasiparticle**. The collection of these quasiparticles which we considered in the introduction, was rather peculiar because they did not interact with each other. This was a result of the harmonic approximation. If we account for the anharmonic effects, the phonons must not only interact with each other but they can also be created and annihilated. This means that the phonon gas, modelled as a continuum, cannot satisfy the usual conservation laws of material media, even though it is supposed to be described by similar fields of **phonon number density** n („the mass density"), the **phonon flux j** („the mass flux"), the **energy density** ε, and the **energy flux q** satisfying some balance laws.

The kinetic theory of a similar gas of photons (light particles) was already anticipated by Boltzmann. The recent kinetic approach for phonons was constructed successfully by W. LARECKI (1991). The thermodynamical counterpart for photons was investigated in part by A. M. ANILE, S. PENNISI AND M. SAMMARTINO (1991). In the following sections we present the results of G. M. KREMER and I. MÜLLER (1992) for phonons. They have used the approach of the extended thermodynamics.

Remark 9.1. *On the Moments of the Boltzmann-Peierls Equation*
The kinetic theory of phonons is based on the notion of the **phase density** function f(ξ,**k**,t) such that f(ξ,**k**,t)$d\xi d\mathbf{k}$ describes the number of phonons in a $d\xi d\mathbf{k}$-volume of the phase space at the instant of time t. In contrast to the usual gas particles (compare the remark 5.1.), the phase space of these quasiparticles is defined by the collection of positions ξ and **quasimomenta** $\hbar\mathbf{k}$ ($\hbar$ - the Planck constant) rather than the positions ξ and the momenta mc. In many interaction processes of phonons the quantity $\hbar\mathbf{k}$ plays the same role as the usual momentum. However, as we already mentioned, the quasimomentum does not satisfy a conservation law, and it may contain an additional term not only connected with the quasiparticle but also with its interaction with the background lattice (e.g. in the so-called Umklapp processes).

The phase density f satisfies an equation similar to (5.108) which is called the **Boltzmann-Peierls equation**

$$\frac{\partial f}{\partial t}+\frac{\partial}{\partial \xi^i}\left(\frac{\partial \omega}{\partial k_i} f\right)=\mathcal{C}(f), \tag{9.1}$$

where $\mathcal{C}$ is the **collision operator for phonons**. Again, in contrast to the Boltzmann collision operator appearing in (5.108), the above operator does not have the conservation properties (5.109).

The simplified kinetic models of phonons are based on the assumption that dispersion relation is **linear**

$$\omega = ck, \quad k^2 \equiv \mathbf{k}\cdot\mathbf{k}, \tag{9.2}$$

where c denotes the speed of longitudinal waves which is assumed to be constant. The assumption (9.2) limits the applicability of the model to the range of small frequencies. In such a case the equation (9.1) becomes

$$\frac{\partial f}{\partial t}+\frac{\partial}{\partial \xi^i}\left(c\frac{k^i}{k} f\right)=\mathcal{C}(f). \tag{9.3}$$

Similarly to the case of the Boltzmann equation, we construct the **moments** of this equation (the equations of transfer) by multiplying (9.3) by 1, ck_i/k, $\hbar\omega$ and $\hbar k_i$ and integrating with respect to **k**. We obtain

$$\begin{aligned}
&\frac{\partial}{\partial t}\int f\, d\mathbf{k} + \frac{\partial}{\partial \xi^i}\int c\frac{k^i}{k} f\, d\mathbf{k} = \int \mathcal{C}\, d\mathbf{k},\\
&\frac{\partial}{\partial t}\int c\frac{k^i}{k} f\, d\mathbf{k} + \frac{\partial}{\partial \xi^i}\int c^2\frac{k^i k^j}{k^2} f\, d\mathbf{k} = \int c\frac{k^i}{k}\mathcal{C}\, d\mathbf{k},\\
&\frac{\partial}{\partial t}\int \hbar c k f\, d\mathbf{k} + \frac{\partial}{\partial \xi^i}\int \hbar c^2 k^i f\, d\mathbf{k} = \int \hbar\omega\mathcal{C}\, d\mathbf{k},\\
&\frac{\partial}{\partial t}\int \hbar k^i f\, d\mathbf{k} + \frac{\partial}{\partial \xi^i}\int \hbar c\frac{k^i k^j}{k} f\, d\mathbf{k} = \int \hbar k^i\mathcal{C}\, d\mathbf{k}.
\end{aligned} \tag{9.4}$$

We introduce the following notation:

$n = \int f\, d\mathbf{k}$	Phonon number density
$\mathbf{j} = \int c\frac{\mathbf{k}}{k} f\, d\mathbf{k}$	Phonon flux
${}_nP = \int \mathcal{C}\, d\mathbf{k}$	Phonon production
$\mathbf{J} = \int c^2 \frac{\mathbf{k}\otimes\mathbf{k}}{k^2} f\, d\mathbf{k}$	Flux of the phonon flux
${}_j\mathbf{P} = \int c\frac{\mathbf{k}}{k}\mathcal{C}\, d\mathbf{k}$	Production of the phonon flux

(9.5)

$\varepsilon = \int \hbar c k f\, d\mathbf{k}$	Energy density
$\mathbf{q} = \int \hbar c^2 \mathbf{k} f\, d\mathbf{k}$	Energy flux
${}_\varepsilon P = \int \hbar\omega\mathcal{C}\, d\mathbf{k}$	Energy production
$\mathbf{p} = \int \hbar\mathbf{k} f\, d\mathbf{k}$	Momentum
$\mathbf{P} = \int \hbar c \frac{\mathbf{k}\otimes\mathbf{k}}{k} f\, d\mathbf{k}$	Flux of momentum
${}_p\mathbf{P} = \int \hbar\mathbf{k}\mathcal{C}\, d\mathbf{k}$	Production of momentum

(9.6)

Then the equations (9.4) have the following form of **balance laws**

$$\frac{\partial n}{\partial t} + \frac{\partial j^i}{\partial \xi^i} = {}_nP, \qquad \text{- the balance of phonon number,}$$

$$\frac{\partial j^i}{\partial t} + \frac{\partial J^{ij}}{\partial \xi^j} = {}_jP^i, \qquad \text{- the balance of phonon flux,} \tag{9.7}$$

and

$$\frac{\partial \varepsilon}{\partial t}+\frac{\partial q^i}{\partial \xi^i}={}_{\varepsilon}P, \qquad \text{- the balance of energy,}$$

(9.8)

$$\frac{\partial p^i}{\partial t}+\frac{\partial P^{ij}}{\partial \xi^j}={}_{p}P^i, \qquad \text{- the balance of momentum.}$$

Let us notice that the tensors **J** and **P** are **symmetric**. Simultaneously, from the definitions $(9.5)_4$, $(9.6)_1$, $(9.6)_2$, and $(9.6)_4$, we obtain

$$\operatorname{tr}\mathbf{J}=nc^2, \qquad \operatorname{tr}\mathbf{P}=\varepsilon, \qquad \mathbf{q}=c^2\mathbf{p}. \tag{9.9}$$

The relations (9.7), (9.8), and (9.9) form the basis of the thermodynamical theory of phonons presented further on in this chapter. It is obvious that the theory can also be constructed directly on the basis of the above relations provided we assume some explicit form of the collision operator $\mathcal{C}$. This kinetic approach shall not be considered in these notes•

9.2 Thermodynamical Model of the Phonon Gas

The aim of the thermodynamical model of the phonon gas is to formulate thermodynamically admissible field equations for the eight fields n, **j**, ε, **q** of phonon number density, phonon flux, energy density per unit volume, and the energy flux, respectively. These equations should follow from the balance laws

$$\begin{aligned}&\frac{\partial n}{\partial t}+\operatorname{div}\mathbf{j}={}_{n}P, \quad \frac{\partial \mathbf{j}}{\partial t}+\operatorname{div}\mathbf{J}={}_{j}\mathbf{P},\\ &\frac{\partial \varepsilon}{\partial t}+\operatorname{div}\mathbf{q}={}_{\varepsilon}P, \quad \frac{\partial \mathbf{p}}{\partial t}+\operatorname{div}\mathbf{P}={}_{p}\mathbf{P},\end{aligned} \tag{9.10}$$

[c.f. (9.7), (9.8)] supplemented with the constitutive relations

$$\begin{aligned}&\mathbf{J}=\mathbf{J}(n,\mathbf{j},\varepsilon,\mathbf{p}), \quad {}_{n}P={}_{n}P(n,\mathbf{j},\varepsilon,\mathbf{p}), \quad {}_{j}\mathbf{P}={}_{j}\mathbf{P}(n,\mathbf{j},\varepsilon,\mathbf{p}),\\ &\mathbf{P}=\mathbf{P}(n,\mathbf{j},\varepsilon,\mathbf{p}), \quad {}_{\varepsilon}P={}_{\varepsilon}P(n,\mathbf{j},\varepsilon,\mathbf{p}), \quad {}_{p}\mathbf{P}={}_{p}\mathbf{P}(n,\mathbf{j},\varepsilon,\mathbf{p}).\end{aligned} \tag{9.11}$$

The relations (9.9) are assumed to be satisfied, i.e.

$$\operatorname{tr}\mathbf{J}=nc^2, \quad \mathbf{J}^T=\mathbf{J}, \quad \operatorname{tr}\mathbf{P}=\varepsilon, \quad \mathbf{P}^T=\mathbf{P}, \quad \mathbf{q}=c^2\mathbf{p}, \tag{9.12}$$

the last relation being already used in (9.11) by means of replacing **q** by **p** as a field variable.

The constitutive relations (9.11) must be chosen in such a way that the following entropy inequality is identically satisfied for all solutions of the field equations

$$\frac{\partial \eta}{\partial t}+\operatorname{div}\mathbf{h}\geq 0, \quad \eta=\eta(n,\mathbf{j},\varepsilon,\mathbf{p}), \quad \mathbf{h}=\mathbf{h}(n,\mathbf{j},\varepsilon,\mathbf{p}). \tag{9.13}$$

In addition, we assume that η, which is the specific entropy per unit volume, is concave (thermodynamical stability), i.e. the Hessian matrix of second derivatives of η must be negative definite.

Bearing Liu's Theorem in mind, we eliminate the constraints on the solutions of the inequality (9.13) by means of the Lagrange multipliers λ, $\boldsymbol{\lambda}$, Λ, $\boldsymbol{\Lambda}$:

$$\forall(n,j,\varepsilon,p): \quad \frac{\partial \eta}{\partial t}+\operatorname{div}h-\lambda\left(\frac{\partial n}{\partial t}+\operatorname{div}j-{}_nP\right)-\boldsymbol{\lambda}\cdot\left(\frac{\partial j}{\partial t}+\operatorname{div}J-{}_jP\right)- \\ -\Lambda\left(\frac{\partial \varepsilon}{\partial t}+\operatorname{div}q-{}_\varepsilon P\right)-\boldsymbol{\Lambda}\cdot\left(\frac{\partial p}{\partial t}+\operatorname{div}P-{}_pP\right)\geq 0. \tag{9.14}$$

According to our constitutive assumptions (9.11), (9.13) the above inequality can be written in the following explicit form:

$$\left(\frac{\partial \eta}{\partial n}-\lambda\right)\frac{\partial n}{\partial t}+\left(\frac{\partial \eta}{\partial j^k}-\lambda_k\right)\frac{\partial j^k}{\partial t}+\left(\frac{\partial h^k}{\partial n}-\lambda_l\frac{\partial J^{kl}}{\partial n}-\Lambda_l\frac{\partial P^{kl}}{\partial n}\right)\frac{\partial n}{\partial x^k}+ \\ +\left(\frac{\partial h^k}{\partial j^l}-\lambda\delta_l^k-\lambda_m\frac{\partial J^{km}}{\partial j^l}-\Lambda_m\frac{\partial P^{km}}{\partial j^l}\right)\frac{\partial j^l}{\partial x^k}+ \\ +\left(\frac{\partial \eta}{\partial \varepsilon}-\Lambda\right)\frac{\partial \varepsilon}{\partial t}+\left(\frac{\partial \eta}{\partial p^k}-\Lambda_k\right)\frac{\partial p^k}{\partial t}+\left(\frac{\partial h^k}{\partial \varepsilon}-\lambda_l\frac{\partial J^{kl}}{\partial \varepsilon}-\Lambda_l\frac{\partial P^{kl}}{\partial \varepsilon}\right)\frac{\partial \varepsilon}{\partial x^k}+ \\ +\left(\frac{\partial h^k}{\partial p^l}-\Lambda c^2\delta_l^k-\lambda_m\frac{\partial J^{km}}{\partial p^l}-\Lambda_m\frac{\partial P^{km}}{\partial p^l}\right)\frac{\partial p^l}{\partial x^k}+ \\ +\lambda\,{}_nP+\lambda_k\,{}_jP^k+\Lambda\,{}_\varepsilon P+\Lambda_k\,{}_pP^k\geq 0, \tag{9.15}$$

for all fields n, **j**, ε, and **p**. Hence, all expressions in the parentheses must vanish independently from each other. These conditions can be written in the following compact form:

$$d\eta=\lambda dn+\boldsymbol{\lambda}\cdot dj+\Lambda d\varepsilon+\boldsymbol{\Lambda}\cdot dp, \\ dh=\lambda dj+\Lambda c^2 dp+dJ\boldsymbol{\lambda}+dP\boldsymbol{\Lambda}. \tag{9.16}$$

Exercise 9.1. Prove the relations (9.16)•

We are left with the residual inequality

$$\lambda\,{}_nP+\boldsymbol{\lambda}\cdot{}_jP+\Lambda\,{}_\varepsilon P+\boldsymbol{\Lambda}\cdot{}_pP\geq 0. \tag{9.17}$$

The state of **thermodynamical equilibrium** is defined as that in which the flux of phonons **j** and the momentum **p** vanish. Due to the isotropy, this means that the multipliers $\boldsymbol{\lambda}$ and $\boldsymbol{\Lambda}$ must vanish in the thermodynamical equilibrium. Therefore, the relations $(9.16)_1$ and (9.17) become

$$d\eta|_E=\lambda|_E\,dn+\Lambda|_E\,d\varepsilon, \quad \lambda|_E\,{}_nP|_E+\Lambda|_E\,{}_\varepsilon P|_E=0, \tag{9.18}$$

where the subscript E denotes the equilibrium state.

It should be kept in mind that the densities ε and η used in the above description refer to the unit volume of the phonon gas. This means that the Gibbs equation describing the equilibrium state has the following form:

$$d\eta|_E = \frac{1}{T}\left(d\varepsilon - \frac{1}{n} g|_E \, dn\right), \quad g|_E \equiv \varepsilon - T\eta|_E + \tfrac{1}{3}\mathrm{tr}\,\mathbf{P}. \tag{9.19}$$

Exercise 9.2. Prove the Gibbs equation (9.19)•

Comparison of the relations (9.18) and (9.19) yields

$$\lambda|_E = -\frac{g|_E}{nT}, \quad \Lambda|_E = \frac{1}{T}. \tag{9.20}$$

The integrability condition implied by (9.18) leads to the following relation:

$$\frac{\partial \lambda|_E}{\partial \varepsilon} = \frac{\partial \Lambda|_E}{\partial n} \quad \Rightarrow \quad -\frac{1}{n}\frac{\partial}{\partial \varepsilon}\left(\frac{g|_E}{T}\right) = \frac{\partial}{\partial n}\left(\frac{1}{T}\right). \tag{9.21}$$

Simultaneously [see: (9.12)]

$$\mathrm{tr}\,\mathbf{P} = \varepsilon, \quad \frac{\partial \eta|_E}{\partial \varepsilon} = \Lambda|_E = \frac{1}{T}. \tag{9.22}$$

Hence,

$$T - 4\varepsilon\frac{\partial T}{\partial \varepsilon} = 3n\frac{\partial T}{\partial n}. \tag{9.23}$$

It is customary to write the energy ε as a function of the temperature T and the number density n. After a change of variables $(\varepsilon,n)\to(T,n)$ the equation (9.23) becomes

$$T\frac{\partial \varepsilon}{\partial T} + 3n\frac{\partial \varepsilon}{\partial n} = 4\varepsilon. \tag{9.24}$$

Integration along the characteristics yields

$$\varepsilon(T,n) = \sigma\left(\tfrac{T^3}{n}\right)T^4, \tag{9.25}$$

where σ is an arbitrary function of the single argument (the parameter constant along the characteristic) T^3/n.

In a particular case of the energy being independent of n we have

$$g|_E = 0 \;\Rightarrow T\eta|_E = \tfrac{4}{3}\varepsilon \;\Rightarrow\; \varepsilon = \sigma T^4, \; \sigma = \mathrm{const.} \tag{9.26}$$

This case was considered for the photon gas by Kirchhoff. In such a case the relation (9.25) (i.e. $(9.26)_3$, σ=const.) is called the **Stefan-Boltzmann law**. It was found empirically by Stefan and derived from the Gibbs equation by Boltzmann.

According to our assumption of the concavity of η, we can perform the Legendre transformation described in Sect. 6.3. The **main fields** λ, $\boldsymbol{\lambda}$, Λ, $\boldsymbol{\Lambda}$ yield the existence of the following **potentials**:

$$\begin{aligned} H^0 &= -\eta + \lambda n + \boldsymbol{\lambda}\cdot j + \Lambda\varepsilon + \boldsymbol{\Lambda}\cdot p, \\ H &= -h + \lambda j + J\boldsymbol{\lambda} + \Lambda c^2 p + P\boldsymbol{\Lambda}, \end{aligned} \tag{9.27}$$

which allow us to write the relations (9.16) in the form

$$\begin{aligned} dH^0 &= n\,d\lambda + j\cdot d\boldsymbol{\lambda} + \varepsilon\, d\Lambda + p\cdot d\boldsymbol{\Lambda}, \\ dH &= j\,d\lambda + J\,d\boldsymbol{\lambda} + pc^2\, d\Lambda + P\,d\boldsymbol{\Lambda}. \end{aligned} \tag{9.28}$$

Hence, the knowledge of H^0 and H as functions of λ, $\boldsymbol{\lambda}$, Λ, and $\boldsymbol{\Lambda}$ suffices to determine n, **j**, ε, **p**, **J** and **P**. The conditions (9.28) have not been fully exploited yet. Further we limit our considerations in this chapter to the simplified case of four fields ε and **p**.

9.3 Four-Field Model

We proceed to present the results of the restricted model of the phonon gas in which only the fields of the energy ε and the momentum **p** are considered. Due to the proportionality of **p** to the flux **q** [the relation $(9.12)_5$] this model is reminiscent of the model of a **rigid heat conductor.** This similarity is, certainly, not accidental. As we explained in Sect. 9.1., the phonon gas in a crystalline lattice should be understood as a propagation of disturbances whose thermal part – roughly speaking, responsible for the thermodynamical relaxation to the equilibrium state of the system – is connected with the heat conduction, the harmonic background reproducing the normal acoustic (elastic) waves.

The relevant balance equations in this simplified case are those of energy and momentum

$$\frac{\partial \varepsilon}{\partial t} + \operatorname{div}\mathbf{q} = {}_{\varepsilon}P, \quad \frac{\partial \mathbf{p}}{\partial t} + \operatorname{div}\mathbf{P} = {}_{p}\mathbf{P}, \tag{9.29}$$

with the following conditions

$$\mathbf{P}^T = \mathbf{P}, \quad \operatorname{tr}\mathbf{P} = \varepsilon, \quad \mathbf{q} = c^2\mathbf{p}. \tag{9.30}$$

The constitutive relations turning the equations (9.29) into field equations for ε and **p** are assumed to have the form

$$\mathbf{P} = \mathbf{P}(\varepsilon, \mathbf{p}), \quad {}_{\varepsilon}P = {}_{\varepsilon}P(\varepsilon, \mathbf{p}), \quad {}_{p}\mathbf{P} = {}_{p}\mathbf{P}(\varepsilon, \mathbf{p}). \tag{9.31}$$

These functions must be isotropic. According to the representations discussed in Chap. 5, we have

$$\begin{aligned} \mathbf{P} &= \pi_1(\varepsilon, p^2)\mathbf{1} + \pi_2(\varepsilon, p^2)\mathbf{p}\otimes\mathbf{p}, \qquad p^2 \equiv \mathbf{p}\cdot\mathbf{p}, \\ {}_{\varepsilon}P &= A(\varepsilon, p^2)p^2, \qquad {}_{p}\mathbf{P} = B(\varepsilon, p^2)\mathbf{p}. \end{aligned} \tag{9.32}$$

The condition $(9.30)_2$ then yields

$$3\pi_1 + \pi_2 p^2 = \varepsilon. \tag{9.33}$$

We introduce the following notation

$$\chi \equiv 1 - \frac{2}{\varepsilon}\pi_1. \tag{9.34}$$

Then

$$\pi_2 = \frac{\varepsilon}{2p^2}(3\chi - 1), \tag{9.35}$$

and we obtain the following relation for the momentum flux **P**:

$$\mathbf{P} = \varepsilon\left(\frac{1-\chi}{2}\mathbf{1} + \frac{3\chi - 1}{2}\frac{\mathbf{p}\otimes\mathbf{p}}{p^2}\right), \qquad \chi = \chi(\varepsilon, p^2). \tag{9.36}$$

The constitutive scalar χ is called the **Eddington factor**. This was introduced in Eddington's considerations of the photon gas to describe radiative stresses in stars.

The production terms $_\varepsilon P$ and $_p\mathbf{P}$ must vanish in the **thermodynamical equilibrium,** which is defined in the present case by the condition **p**=0. This can be accounted for in the relations (9.32) with the understanding that the functions A and B are non-singular in the equilibrium.

The entropy principle, restricting the relations (9.32) to those which are thermodynamically admissible, takes the following form in the case under consideration:

$$\frac{\partial \eta}{\partial t} + \operatorname{div}\mathbf{h} \geq 0, \qquad \eta = \eta(\varepsilon, p^2), \quad \mathbf{h} = \kappa(\varepsilon, p^2)\mathbf{p}, \tag{9.37}$$

where the entropy function, η, is assumed to be **concave**.

Now we can repeat the derivations of the previous section. After simple manipulations, we arrive at the following potentials:

$$\begin{aligned} H^0 &= -\eta + \Lambda\varepsilon + \mathbf{\Lambda}\cdot p, \\ H &= -h + \Lambda c^2 p + P\mathbf{\Lambda}, \end{aligned} \tag{9.38}$$

with

$$\begin{aligned} dH^0 &= \varepsilon\, d\Lambda + p\cdot d\mathbf{\Lambda}, \\ dH &= pc^2\, d\Lambda + P\, d\mathbf{\Lambda}, \end{aligned} \tag{9.39}$$

and the following residual inequality

$$\Lambda\, {}_\varepsilon P + \mathbf{\Lambda}\cdot {}_p P \geq 0. \tag{9.40}$$

The potentials H^0, H are isotropic functions of the multipliers (main fields) Λ, $\mathbf{\Lambda}$, i.e.

$$H^0 = H^0(\Lambda, L), \qquad \mathbf{H} = H(\Lambda, L)\,\mathbf{\Lambda}, \qquad L \equiv \mathbf{\Lambda}\cdot\mathbf{\Lambda}. \tag{9.41}$$

The relations (9.39) read in the explicit form

$$\frac{\partial H^0}{\partial \Lambda} = \varepsilon, \qquad \frac{\partial H^0}{\partial L} 2\Lambda = p,$$
$$\frac{\partial H}{\partial \Lambda}\Lambda = c^2 p, \quad \frac{\partial H}{\partial L} 2\Lambda \otimes \Lambda + H1 = P. \tag{9.42}$$

We proceed to exploit these relations. Firstly we find the form of potentials as functions of the main fields. Bearing the condition $(9.30)_2$ in mind, the trace of $(9.42)_4$ leads to

$$2L\frac{\partial H}{\partial L} + 3H = \varepsilon. \tag{9.43}$$

Simultaneously, substitution of $(9.42)_2$ in $(9.42)_3$ yields

$$2c^2 \frac{\partial H^0}{\partial L} = \frac{\partial H}{\partial \Lambda}. \tag{9.44}$$

Consequently, if we account for the relation $(9.42)_1$, we arrive at the following set of partial differential equations for H^0 and H:

$$\frac{\partial H^0}{\partial L} - \frac{1}{2c^2}\frac{\partial H}{\partial \Lambda} = 0,$$
$$\frac{\partial H}{\partial L} - \frac{1}{2L}\frac{\partial H^0}{\partial \Lambda} = -\frac{3}{2L}H. \tag{9.45}$$

In order to find the general solution of this set, let us change the independent variables to variables along the characteristics. The eigenvalues μ for the system (9.45) follow from the relation

$$\begin{vmatrix} \mu & \frac{1}{2c^2} \\ \frac{1}{2L} & \mu \end{vmatrix} = 0 \quad \Rightarrow \quad \mu^{(1),(2)} = \mp \frac{1}{2c\sqrt{L}}. \tag{9.46}$$

Hence, the parameters ξ_1, ξ_2, describing the characteristics, are given by the equations

$$\frac{d\Lambda}{dL} = \mp \frac{1}{2c\sqrt{L}} \quad \Rightarrow \quad \begin{cases} \xi_1 = \frac{1}{c}\sqrt{L} + \Lambda, \\ \xi_2 = \frac{1}{c}\sqrt{L} - \Lambda. \end{cases} \tag{9.47}$$

Let us perform the transformation of variables $(L,\Lambda) \rightarrow (\xi_1,\xi_2)$ in the set (9.45). We obtain

$$\tfrac{1}{2}(\xi_1+\xi_2)\left(\frac{\partial H}{\partial \xi_1}-\frac{\partial H}{\partial \xi_2}\right)-\left(\frac{\partial H^0}{\partial \xi_1}+\frac{\partial H^0}{\partial \xi_2}\right)=0,$$
$$\tfrac{1}{2}(\xi_1+\xi_2)\left(\frac{\partial H}{\partial \xi_1}+\frac{\partial H}{\partial \xi_2}\right)-\left(\frac{\partial H^0}{\partial \xi_1}-\frac{\partial H^0}{\partial \xi_2}\right)=-3H. \tag{9.48}$$

If we eliminate H^0 from this set, the following equation for H can be deduced

$$\frac{\partial^2 H}{\partial \xi_1 \partial \xi_2}+\frac{2}{\xi_1+\xi_2}\left(\frac{\partial H}{\partial \xi_1}+\frac{\partial H}{\partial \xi_2}\right)=0. \tag{9.49}$$

This equation is of the **Poisson-Euler-Darboux type.**

Equations of this type can be solved by the Riemann method*). In our case we use the following C^2-functions

$$\Phi(\xi_1,\xi_2)\equiv\frac{\partial}{\partial \xi_1}\frac{f(\xi_1)}{(\xi_1+\xi_2)^2},\quad \Psi(\xi_1,\xi_2)\equiv\frac{\partial}{\partial \xi_2}\frac{g(\xi_2)}{(\xi_1+\xi_2)^2}, \tag{9.50}$$

which satisfy the equation (9.49) for **arbitrary** C^3-functions f and g of a single variable. Hence, the general solution has the form

$$H(\xi_1,\xi_2)=\text{const.}+\frac{\partial}{\partial \xi_1}\frac{f(\xi_1)}{(\xi_1+\xi_2)^2}+\frac{\partial}{\partial \xi_2}\frac{g(\xi_2)}{(\xi_1+\xi_2)^2}. \tag{9.51}$$

Exercise 9.3. Check that Φ and Ψ defined by (9.50) are solutions of (9.49)•

For our purposes we can choose an arbitrary solution of (9.49). We select the simplest one by substituting

$$f(\xi_1)=-\frac{\sigma}{3}\xi_1^{-1},\quad g(\xi_2)=-\frac{\sigma}{3}\xi_2^{-1}, \tag{9.52}$$

where σ is a constant and the constant in (9.51) has been chosen to be zero. Then

$$H(\xi_1,\xi_2)=\frac{\sigma}{3}\frac{1}{\xi_1^2\xi_2^2}. \tag{9.53}$$

Returning to the original variables (Λ, L) we finally obtain

$$H(\Lambda,L)=\frac{\sigma}{3}\frac{1}{\left(\Lambda^2-\frac{1}{c^2}\right)^2}. \tag{9.54}$$

*) see: R. COURANT, Partial Differential Equations, Chap.5, (1962).

This result allows us to calculate ε and p^2 as functions of Λ and L. Namely, making use of (9.43) and $(9.42)_3$, we get

$$\varepsilon = \frac{\sigma}{3}\frac{3\Lambda^2 + \frac{L}{c^2}}{\left(\Lambda^2 - \frac{L}{c^2}\right)^3}, \qquad c^2 p^2 = \frac{\sigma^2}{9}\frac{16\Lambda^2 \frac{L}{c^2}}{\left(\Lambda^2 - \frac{L}{c^2}\right)^6}. \tag{9.55}$$

To obtain physically relevant results we have to invert the variables again. In the case considered in this section this happens to be possible without any approximations. Up to now this is the only case in extended thermodynamics where such an inversion is possible. The main fields Λ and $\mathbf{\Lambda}$ can be found as functions of ε and $\mathbf{p}$ if we solve (9.55) for Λ and L. Dividing ε^2 by c^2p^2 we obtain the following equation

$$\left(\frac{1}{\Lambda^2 c^2}\right)^2 + \left(6 - 16\frac{\varepsilon^2}{c^2 p^2}\right)\left(\frac{L}{\Lambda^2 c^2}\right) + 9 = 0, \tag{9.56}$$

which yields

$$\frac{L}{\Lambda^2 c^2} = \tfrac{9}{4}\frac{c^2 p^2}{\varepsilon^2}\left(1 + \sqrt{\Delta}\right)^{-2}, \qquad \Delta \equiv 1 - \tfrac{3}{4}\frac{c^2 p^2}{\varepsilon^2}. \tag{9.57}$$

Again the use of $(9.55)_1$ and of the above result leads to relations for Λ and $L=\mathbf{\Lambda}\cdot\mathbf{\Lambda}$

$$\Lambda = \tfrac{1}{\sqrt{2}}\left(\frac{\sigma}{\varepsilon}\right)^{\frac{1}{4}} \frac{\left(1+\sqrt{\Delta}\right)^{\frac{5}{4}}}{\left(1 + \sqrt{\Delta} - \frac{3}{2}\frac{c^2 p^2}{\varepsilon^2}\right)^{\frac{3}{4}}},$$

$$\frac{L}{c^2} = \tfrac{9}{8}\left(\frac{\sigma}{\varepsilon}\right)^{\frac{1}{2}} \frac{c^2 p^2}{\varepsilon^2} \frac{\left(1+\sqrt{\Delta}\right)^{\frac{1}{2}}}{\left(1 + \sqrt{\Delta} - \frac{3}{2}\frac{c^2 p^2}{\varepsilon^2}\right)^{\frac{3}{2}}}. \tag{9.58}$$

Substitution in (9.54) yields immediately

$$H = \frac{\varepsilon}{3}\left(2\sqrt{\Delta} - 1\right) = \tfrac{1}{3}\left(\sqrt{4\varepsilon^2 - 3c^2 p^2} - \varepsilon\right). \tag{9.59}$$

By means of the relations (9.42) and of the above result, we can find the potentials H^0 and H, and consequently, the thermodynamically admissible constitutive relations of the present model. Bearing the colinearity of $\mathbf{\Lambda}$ and $\mathbf{p}$ in mind [compare $(9.42)_3$], we can write the flux of momentum in the form

$$\mathbf{P} = H\mathbf{1} + 2L\frac{\partial H}{\partial L}\frac{\mathbf{p}\otimes\mathbf{p}}{p^2}. \tag{9.60}$$

Simultaneously,

$$L\frac{\partial H}{\partial L}=\tfrac{2}{3}\sigma\frac{\frac{L}{c^2}}{\left(\Lambda^2-\frac{L}{c^2}\right)^3}=2\varepsilon\frac{\frac{L}{\Lambda^2c^2}}{3+\frac{L}{\Lambda^2c^2}}, \tag{9.61}$$

where $(9.55)_1$ was used. Hence

$$L\frac{\partial H}{\partial L}=\varepsilon\left(1-\sqrt{\Delta}\right)=\varepsilon-\tfrac{1}{2}\sqrt{4\varepsilon^2-3c^2p^2}. \tag{9.62}$$

Finally, we obtain an explicit relation for the momentum flux

$$\mathbf{P}=\tfrac{1}{3}\left(\sqrt{4\varepsilon^2-3c^2p^2}-\varepsilon\right)\mathbf{1}+\left(2\varepsilon-\sqrt{4\varepsilon^2-3c^2p^2}\right)\frac{\mathbf{p}\otimes\mathbf{p}}{p^2}. \tag{9.63}$$

Comparison with (9.36) yields the following relation for the Eddington factor - the most important result of the present model (Fig. 9.1),

$$\chi=\tfrac{1}{3}\left(5-4\sqrt{\Delta}\right)=\tfrac{5}{3}-\tfrac{2}{3}\sqrt{4-3\frac{c^2p^2}{\varepsilon^2}}. \tag{9.64}$$

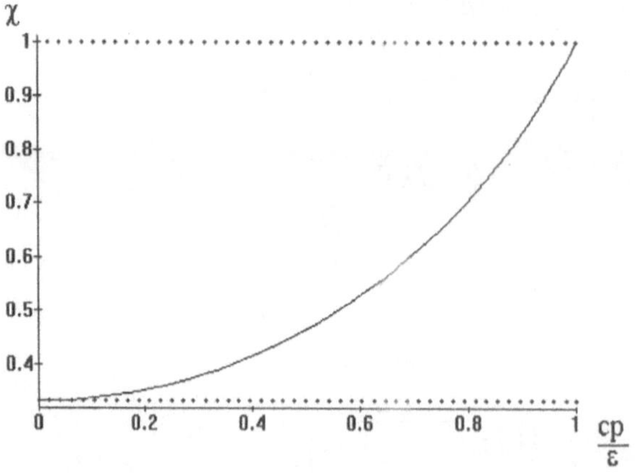

Fig. 9.1. *Eddington factor as a function of* cp/ε.

Now we are in the position to write the results in terms of the Eddington factor. It is easy to prove the following useful identities:

$$\frac{c^2p^2}{\varepsilon^2}=\tfrac{3}{4}(3-\chi)\left(\chi-\tfrac{1}{3}\right),\quad \Lambda=\frac{1}{2^{\frac{3}{4}}}\frac{1}{3^{\frac{1}{4}}}\left(\frac{\sigma}{\varepsilon}\right)^{\frac{1}{4}}\frac{(3-\chi)^{\frac{1}{2}}}{(1-\chi)^{\frac{3}{4}}},$$
$$\Lambda^2-\frac{L}{c^2}=\frac{2^{\frac{1}{2}}}{3^{\frac{1}{2}}}\left(\frac{\sigma}{\varepsilon}\right)^{\frac{1}{2}}\frac{1}{(1-\chi)^{\frac{1}{2}}}. \tag{9.65}$$

The entropy density η and the entropy flux $\mathbf{h}$ follow if we use (9.38). First of all, according to (9.44) we have

$$H^0=-\frac{\sigma}{3}\frac{\Lambda}{\left(\Lambda^2-\frac{L}{c^2}\right)^2}+c, \tag{9.66}$$

where c is the constant of integration. Then

$$\eta=-H^0+\Lambda\varepsilon+\boldsymbol{\Lambda}\cdot\mathbf{p}=-H^0+\Lambda\varepsilon+L^{\frac{1}{2}}\frac{\boldsymbol{\Lambda}}{L^{\frac{1}{2}}}\cdot\mathbf{p}=$$
$$=-H^0+\Lambda\varepsilon+L^{\frac{1}{2}}p, \tag{9.67}$$

where again the colinearity of $\boldsymbol{\Lambda}$ and $\mathbf{p}$ has been used. Consequently,

$$\eta=\tfrac{4}{3}\sigma\frac{\Lambda}{\left(\Lambda^2-\frac{L}{c^2}\right)^2}-c=\tfrac{4}{3}\sigma^{\frac{1}{4}}\varepsilon^{\frac{3}{4}}\frac{3^{\frac{3}{4}}}{2^{\frac{7}{4}}}(3-\chi)^{\frac{1}{2}}(1-\chi)^{\frac{1}{4}}-c. \tag{9.68}$$

For the flux H we have

$$\mathbf{H}=H\boldsymbol{\Lambda}=HL^{\frac{1}{2}}\frac{\mathbf{p}}{p}, \tag{9.69}$$

and $(9.38)_2$ yields

$$\mathbf{h}=-\mathbf{H}+\Lambda c^2\mathbf{p}+P\boldsymbol{\Lambda}=-HL^{\frac{1}{2}}\frac{1}{p}\mathbf{p}+\Lambda c^2\mathbf{p}+L^{\frac{1}{2}}\frac{1}{p}P\mathbf{p}=$$
$$=\frac{\left(\Lambda^2-\frac{L}{c^2}\right)}{\Lambda}c^2\mathbf{p}=\frac{2^{\frac{5}{4}}}{3^{\frac{1}{4}}}\frac{(1-\chi)^{\frac{1}{4}}}{(3-\chi)^{\frac{1}{2}}}\left(\frac{\sigma}{\varepsilon}\right)^{\frac{1}{4}}c^2\mathbf{p}. \tag{9.70}$$

Let us notice that the factor of proportionality for the heat flux $\mathbf{q}$ ($=c^2\mathbf{p}$) and the entropy flux $\mathbf{h}$ is not equal, in general, to 1/T. Even if we assumed the fraction σ/ε to be given approximately by the equilibrium relation (9.25) with σ being a constant, the coefficient would still be dependent on the Eddington factor. We return to this point in the following.

Yet the residual inequality (9.40), which according to the relations (9.32) can be written as follows

$$\Lambda p^2\left(A + B\frac{c}{p}\left(\frac{L}{\Lambda^2 c^2}\right)^{\frac{1}{2}}\right) \geq 0, \tag{9.71}$$

remains unexploited.

Substitution of the above results yields the following restriction for the functions A and B

$$\frac{1}{(1-\chi)^{\frac{3}{4}}}\left(A - \frac{2Bc^2}{\varepsilon}\frac{1}{3-\chi}\right) \geq 0. \tag{9.72}$$

9.4 Thermal Waves

Let us summarize the results of the previous section. For the four fields of the phonon energy density ε and the phonon momentum $\mathbf{p}$ we have the following set of field equations

$$\frac{\partial \varepsilon}{\partial t} + c^2 \operatorname{div} \mathbf{p} = Ap^2, \qquad \frac{\partial \mathbf{p}}{\partial t} + \operatorname{div} \mathbf{P} = B\mathbf{p}, \tag{9.73}$$

where the momentum flux $\mathbf{P}$ is given by

$$\mathbf{P} = \varepsilon\left(\frac{1-\chi}{2}\mathbf{1} + \frac{3\chi - 1}{2}\frac{\mathbf{p}\otimes\mathbf{p}}{p^2}\right), \tag{9.74}$$

and

$$\chi = \tfrac{5}{3} - \tfrac{2}{3}\sqrt{4 - 3\frac{c^2 p^2}{\varepsilon^2}}, \tag{9.75}$$

is the Eddington factor.

It can be seen that there is no ambiguity on the left-hand side of these equations due to the entropy inequality. Without any approximation, the extended thermodynamics specifies in this case the field equations where only the production terms can be chosen.

For the entropy density η and the entropy flux $\mathbf{h}$ we have also the explicit relations (9.68) and (9.70), respectively.

The functions A and B defining the productions are arbitrary within the restrictions of inequality (9.72).

To demonstrate the physical meaning of the above results, we proceed to discuss two particular cases.

1. **Near-equilibrium case:** $cp \ll \varepsilon$.

According to this condition the Eddington factor χ tends to the limit value $^1/_3$. Therefore, the momentum flux becomes isotropic

$$\mathbf{P} = \tfrac{1}{3}\varepsilon\mathbf{1}. \tag{9.76}$$

The entropy and the entropy flux follow in the form of

$$\eta = \tfrac{4}{3}\sigma^{\frac{1}{4}}\varepsilon^{\frac{1}{4}} - c, \qquad \mathbf{h} = \left(\frac{\sigma}{\varepsilon}\right)^{\frac{1}{4}} c^2\mathbf{p} \equiv \left(\frac{\sigma}{\varepsilon}\right)^{\frac{1}{4}} \mathbf{q}, \tag{9.77}$$

$\mathbf{q}$ being the heat flux. Making use of the equilibrium relation (9.25) for σ, we obtain

$$\eta = \tfrac{4}{3}\sigma T^3 - c, \qquad \mathbf{h} = \frac{1}{T}\mathbf{q}. \tag{9.78}$$

The T^3-dependence agrees with the classical results for the low-temperature limits of the second sound models of liquid helium*). The second relation shows that the near-equilibrium approximation renders the classical relation between heat and entropy fluxes, as we already indicated in the previous section.

The field equations (9.73) have, in this case, the following form:

$$\frac{\partial\varepsilon}{\partial t} + c^2 \operatorname{div}\mathbf{p} = 0, \qquad \frac{\partial\mathbf{p}}{\partial t} + \tfrac{1}{3}\operatorname{grad}\varepsilon = B|_E\,\mathbf{p}, \tag{9.79}$$

where $B|_E$ denotes the equilibrium value of the function B. Let us notice the similarity of the second equation to the Cattaneo equation $(5.127)_2$ for the heat flux $\mathbf{q}$. If ε was solely a function of T, we would have the heat conductivity K equal to $-\dfrac{c_v c^2}{3B|_E}$, $c_v \equiv \dfrac{\partial\varepsilon}{\partial T}$ and the relaxation time $\tau = -{}^1/_{B|E}$ with $B|_E < 0$.

It is instructive to eliminate $\mathbf{p}$ from these equations. We arrive easily at the **telegraph equation** for the energy ε

$$\frac{\partial^2\varepsilon}{\partial t^2} - B|_E\frac{\partial\varepsilon}{\partial t} - \frac{c^2}{3}\Delta^2\varepsilon = 0, \tag{9.80}$$

Δ^2 being the Laplace operator. Hence, the disturbances propagate in all directions with finite speed

$$c_{II} = \frac{c}{\sqrt{3}}, \tag{9.81}$$

and are damped due to the production $B|_E$. These **thermal waves** are commonly known as the **second sound**.

*) e.g. see: The review article of J. DE BOER, *Phonons in Liquids*, in T. A. BAK (ed.), *Phonons and Phonon Interactions*, Aarhus Summer School Lectures (1963), W. A. Benjamin, N.Y., (1964).

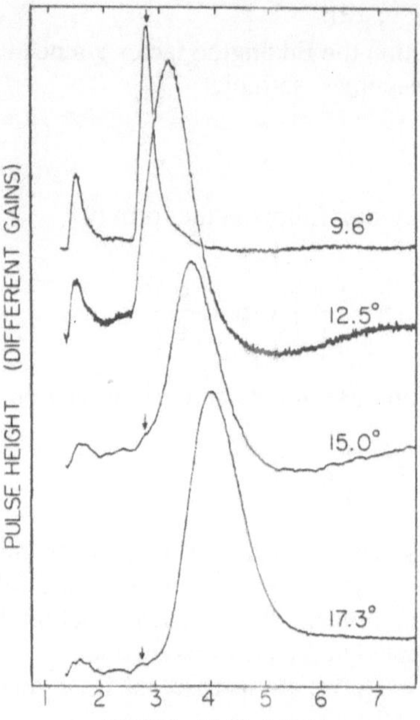

Fig. 9.2. *Arrival time measured in* NaF-*single crystals. An indication of the existence of the second sound.*

The above result is well supported by experiments on liquid and solid helium, natrium fluoride (NaF)-crystals and some other substances. In Fig. 9.2. we reproduce the result of H. E. JACKSON AND C. T. WALKER (1971) for a NaF-crystal obtained by the so-called heat-pulse propagation technique. Roughly speaking, the single crystal sample in the form of a bar of length 7.9 mm was loaded on one end, and the time of arrival of impulses on the other end was measured. The first peak corresponds to the arrival of the fastest (longitudinal) wave followed by the transversal wave, and subsequently, by a less prominent peak which is prescribed to the second sound wave. The tail of the curve in Fig. 9.2. corresponds to the thermal conductivity of the diffusive character.

Apart from the above-mentioned experiments, at relatively low temperatures it is speculated that some effects in layers of insulators at room temperatures can also be connected with the second sound. However, I am not aware of any reliable experimental data supporting these speculations.

2. **Free-streaming case:** $c\mathbf{p} \lesssim \varepsilon\mathbf{n}$, where $\mathbf{n}$ is a unit vector in the streaming direction.

In this case the Eddington factor χ tends to the value 1. The formula (9.74) becomes

$$\mathbf{P} = \varepsilon\,\mathbf{n} \otimes \mathbf{n}, \tag{9.82}$$

the entropy flux $\mathbf{h}$ given by (9.70) vanishes identically and the entropy η is constant. It means that the entropy production defining the left-hand side of the inequality (9.72) must vanish. Inspection of this inequality shows that the following relation must hold in this case

$$\lim_{\chi\to 1}\frac{1}{(1-\chi)^{\frac{3}{4}}}\left[A-\frac{2Bc^2}{\varepsilon}\frac{1}{3-\chi}\right]=0. \tag{9.83}$$

We assume that both A and B tend to zero.

Under this assumption the field equations (9.73) become

$$\frac{\partial \varepsilon}{\partial t}+c(\mathbf{n}\cdot \operatorname{grad})\varepsilon=0, \qquad \frac{1}{c}\mathbf{n}\left(\frac{\partial \varepsilon}{\partial t}+c(\mathbf{n}\cdot \operatorname{grad})\varepsilon\right)=0. \tag{9.84}$$

Hence, we have obtained a single equation for undamped waves in the **n**-direction whose speed is equal to c. In such a case the phonons do not interact with each other (harmonic approximation!) and we deal with the usual acoustic (longitudinal) waves. This mode of propagation is called **ballistic** because all phonons move with the same velocity without mutual interference.

The above limit cases justify the choice of the interval for values of the Eddington factor indicated in Fig. 9.1. as well as the names of different regions shown in this figure.

Example 9.1. *Characteristic Speeds for the Four-Field Model*

The above limit cases demonstrate the most appealing features of the set (9.73) – the existence of two types of waves in different ranges of the phonon flux. This can be appreciated even better if we consider the characteristic form of this set in the one-dimensional case

$$\mathbf{p}=(p,0,0)^T. \tag{9.85}$$

Then the field equations have the following form:

$$\begin{aligned}&\frac{\partial \varepsilon}{\partial t}+c^2\frac{\partial p}{\partial x}=Ap^2, \qquad x\equiv x^1,\\&\frac{\partial p}{\partial t}+\frac{\partial}{\partial x}(\varepsilon\chi)=Bp.\end{aligned} \tag{9.86}$$

Substitution of formula (9.75) for the Eddington factor yields easily the following characteristic speeds (eigenvalues) for (9.86)

$$\frac{1}{c}\mu_{1,2}=\frac{2}{3}\frac{cp}{\varepsilon}\left(\frac{5}{3}-\chi\right)^{-1}\left\{1\pm\left[\frac{1}{4}\chi(5-3\chi)^2\frac{\varepsilon^2}{c^2p^2}+3\chi-4\right]^{\frac{1}{2}}\right\}. \tag{9.87}$$

These are shown in Fig. 9.3. As expected from the limit analysis, the propagation for low values of cp/ε proceeds along two characteristics corresponding to the characteristic speeds $\mu=\pm c/\sqrt{3}$. For the value cp/ε close to 1 the characteristics approach each other until, for $cp/\varepsilon=1$, the eigenvalue $\mu=c$ becomes common for both of them ■

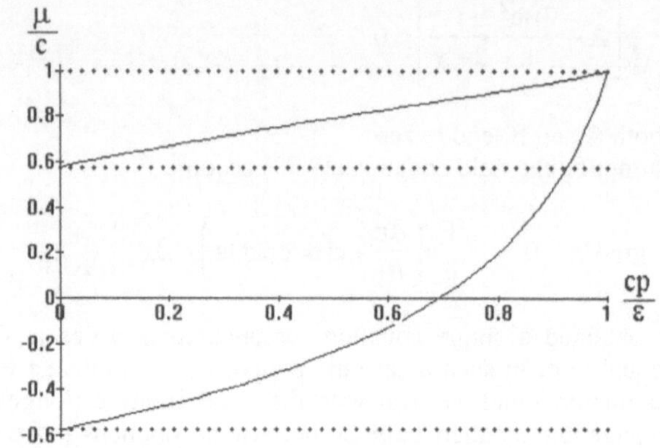

Fig. 9.3. *Characteristic speeds for the four-field model*

Let us complete this chapter with two remarks.

First of all, the results of Kremer and Müller, quoted in this chapter, concern to some extent also the model of the photon gas. The discussion of similarities and differences can be found in the original papers. The problem of the photon gas is strongly related to the problem of radiation which still yields many controversies in thermodynamics. Its presentation lies beyond the scope of these notes.

Secondly, it should be mentioned that an extensive research work is currently being carried out on the theory of the second sound based on a semi-kinetic approach which relies on the kinetic transfer equations closed by means of Dreyer's principle of the maximum entropy. Due to the purely phenomenological character of the material contained in this book, this approach could not be included in spite of many interesting results.

10 Some Multicomponent Systems

10.1 Introduction

The models of continua, which have been discussed in this book, are based on the assumption that the kinematics of a body is described by a single mapping of the manifold of a body into the space of motion, i.e. by the deformation function (2.1).

In many cases of practical importance this assumption is too restrictive. Systems such as mixtures of fluids, suspensions, granular materials, porous materials, etc. require a more sophisticated description of the kinematics. This kinematics should account for the *microstructure*, which can be understood in many cases as the extension of the kinematics of a single body to a few kinematical descriptions of different *constituent bodies*. Then the number of constituents determines the number of deformation functions needed to obtain the kinematics of such a *multicomponent body*.

A continuous description of a multicomponent body with A fluid components was proposed by C. TRUESDELL (1957) [see also: C. TRUESDELL, R. A. TOUPIN (1960)]. In his model, Truesdell assumes that each of the A components possesses its own velocity field defined on the common domains of the space of motion (Eulerian description)

$$\mathbf{v}^\alpha = \mathbf{v}^\alpha(\mathbf{x}, t), \qquad \mathbf{x} \in \mathcal{B}_t, \quad t \in \mathcal{T}, \quad \alpha = 1, \ldots, A. \tag{10.1}$$

As usual in the theory of fluids the current mass densities of components ρ_t^α and the temperature fields T^α complete the list of fields. It is also assumed that each component must satisfy the *partial balance equations* of the following form:

- mass balance

$$\frac{\partial \rho_t^\alpha}{\partial t} + \operatorname{div} \rho_t^\alpha \mathbf{v}^\alpha = \hat{\rho}_t^\alpha, \tag{10.2}$$

- momentum balance

$$\frac{\partial \rho_t^\alpha \mathbf{v}^\alpha}{\partial t} + \operatorname{div}\left(\rho_t^\alpha \mathbf{v}^\alpha \otimes \mathbf{v}^\alpha - \mathbf{T}^\alpha\right) = \hat{\mathbf{p}}_t^\alpha + \rho_t^\alpha \mathbf{b}^\alpha, \tag{10.3}$$

- energy balance

$$\begin{aligned}\frac{\partial}{\partial t}\left[\rho_t^\alpha\left(\varepsilon^\alpha + \tfrac{1}{2} v^{\alpha 2}\right)\right] + \operatorname{div}\left[\rho_t^\alpha\left(\varepsilon^\alpha + \tfrac{1}{2} v^{\alpha 2}\right)\mathbf{v}^\alpha + \mathbf{q}^\alpha - \mathbf{T}^\alpha \mathbf{v}^\alpha\right] = \\ = \hat{\varepsilon}_t^\alpha + \rho_t^\alpha \mathbf{b}^\alpha \cdot \mathbf{v}^\alpha + \rho_t^\alpha r^\alpha,\end{aligned} \tag{10.4}$$

where ρ_t^α, $\mathbf{T}^\alpha$, $\hat{\mathbf{p}}_t^\alpha$, $\mathbf{b}^\alpha$, ε^α, $\mathbf{q}^\alpha$, $\hat{\varepsilon}_t^\alpha$, r^α, denote the partial mass densities of the components per unit (common for all components!) volume in the current configuration, partial Cauchy stress tensors, partial momentum sources, partial body forces, partial specific internal energies, partial heat fluxes, partial sources of energies, partial energy radiations, respectively.

These equations are not assumed to be conservation laws because, apart from the exchange with the external world, the components may exchange mass, momentum and energy locally among themselves. The first type of an exchange happens, for example, in the case of chemical reactions if the components are defined as systems of different chemical composition or in the case of phase transformations if the components are understood as different phases of the same material. The second type of an exchange appears due to the relative motion of components and the third one is due to different temperatures of components.

Truesdell's theory of mixtures imposes additional restrictions on the source terms which, in a sense, indicate that the multicomponent model is a „better" description of a system than the single component model due to the fact that it accounts for the simplest and most important features of the microstructure. Namely, it is assumed that the summation of the partial balance equations must yield the corresponding conservation laws of the single component model with an appropriate definition of the fields of such a single component model. We then have

$$
\begin{aligned}
&\frac{\partial \rho}{\partial t} + \operatorname{div} \rho \mathbf{v} = 0, \quad \rho \equiv \sum_{\alpha=1}^{A} \rho_t^\alpha, \quad \rho \mathbf{v} \equiv \sum_{\alpha=1}^{A} \rho_t^\alpha \mathbf{v}^\alpha, \quad \underline{\sum_{\alpha=1}^{A} \hat{\rho}_t^\alpha = 0}, \\
&\frac{\partial \rho \mathbf{v}}{\partial t} + \operatorname{div}\left(\rho \mathbf{v} \otimes \mathbf{v} - \mathbf{T}\right) = \rho \mathbf{b}, \qquad \underline{\sum_{\alpha=1}^{A} \hat{\mathbf{p}}_t^\alpha = 0}, \\
&\qquad \mathbf{T} \equiv \sum_{\alpha=1}^{A} \left(\mathbf{T}^\alpha - \rho_t^\alpha \mathbf{v}^\alpha \otimes \mathbf{v}^\alpha\right) + \rho \mathbf{v} \otimes \mathbf{v}, \quad \rho \mathbf{b} \equiv \sum_{\alpha=1}^{A} \rho_t^\alpha \mathbf{b}^\alpha, \\
&\frac{\partial}{\partial t}\left[\rho\left(\varepsilon + \tfrac{1}{2} v^2\right)\right] + \operatorname{div}\left[\rho\left(\varepsilon + \tfrac{1}{2} v^2\right)\mathbf{v} + \mathbf{q} - \mathbf{T}\mathbf{v}\right] = \rho \mathbf{b} \cdot \mathbf{v} + \rho r, \\
&\qquad \rho \varepsilon \equiv \sum_{\alpha=1}^{A} \rho_t^\alpha \left(\varepsilon^\alpha + \tfrac{1}{2} v^{\alpha 2}\right) - \tfrac{1}{2} \rho v^2, \\
&\qquad \mathbf{q} \equiv \sum_{\alpha=1}^{A} \left[\mathbf{q}^\alpha + \rho_t^\alpha \left(\varepsilon^\alpha + \tfrac{1}{2} v^{\alpha 2}\right)\mathbf{v}^\alpha - \mathbf{T}^\alpha \mathbf{v}^\alpha\right] - \\
&\qquad\qquad - \rho\left(\varepsilon + \tfrac{1}{2} v^2\right)\mathbf{v} + \mathbf{T}\mathbf{v}, \\
&\qquad \rho r \equiv \sum_{\alpha=1}^{A} \left(\rho_t^\alpha r^\alpha + \rho_t^\alpha \mathbf{b}^\alpha \cdot \mathbf{v}^\alpha\right) - \rho \mathbf{b} \cdot \mathbf{v}, \qquad \underline{\sum_{\alpha=1}^{A} \hat{\varepsilon}^\alpha = 0.}
\end{aligned}
\tag{10.5}
$$

The underlined relations among sources restrict the partial balance equations of such a mixture theory.

Obviously, if we add constitutive relations for $\{\mathbf{T}^\alpha, \varepsilon^\alpha, \mathbf{q}^\alpha, \hat{\rho}_t^\alpha, \hat{\mathbf{p}}_t^\alpha, \hat{\varepsilon}^\alpha\}$, the balance equations (10.2–10.4) transform into field equations for the fields $\{\rho_t^\alpha, \mathbf{v}^\alpha, T^\alpha\}$.

We shall not discuss this model any further in this book. A presentation of the classical thermodynamical properties of mixtures of fluids can be found in the books by C. TRUESDELL (1984) and of I. MÜLLER (1985) where the original sources are also quoted. The extended thermodynamics of fluid mixtures is presented in the book by I. MÜLLER and T. RUGGERI (1993).

In this chapter, we briefly present the thermodynamical restrictions of the multicomponent model whose microstructure is further extended onto a single scalar field of **po-**

rosity. Such a model is the simplest description of the so-called **immiscible mixtures** [see: C. TRUESDELL (1984)] in which the components do not form a mixture on the molecular level as it is the case for fluids but remain separated to the sub-macroscopical level of observations. Systems such as rocks, clays, solid filters, biological tissues (lungs, liver, etc.) belong to this class.

The model which we are going to discuss has two novel features connected with a new „philosophy“ of the porous material. Namely, the motion of the single-solid phase, the so-called **skeleton**, is considered to be described as for a single body and, simultaneously, the microscopical channels of the skeleton are considered to be the *confinement* for the motion of A fluid components. The geometry of these microscopical channels is reflected on the macroscopical level solely by a scalar field n which is called the **porosity**.

This type of description of porous materials yields major differences in comparison with other models appearing in the literature. First of all, it is natural to introduce a consistent Lagrangian description of **all** components with respect to the **same reference configuration** of the skeleton. We will present this in the next section. Secondly, the additional microstructural variables are reduced to the **single field of porosity** instead of A+1 volume fractions of all components satisfying the *saturation condition* and appearing in almost all other models. These additional microstructural fields require field equations which are not easy to construct and, in addition, yield the constraint (the saturation condition) connected with certain unsolved thermodynamical problems. Moreover, these fields do not seem to be physically well motivated, at least in the case of a single solid component, in spite of an immense advertisement for the „volume fraction concept“ in some papers on the theory of porous materials. They extend unnecessarily the number of microstructural fields violating the old (the XIVth century!) philosophical principle of the Ockham razor [see: C. TRUESDELL (1966)]. Apart from this, the constraint eliminates one of the modes of propagation of sound waves (the so-called P2-wave) which has a very important practical bearing in porous materials.

The porosity appearing in the model discussed in this section has the same thermodynamical status as all other thermodynamical fields. Namely, it is assumed to satisfy its own balance equation and, due to the source term in this equation, it possesses the property of spontaneous relaxation to the thermodynamical equilibrium.

Models with the porosity as a thermodynamical field were already proposed earlier in a different framework. R. M. BOWEN (1982) considered the porosity as a microstructural variable with a corresponding *evolution equation* (i.e. without the flux term). M. A. GOODMAN, S. C. COWIN (1972) proposed a balance equation following from some heuristic considerations of the microscopical interactions of components. They called it the *balance of equilibrated pressures*. Both models were used to describe dynamical processes in porous materials, such as sound waves and combustion of granular materials. The results seem to match well with experiments, which certainly indicates that the model with an additional equation for the porosity is adequate for porous materials. However, the motivation of both approaches does not seem to be very convincing.

10.2 Lagrangian Description

Let us begin with the description of the motion of a skeleton. We assume that it is given by the mapping

$$\mathbf{x} = \mathbf{f}^S(\mathbf{X}, t), \qquad \mathbf{X} \in \mathcal{B}_0, \quad t \in \mathcal{T}, \tag{10.6}$$

which has the same properties as the mapping (2.1) for a single component body. Certainly, the function (10.6) does not describe the real microscopical motion of a skeleton. It reflects solely the **smeared-out kinematics**. By means of such a description we loose most details of the real motion. The porosity delivers certain additional information about this motion which we will show in the next section.

Even though it is possible to introduce some averaging procedures in the opposite direction, i.e. to describe the macroscopical motion by some averages of micromotions, we shall not discuss this problem any further in this book, with the sole of porosity*).

The velocity and the deformation gradient of a skeleton are given by the following relations:

$$\mathbf{x}'^S = \frac{\partial \mathbf{f}^S}{\partial t}(\mathbf{X}, t), \quad \mathbf{F}^S = \operatorname{Grad} \mathbf{f}^S(\mathbf{X}, t). \tag{10.7}$$

This is, of course, the Lagrangian description.

The fluid components are at first assumed to be described in the Eulerian way. This means that their velocity fields are defined on the current configuration of a skeleton

$$\mathbf{v}^\alpha = \mathbf{v}^\alpha(\mathbf{x}, t), \quad \alpha = 1, \cdots, A, \qquad \mathbf{x} \in \mathbf{f}^S(\mathcal{B}_0, t), \quad t \in \mathcal{T}. \tag{10.8}$$

Again, these are smeared-out velocities. In most cases of practical interest they cannot be transformed into real velocities of fluid components in the pores of a skeleton. Such a transformation may be possible in exceptional cases of very slow, almost uniform motions or for very large porosities (e.g. in the case of a skeleton in the form of thin membranes, like lungs and some other biological tissues).

By means of the mapping (10.6) we can transform the fields (10.8) into the following fields on the reference configuration of the skeleton:

$$\mathbf{x}'^\alpha = \mathbf{v}^\alpha(\mathbf{f}^S(\mathbf{X}, t), t) \equiv \mathbf{x}'^\alpha(\mathbf{X}, t), \quad \alpha = 1, \cdots, A, \qquad \mathbf{X} \in \mathcal{B}_0, \quad t \in \mathcal{T}. \tag{10.9}$$

Apart from the above natural velocity fields, we need as well the kinematics of the material regions of fluid components projected onto the reference configuration $\mathcal{B}_0$ of the skeleton. At a generic point **x** of the space of motion, the velocity of a fluid particle of the α-component with respect to the particle of the skeleton, which is instantaneously located at this point, is $\mathbf{x}'^\alpha(\mathbf{X}, t) - \mathbf{x}'^S(\mathbf{X}, t)$, where $\mathbf{X} = \mathbf{f}^{S-1}(\mathbf{x}, t)$. Consequently, the instantaneous image of this fluid particle in the reference configuration moves with the velocity

$$\mathbf{X}'^\alpha \equiv \mathbf{F}^{S-1}(\mathbf{x}'^\alpha - \mathbf{x}'^S) = \mathbf{X}'^\alpha(\mathbf{X}, t). \tag{10.10}$$

These are the **Lagrangian velocities** of fluid components [see: K. WILMANSKI (1995, 1,2), (1996,1,2) for details]. They define the motion of the images of fluid components throughout the reference configuration $\mathcal{B}_0$. In the Lagrangian description of this chapter, they are counterparts of the **diffusion velocity** of fluid components with respect to the skeleton. In such a description we do not need the *barycentric velocity* which is usually introduced in the classical theory of mixtures and which is used to define the diffusion velocities.

We are now in the position to specify the classical balance equations for the components of a porous medium. We have for the skeleton in the inertial frame of reference

*) e.g. see: J. BEAR (1972), J. BEAR, Y. BACHMAT (1984), F. DOBRAN (1991) for some details of such averages.

Global form for any subbody $\mathcal{P}^S \subset \mathcal{B}_0$	Local form
Mass balance	
$\frac{d}{dt}\int\limits_{\mathcal{P}^S} \rho^S dV = \int\limits_{\mathcal{P}^S} \hat{\rho}^S dV,$	$\frac{\partial \rho^S}{\partial t} = \hat{\rho}^S,$
Momentum balance	
$\frac{d}{dt}\int\limits_{\mathcal{P}^S} \rho^S \mathbf{x}'^S dV = \oint\limits_{\partial\mathcal{P}^S} \mathbf{P}^S \mathbf{N} dS +$ $+ \int\limits_{\mathcal{P}^S} \rho^S \mathbf{b}^S + \hat{\mathbf{p}}^S dV,$	$\frac{\partial}{\partial t}(\rho^S \mathbf{x}'^S) = \mathrm{Div}\,\mathbf{P}^S + \rho^S \mathbf{b}^S + \hat{\mathbf{p}}^S,$
Energy balance	
$\frac{d}{dt}\int\limits_{\mathcal{P}^S} \rho^S\left(\varepsilon^S + \tfrac{1}{2}x'^{S2}\right) dV =$ $= \oint\limits_{\partial\mathcal{P}^S} \left(\mathbf{Q}^S + \mathbf{P}^{ST}\mathbf{x}'^S\right)\cdot \mathbf{N} dS +$ $+ \int\limits_{\mathcal{P}^S} \rho^S \mathbf{b}^S \cdot \mathbf{x}'^S + \rho^S r^S + \hat{\varepsilon}^S dV,$	$\frac{\partial}{\partial t}\left[\rho^S\left(\varepsilon^S + \tfrac{1}{2}x'^{S2}\right)\right] + \mathrm{Div}\,\mathbf{Q}^S =$ $= \mathrm{Div}\left(\mathbf{P}^{ST}\mathbf{x}'^S\right) + \rho^S \mathbf{b}^S \cdot \mathbf{x}'^S + \rho^S r^S + \hat{\varepsilon}^S$

(10.11)

For the α-fluid component we have correspondingly

Global form for any subbody $\mathcal{P}^\alpha \subset \mathcal{B}_0$ Material with respect to the α-kinematics	Local form
Mass balance	
$\frac{d}{dt}\int\limits_{\mathcal{P}^\alpha} \rho^\alpha dV = \int\limits_{\mathcal{P}^\alpha} \hat{\rho}^\alpha dV,$	$\frac{\partial \rho^\alpha}{\partial t} + \mathrm{Div}\left(\rho^\alpha \mathbf{X}'^\alpha\right) = \hat{\rho}^\alpha,$
Momentum balance	
$\frac{d}{dt}\int\limits_{\mathcal{P}^\alpha} \rho^\alpha \mathbf{x}'^\alpha dV = \oint\limits_{\partial\mathcal{P}^\alpha} \mathbf{P}^\alpha \mathbf{N} dS +$ $+ \int\limits_{\mathcal{P}^\alpha} \rho^\alpha \mathbf{b}^\alpha + \hat{\mathbf{p}}^\alpha dV,$	$\frac{\partial}{\partial t}(\rho^\alpha \mathbf{x}'^\alpha) + \mathrm{Div}\left(\rho^\alpha \mathbf{x}'^\alpha \otimes \mathbf{X}'^\alpha - \mathbf{P}^\alpha\right) =$ $= \rho^\alpha \mathbf{b}^\alpha + \hat{\mathbf{p}}^\alpha,$
Energy balance	
$\frac{d}{dt}\int\limits_{\mathcal{P}^\alpha} \rho^\alpha\left(\varepsilon^\alpha + \tfrac{1}{2}x'^{\alpha 2}\right) dV =$ $= \oint\limits_{\partial\mathcal{P}^\alpha} \left(-\mathbf{Q}^\alpha + \mathbf{P}^{\alpha T}\mathbf{x}'^\alpha\right)\cdot \mathbf{N} dS +$ $+ \int\limits_{\mathcal{P}^\alpha} \rho^\alpha \mathbf{b}^\alpha \cdot \mathbf{x}'^\alpha + \rho^\alpha r^\alpha + \hat{\varepsilon}^\alpha dV,$	$\frac{\partial}{\partial t}\left[\rho^\alpha\left(\varepsilon^\alpha + \tfrac{1}{2}x'^{\alpha 2}\right)\right] +$ $+ \mathrm{Div}\left[\rho^\alpha\left(\varepsilon^\alpha + \tfrac{1}{2}x'^{\alpha 2}\right)\mathbf{X}'^\alpha + \mathbf{Q}^\alpha\right] =$ $= \mathrm{Div}\left(\mathbf{P}^{\alpha T}\mathbf{x}'^\alpha\right) + \rho^\alpha \mathbf{b}^\alpha \cdot \mathbf{x}'^\alpha + \rho^\alpha r^\alpha + \hat{\varepsilon}^\alpha$

(10.12)

The notation in these equations is as follows:

ρ^S - the partial mass density of the skeleton per unit volume of the reference configuration;

$\hat{\rho}^S$ - the mass source of the skeleton per unit volume of the reference configuration;

$\mathbf{P}^S$ - the partial Piola-Kirchhoff stress tensor of the skeleton;

$\mathbf{b}^S$ - the partial body force of the skeleton per unit mass of the skeleton;

$\hat{\mathbf{p}}^S$ - the source of momentum in the skeleton;

ε^S - the specific internal energy of the skeleton per unit mass of the skeleton;

$\mathbf{Q}^S$ - the partial heat flux in the skeleton per unit reference surface;

r^S - the partial energy radiation in the skeleton per unit mass of the skeleton;

$\hat{\varepsilon}^S$ - the energy source in the skeleton per unit reference volume.

ρ^α - the partial mass density of the α-component per unit volume of the reference configuration of the skeleton;

$\hat{\rho}^\alpha$ - the mass source of the α-component per unit volume of the reference configuration of the skeleton;

$\mathbf{P}^\alpha$ - the partial Piola-Kirchhoff stress tensor of the α-component related to the reference configuration of the skeleton;

$\mathbf{b}^\alpha$ - the partial body force of the α-component per unit mass in the reference configuration of the skeleton;

$\hat{\mathbf{p}}^\alpha$ - the source of momentum in the α-component;

ε^α - the specific internal energy of the α-component per unit mass in the reference configuration of the skeleton;

$\mathbf{Q}^\alpha$ - the partial heat flux in the α-component per unit reference surface;

r^α - the partial energy radiation in the α-component per unit mass in the reference configuration of the skeleton;

$\hat{\varepsilon}^\alpha$ - the energy source in the α-component per unit reference volume.

Once again, all quantities in these equations are related to the reference configuration $\mathcal{B}_0$ of the skeleton.

Exercise 10.1. Prove the local equations (10.12) under the assumption of the global form given in the left part of the above table•

It is easy to show that the above equations transform into the usual balance equations of the theory of mixtures in the Eulerian description. We show this in the example of the mass balance for the α-component.

The partial time derivative in the above equations is calculated for **X**=const. Consequently, it is the material time derivative with respect to the skeleton, i.e.

$$\begin{aligned}\frac{\partial\rho^{\alpha}}{\partial t}(\mathbf{X},t) &= \frac{\partial J^{S}\rho^{\alpha}J^{S-1}}{\partial t} = J^{S}\frac{\partial\rho^{\alpha}J^{S-1}}{\partial t} + \rho^{\alpha}\mathbf{F}^{S-T}\cdot\frac{\partial\mathbf{F}^{S}}{\partial t} = \\ &= J^{S}\left(\frac{\partial\rho^{\alpha}J^{S-1}}{\partial t} + \rho^{\alpha}J^{S-1}\mathbf{F}^{S-T}\cdot\operatorname{Grad}\mathbf{x}'^{S}\right)\overset{*}{=} \\ &\overset{*}{=} J^{S}\left(\frac{\partial\rho^{\alpha}J^{S-1}}{\partial t} + \mathbf{v}^{S}\cdot\operatorname{grad}\left(\rho^{\alpha}J^{S-1}\right) + \rho^{\alpha}J^{S-1}\operatorname{div}\mathbf{v}^{S}\right) \\ &= J^{S}\left(\frac{\partial\rho^{\alpha}J^{S-1}}{\partial t} + \operatorname{div}\left(\rho^{\alpha}J^{S-1}\mathbf{v}^{S}\right)\right), \quad \mathbf{v}^{S} \equiv \mathbf{x}'^{S}\left(\mathbf{f}^{S-1}(\mathbf{x},t),t\right).\end{aligned} \tag{10.13}$$

The asterix means that we transform the variables $(\mathbf{X},t)\to(\mathbf{x},t)$ by means of the function of motion (10.6). Simultaneously,

$$\begin{aligned}\operatorname{Div}\left(\rho^{\alpha}\mathbf{X}'^{\alpha}\right)(\mathbf{X},t) &= \operatorname{Div}\left(\rho^{\alpha}\mathbf{F}^{S-1}\left(\mathbf{x}'^{\alpha}-\mathbf{x}'^{S}\right)\right) = \\ &= J^{S}\mathbf{F}^{S-T}\cdot\operatorname{Grad}\left(\rho^{\alpha}J^{S-1}\left(\mathbf{x}'^{\alpha}-\mathbf{x}'^{S}\right)\right)\overset{*}{=} \\ &\overset{*}{=} J^{S}\operatorname{div}\left(\rho^{\alpha}J^{S-1}\left(\mathbf{v}^{\alpha}-\mathbf{v}^{S}\right)\right).\end{aligned} \tag{10.14}$$

Consequently, we have

$$\frac{\partial\rho^{\alpha}J^{S-1}}{\partial t} + \operatorname{div}\left(\rho^{\alpha}J^{S-1}\mathbf{v}^{S}\right) + \operatorname{div}\left(\rho^{\alpha}J^{S-1}\left(\mathbf{v}^{\alpha}-\mathbf{v}^{S}\right)\right) = \hat{\rho}^{\alpha}J^{S-1}. \tag{10.15}$$

After simplification, we obtain the mass balance equation (10.2) if we identify

$$\rho_t^{\alpha} = \rho^{\alpha}J^{S-1}, \qquad \hat{\rho}_t^{\alpha} = \hat{\rho}^{\alpha}J^{S-1}. \tag{10.16}$$

Exercise 10.2. Show that the transformation rules for the remaining fields have the following form:

$$\mathbf{T}^{\alpha} \overset{*}{=} J^{S-1}\mathbf{P}^{\alpha}\mathbf{F}^{ST}, \quad \hat{\mathbf{p}}_t^{\alpha} \overset{*}{=} J^{S-1}\hat{\mathbf{p}}^{\alpha}, \quad \mathbf{q}^{\alpha} \overset{*}{=} J^{S-1}\mathbf{F}^{S}\mathbf{Q}^{\alpha}, \quad \hat{\varepsilon}_t^{\alpha} \overset{*}{=} J^{S-1}\hat{\varepsilon}^{\alpha}, \tag{10.17}$$

where $\mathbf{T}^{\alpha}$ denotes the partial Cauchy stress tensors, and $\mathbf{q}^{\alpha}$ denotes the vectors of the heat flux in the current configuration•

As usual, it is required that the sums of sources vanish

$$\sum_{\alpha=1}^{A}\hat{\rho}^{\alpha} + \hat{\rho}^{S} = 0, \quad \sum_{\alpha=1}^{A}\hat{\mathbf{p}}^{\alpha} + \hat{\mathbf{p}}^{S} = 0, \quad \sum_{\alpha=1}^{A}\hat{\varepsilon}^{\alpha} + \hat{\varepsilon}^{S} = 0. \tag{10.18}$$

The above equations form the basis for the construction of the field equations. The fields essential for the present model of porous materials are as follows:

$$\left\{\rho^S,\rho^1,\cdots,\rho^A,\mathbf{f}^S,\mathbf{X}'^1,\cdots,\mathbf{X}'^A,T^S,T^1,\cdots,T^A,n\right\}. \tag{10.19}$$

These are 5(A+1)+1 fields. If we close the balance equations by means of appropriate constitutive relations then we, obviously, omit one equation for the case where the porosity n is a thermodynamical field such as all other fields of the model.

The balance equations yield the dynamical compatibility conditions on singular surfaces as was already the case for a single component media. However, in the case of relative motions of components these relations have a special bearing. Namely, the boundary for which we define a boundary value problem for the field equations is usually identical with the boundary of one of the components. For porous materials it is in most cases the boundary of the skeleton. Then the other components flow through this boundary except in the case when the boundary is simultaneously not permeable for these components. This requires a modification of the classical boundary conditions in the continuum mechanics and it is, simultaneously, coupled with certain mathematical problems concerning the admissible classes of solutions (the smoothness of fields). In addition, the dynamical compatibility conditions play an important role in the analysis of waves, in such contact problems as seepage in soil mechanics [e.g. see: J. BEAR (1972)] or interfaces between layers of different porous materials (stratified structures), etc. Here we present only the most important basic relations connected with the dynamical compatibility which have a bearing in thermodynamical considerations. The conditions shall be formulated in the Lagrangian description. We follow the work by K. WILMANSKI (1995, 2).

We consider a smooth orientable surface $\mathcal{S}_t$ given for the instant of time t in the configuration space by the relation [see: (4.52) and (4.53)]

$$g(\mathbf{x},t)=0 \quad\Rightarrow\quad c=-\frac{\frac{\partial g}{\partial t}}{|\operatorname{grad} g|},\quad \mathbf{n}=\frac{\operatorname{grad} g}{|\operatorname{grad} g|}, \tag{10.20}$$

where c and **n** are the **speed of propagation** and the unit normal vector of this surface, respectively. The Lagrangian image $\mathcal{S}_0$ of this surface is then [see: (4.21)] given by

$$G(\mathbf{X},t)\equiv g\left(\mathbf{f}^S(\mathbf{X},t),t\right)=0,\qquad \mathbf{X}\in\mathcal{B}_0,\quad t\in\mathcal{T}. \tag{10.21}$$

The **Lagrangian speed of propagation** U and the unit normal vector **N** of this image are then defined as follows (see:(4.22)-(4.23)):

$$U=-\frac{\frac{\partial G}{\partial t}}{|\operatorname{Grad} G|}=\left|\mathbf{F}^{S-T}\mathbf{N}\right|\left(u-\mathbf{n}\cdot\mathbf{x}'^S\right),\quad \mathbf{N}=\frac{\operatorname{Grad} G}{|\operatorname{Grad} G|},\quad \mathbf{n}=\frac{\mathbf{F}^{S-T}\mathbf{N}}{\left|\mathbf{F}^{S-T}\mathbf{N}\right|}, \tag{10.22}$$

where the formula (2.20) for the current image of the normal vector was used.

These are, certainly, the same relations which we discussed for a single component solid. We can use them due to the Lagrangian description of motion for a multicomponent body.

We proceed to present the dynamical compatibility conditions for a multicomponent system. The dynamical compatibility condition for the mass transport of a skeleton has the following form (see: $(4.34)_1$):

$$U[[\rho^S]]=0. \tag{10.23}$$

This relation is, of course, identically satisfied on the boundary of a porous material because this boundary coincides with the boundary of the skeleton for which U≡0.

The remaining compatibility conditions for a skeleton follow in a similar manner as the conditions (4.34) for the single component body. We have

- momentum balance

$$\rho^S U[[\mathbf{x}'^S]] + [[\mathbf{P}^S]]\mathbf{N} = 0, \tag{10.24}$$

- energy balance

$$\rho^S U[[\varepsilon^S + \tfrac{1}{2}x'^{S2}]] + [[\mathbf{P}^{ST}\mathbf{x}'^S - \mathbf{Q}^S]] \cdot \mathbf{N} = 0, \tag{10.25}$$

- entropy balance

$$\rho^S U[[\eta^S]] - [[\mathbf{H}^S]] \cdot \mathbf{N} = 0, \tag{10.26}$$

where η^S is the specific partial entropy and $\mathbf{H}^S$ denotes its flux for a skeleton. It has been assumed that the singular surface is non-dissipative, i.e. the surface entropy source is identically zero. If it were not the case, we would have to correct the right-hand side of relation (10.26).

On the boundary (U≡0) we have

$$[[\mathbf{P}^S]]\mathbf{N} = 0, \quad [[\mathbf{Q}^S]] \cdot \mathbf{N} = 0, \quad [[\mathbf{H}^S]] \cdot \mathbf{N} = 0. \tag{10.27}$$

The compatibility conditions for the fluid components are more complicated because the Lagrangian images of those components move with the Lagrangian velocities $\mathbf{X}'^\alpha$, in contrast to the skeleton whose Lagrangian image (the reference configuration) does not move at all. Consequently, easy calculations similar to those for the Eulerian description of a single component continuum (see: Sect. 4.5.) yield

- mass balance

$$[[\rho^\alpha(\mathbf{X}'^\alpha \cdot \mathbf{N} - U)]] = 0, \quad \alpha = 1, \cdots, A, \tag{10.28}$$

- momentum balance

$$\rho^\alpha(\mathbf{X}'^\alpha \cdot \mathbf{N} - U)[[\mathbf{x}'^\alpha]] - [[\mathbf{P}^\alpha]]\mathbf{N} = 0, \quad \alpha = 1, \cdots, A, \tag{10.29}$$

- energy balance

$$\rho^\alpha(\mathbf{X}'^\alpha \cdot \mathbf{N} - U)[[\varepsilon^\alpha + \tfrac{1}{2}x'^{\alpha 2}]] - [[\mathbf{P}^{\alpha T}\mathbf{x}'^\alpha - \mathbf{Q}^\alpha]] \cdot \mathbf{N} = 0, \quad \alpha = 1, \cdots, A, \tag{10.30}$$

- entropy balance

$$\rho^\alpha(\mathbf{X}'^\alpha \cdot \mathbf{N} - U)[[\eta^\alpha]] + [[\mathbf{H}^\alpha]] \cdot \mathbf{N} = 0, \quad \alpha = 1, \cdots, A, \tag{10.31}$$

where η^α is the specific partial entropy and $\mathbf{H}^\alpha$ denotes its flux for the α-component. Again it has been assumed that the surface is non-dissipative.

Exercise 10.3. Prove the above dynamical compatibility conditions for the fluid components•

In the particularly important case of the surface being the boundary of a skeleton we have

$$\begin{aligned} &m^{\alpha}[[\mathbf{x}'^{\alpha}]]-[[\mathbf{P}^{\alpha}]]\mathbf{N}=0, \qquad m^{\alpha}\equiv\rho^{\alpha-}\mathbf{X}'^{\alpha-}\cdot\mathbf{N},\\ &m^{\alpha}[[\varepsilon^{\alpha}+\tfrac{1}{2}x'^{\alpha 2}]]-[[\mathbf{P}^{\alpha T}\mathbf{x}'^{\alpha}-\mathbf{Q}^{\alpha}]]\cdot\mathbf{N}=0, \quad \alpha=1,\cdots,A,\\ &m^{\alpha}[[\eta^{\alpha}]]+[[\mathbf{H}^{\alpha}]]\cdot\mathbf{N}=0. \end{aligned} \tag{10.32}$$

Due to the jump of porosity on the boundary of a skeleton, we cannot expect that the velocity fields $\mathbf{x}'^{\alpha}$ are continuous through this surface. Consequently, even in the simplest case of a surface loading we do not obtain the continuity of the partial stress vectors. Hence, we are usually not able to formulate the stress boundary conditions for porous materials in the same way as we do for single component materials. We shall return to this point in the following.

The same difficulty appears for the thermal boundary conditions as an inspection of the relation (10.30) shows.

Remark 10.1. *On the Measurability of the Temperature*
The relations for the partial entropies indicate that the notion of temperature is almost useless in thermomechanical models of porous materials, at least in cases of real dynamical processes. To see this let us consider a simple example. Let us assume that all components have the same temperature T and each partial heat flux is related to the corresponding partial entropy flux by the Fourier relation

$$\mathbf{H}^{S}=\frac{1}{T}\mathbf{Q}^{S}, \quad \mathbf{H}^{\alpha}=\frac{1}{T}\mathbf{Q}^{\alpha}, \quad \alpha=1,\cdots,A. \tag{10.33}$$

Then we have

$$\begin{aligned} &[[\mathbf{Q}_{I}]]\cdot\mathbf{N}=\sum_{\alpha=1}^{A}\left\{-m^{\alpha}[[\varepsilon^{\alpha}+\tfrac{1}{2}x'^{\alpha 2}]]+[[\mathbf{P}^{\alpha}\cdot(\mathbf{x}'^{\alpha}\otimes\mathbf{N})]]\right\},\\ &\left[\left[\frac{1}{T}\mathbf{Q}_{I}\right]\right]\cdot\mathbf{N}=-\sum_{\alpha=1}^{A}m^{\alpha}[[\eta^{\alpha}]], \end{aligned} \tag{10.34}$$

where

$$\mathbf{Q}_{I}\equiv\mathbf{Q}^{S}+\sum_{\alpha=1}^{A}\mathbf{Q}^{\alpha}, \tag{10.35}$$

is the **intrinsic heat flux.**

Consequently, if the right-hand sides of relations (10.34) are different from zero we do not obtain the continuity of the temperature on the boundary

$$[[T]]\neq 0, \tag{10.36}$$

which would be the case in a single component model. Hence, the boundary values of the temperature of porous materials cannot be measured with a thermometer, and the usefulness of this notion becomes rather exceptional [see: K. WILMANSKI (1995, 2)]•

Remark 10.2. *On the Natural Mechanical Boundary Conditions*
The above considerations concerning the compatibility conditions show that the permeability of the boundary for fluid components yields serious difficulties in the formulation of physically justified relations for such a boundary. We shall not discuss this problem in any details in this book. However, it is worthwhile mentioning that in accordance with the permeability the boundary conditions can be divided into two classes.

The first class appears for impermeable boundaries. For such a boundary, the velocity of all components on the boundary is identical, and the external loading is taken over by the whole resultant stress vector of all components.

$$\mathbf{P}_I \mathbf{N}\Big|_{\partial\mathcal{B}_0} = J^S \sqrt{\mathbf{C}^{S-1} \cdot \mathbf{N} \otimes \mathbf{N}}\, \mathbf{t}_{ext}, \quad \mathbf{X}'^{\alpha}\Big|_{\partial\mathcal{B}_0} = 0, \quad \alpha = 1, \cdots, A,$$
$$\mathbf{P}_I \equiv \mathbf{P}^S + \sum_{\alpha=1}^{A} \mathbf{P}^{\alpha}, \tag{10.36}$$

where $\mathbf{t}_{ext}$ denotes the vector of external loading per unit surface in the current configuration, and we use relation $(10.22)_3$ for the normal vector, and the transformation rules (10.17) for the Cauchy stress tensors.

Exercise 10.4. Prove the above relation for the intrinsic stress vector•

There is no need for additional conditions because the motion of the boundary is determined by the motion of the skeleton and there is no free outflow of the fluid components.

The second class appears if the boundary is permeable for one fluid, say, the α-component. Then the external loading is not only supported by the resultant stress vector, but also by the kinematical part connected with the presence of the outflow m^α [see: the formula $(10.32)_1$]. In addition, the outflow m^α is needed to determine the motion of the boundary of the α-component. It is a **problem with a free boundary**. Many questions connected with such boundary value problems are still open. In some practical applications, one can approximate these problems neglecting the influence of the outflow because m^α is small in many porous materials. Mathematically speaking, it yields a perturbation problem with respect to a small parameter appearing in the boundary conditions•

10.3 Balance Equation for Porosity

In this section, we aim for a derivation of those macroscopical relations which reflect the existence of semi-microscopical geometrical properties of a porous material. In contrast to a mixture of fluid components, whose microstructure lies at the molecular level of observation, the porous materials are characterized by a microstructure of almost macroscopical dimensions. This is sometimes called a meso-domain [see: D. R. AXELRAD (1993) for a discussion of all three levels of observation]. For this reason, the macroscopical continuous descriptions of these materials, the theories of immiscible mixtures, must be severely limited in their range of applicability. For instance, we cannot expect that such descriptions would properly reflect the real response of porous materials to short monochromatic wave propagations. If the wavelengths are smaller or of the order of the characteristic dimensions of the semi-microscopical channels of the pore space (high frequency waves), then their continuous description demands special caution. Effects such as critical damping due to the real geometry of a porous body would not appear in the continuous theory at all. Furthermore, even geometrical properties beyond those accounting for porosity alone, such as a geometrical anisotropy of the microstructure or a tortuosity of channels, have been investigated as yet to a rather limited extent.

The basis for a macroscopical description of the semi-microscopical geometry is the assumption that solely the volume contribution of pores to the typical microscopical domain, has a bearing on the distinction between miscible and immiscible mixtures. This is in the spirit of the confinement of the space of motion for the fluid components by the skeleton. In order to construct any macroscopical continuous model of a real porous material, we have to smear out its properties over the volumes whose dimensions are large when compared with the characteristic dimensions of this real structure, such as the dimensions of pores, and which are simultaneously small when compared with the characteristic macroscopical dimensions such as the length of macroscopical sound waves. Once we have chosen such a volume, we can construct the volume averages at every point of a real material over *domains of real observations* which possess this volume. We consider the true instantaneous configurations of a real porous material to be embedded in the three-dimensional Euclidean space. In such a case, we can use the „identical" (isomorphic) domains of real observations at each point of these true configurations which can be obtained, for instance, by shifting a standard three-dimensional domain from point to point in the configuration space. Macroscopical properties obtained by averaging over such domains are the fields on the configuration space. If we use the Lagrangian description, as we do in this chapter, it is more convenient to map this whole construction onto the reference configuration $\mathcal{B}_0$ of the skeleton by means of the deformation function $\mathbf{f}^S(\mathbf{X},t)$. At a chosen point $\mathbf{x}$ in the configuration space, in which we want to locate the domain of real observations, we project all properties within this domain by means of the mappings $\mathbf{f}^{S-1}(\mathbf{x},t)$ and $\mathbf{F}^{S-1}(\mathbf{x},t)$ (constant over the domain located at $\mathbf{x}$!). Then we observe , for instance, the difference (transformed by $\mathbf{F}^{S-1}$) of the real microscopical velocity of points of a real material and of the macroscopical velocity of the skeleton $\mathbf{x}'^S(\mathbf{X},t)$, rather than the real microscopical velocity itself. These images of the domains of real observations are called **control domains** of the microstructure connected with a chosen material point $\mathbf{X}$ of the skeleton. We denote them by $\mathcal{M}_X$.

We show schematically the image of a domain of observation $\mathcal{M}_X$ (the control domain) in Fig. 10.1. It is a part of the space limited by the boundary $\partial\mathcal{M}_X$ of a magnifying glass. In this figure, the „height" o(L) implicates that the characteristic dimension of the control domain is much smaller than the characteristic macroscopical dimension L. The position vector $\mathbf{X}$ indicates that the image $\mathcal{M}_X$ of a real domain is obtained by the mapping $\mathbf{f}^S(\mathbf{X},t)$. It is symbolically placed at the centre of the magnifying glass. Simultaneously, the position vector $\mathbf{Y}$ of a generic point of the microstructure is the image (with respect to the same mapping) of the position vector of a real porous material in its current configuration. This vector runs, certainly, through the entire body and not only through the domain of observation. Consequently, it is characterized by two different types of changes relative to the control domain. It may change „a little" going through the points within $\mathcal{M}_X$. It may change, however, „quite a bit" when shifted on distances of macroscopical dimensions. We shall use this property further in order to introduce the so-called multiscaling technique.

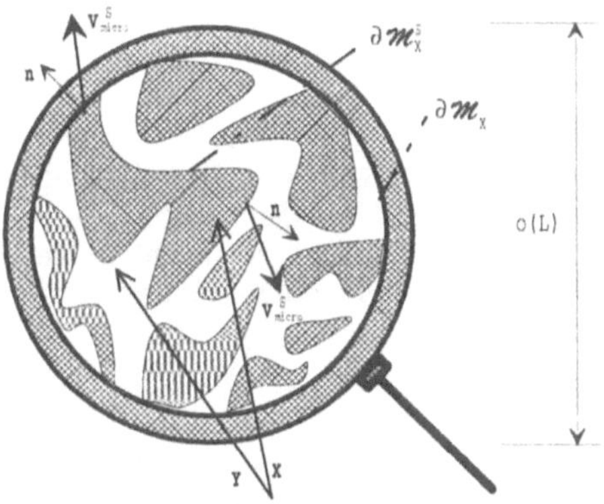

Fig. 10.1 *The geometrical notions within a microstructure* $\mathcal{M}_X$ *at the point* **X** *of a porous body.* The parts with the pattern correspond to the microscopical domain $\mathcal{M}^S_X$ of a skeleton.

We proceed to discuss the distribution of a real skeleton within the control domain. It is assumed that the geometrical properties of a real skeleton and their time changes can be described in the same way as it is done in the continuum mechanics.

We are interested in the volume fraction of a control domain which is occupied by the skeleton. This is obviously given by the following relation:

$$1 - n(\mathbf{X}, t) = \frac{1}{V_c} \int_{\mathcal{M}_X} \mathcal{H}(\mathbf{Y}, t)\, dV, \quad V_c \equiv \int_{\mathcal{M}_X} dV = \text{const.},$$
$$\mathcal{H}(\mathbf{Y}, t) = \begin{cases} 1 \text{ for } \mathbf{Y} \in \mathcal{M}^S_X, \\ 0 \text{ for } \mathbf{Y} \notin \mathcal{M}^S_X, \end{cases} \tag{10.37}$$

where $\mathcal{M}^S_X$ is the part of a control domain occupied by the skeleton. The scalar quantity n shall be, of course, identified with the porosity.

We proceed to investigate the time changes of the porosity. The most important observation is that the time changes of the image $\mathcal{M}^S_X$ are determined solely by the velocity of the points of the boundary $\partial \mathcal{M}^S_X$. The relative velocity of these points with respect to a chosen macroscopical point **X**, i.e. the difference of the microscopical velocity and the velocity $\mathbf{x}'^S(\mathbf{X}, t)$, is denoted by $\mathbf{V}^S{}_{micro}$. Let us notice that this is a true relative velocity of the **material points of a true skeleton** because we made the continuity assumption on the semimicroscopical level of observation. This means that the material points of a skeleton, which happen to be located on the boundary $\partial \mathcal{M}^S_X$ at the initial instant of time must remain boundary points forever. Consequently, if we were able to solve the full boundary value problems on the semi-microscopical level, the velocity $\mathbf{V}^S{}_{micro}$ and the changes of the image of $\mathcal{M}^S_X$ would be determined.

Bearing this in mind, and denoting by **n** the unit outward normal vector to the boundary surface $\partial \mathcal{M}^S_X$, we can write the time changes of the porosity n in the following way:

$$\frac{\partial n}{\partial t} \equiv -\frac{1}{V_c}\frac{\partial}{\partial t}\int_{\mathcal{M}_X} \mathcal{H}(\mathbf{Y},t)\, d\,V_Y = -\frac{1}{V_c}\int_{\partial \mathcal{M}_X^S} \mathbf{V}^S_{\text{micro}} \cdot \mathbf{n}\, d\,S_Y. \tag{10.38}$$

The surface $\partial \mathcal{M}^S_X$ does not contain points which coincide instantaneously with points of the boundary $\partial \mathcal{M}_X$ (the boundary of a magnifying glass in Fig. 10.1) because these do not move into the pore spaces and their neighbourhoods lie inside the skeleton. Consequently, the integral on the right-hand side of the above relation is not evaluated over a closed surface. However, it can be *approximately* written as an integral over a closed surface. Namely,

$$\int_{\partial \mathcal{M}_X^S} d\,S_Y = \oint_{A_{\text{micro}}} d\,S_Y - \int_{\mathcal{M}_X^S \cap \partial \mathcal{M}_X} d\,S_Y, \tag{10.39}$$

where

$$A_{\text{micro}} \equiv \partial \mathcal{M}_X^S \cup \left(\mathcal{M}_X^S \cap \partial \mathcal{M}_X\right). \tag{10.40}$$

The surface $\mathcal{M}_X^S \cap \partial \mathcal{M}_X$ contains, certainly, the points of a true skeleton which instantaneously coincide with the boundary of a control domain $\mathcal{M}_X$. We assume that the area of this surface, i.e. the second integral on the right-hand side of (10.39) is much smaller than the area of $\partial \mathcal{M}_X^S$. Physically, this means that the microscopical channels have a small diameter compared with the length within the control domain. There is also a positive contribution of the tortuosity to this assumption.

Simultaneously, the velocity of points on the surface $\mathcal{M}_X^S \cap \partial \mathcal{M}_X$ is of the same order of magnitude as $\mathbf{V}^s{}_{\text{micro}}$ because this is also the velocity of the material points of a true skeleton. Hence,

$$\begin{aligned}\frac{\partial n}{\partial t} &\approx -\frac{1}{V_c}\oint_{A_{\text{micro}}} \mathbf{V}^S_{\text{micro}} \cdot \mathbf{n}\, d\,S_Y = -\frac{1}{V_c}\int_{\mathcal{M}_X^S} \text{div}_Y \mathbf{V}^S_{\text{micro}}\, d\,V_Y = \\ &= -\frac{1}{V_c}\int_{\mathcal{M}_X} \mathcal{H}(\mathbf{Y},t)\, \text{div}_Y \mathbf{V}^S_{\text{micro}}\, d\,V_Y.\end{aligned} \tag{10.41}$$

The divergence in the above formula is calculated with respect to the variable $\mathbf{Y}$. We can write the right-hand side in the following form:

$$\begin{aligned}-\frac{1}{V_c}\int_{\mathcal{M}_X} \mathcal{H}(\mathbf{Y},t)\, \text{div}_Y \mathbf{V}^S_{\text{micro}}\, d\,V_Y &= -\frac{1}{V_c}\int_{\mathcal{M}_X} \text{div}_Y\left[\mathcal{H}(\mathbf{Y},t)\mathbf{V}^S_{\text{micro}}\right] d\,V_Y + \\ &+ \frac{1}{V_c}\int_{\mathcal{M}_X} \mathbf{V}^S_{\text{micro}} \cdot \text{grad}_Y \mathcal{H}(\mathbf{Y},t)\, d\,V_Y.\end{aligned} \tag{10.42}$$

The gradient in (10.42) describes, of course, the Dirac δ-distribution with the support being identical with $\partial \mathcal{M}_X^S$. It reduces to $\mathbf{n}\delta(z)$ in the local coordinates with the variable z measuring the distance from this surface. Consequently,

$$\frac{\partial n}{\partial t} \approx -\frac{1}{V_c}\int_{\mathcal{M}_X} \text{div}_Y\left[\mathcal{H}(\mathbf{Y},t)\mathbf{V}^S_{\text{micro}}\right] d\,V_Y + \frac{1}{V_c}\oint_{A_{\text{micro}}} \mathbf{V}^S_{\text{micro}} \cdot \mathbf{n}\, d\,S_Y. \tag{10.43}$$

Substitution in (10.41) yields finally

$$\frac{\partial n}{\partial t} = -\frac{1}{2}\frac{1}{V_c}\int_{\mathcal{M}_X} \mathrm{div}_Y\left[\mathcal{H}(\mathbf{Y},t)\mathbf{V}^S_{micro}\right] d\,V_Y. \tag{10.44}$$

The velocity field in the square brackets is defined on the entire control domain, in contrast to the field $\mathbf{V}^S{}_{micro}$. It depends on the location of the point $\mathbf{Y}$ within a control domain as well as on the choice of the point $\mathbf{X} \in \mathcal{B}_0$. This dependence can be used to approximate semi-microscopical properties because the microstructure introduces a small parameter $\varepsilon = d_m/L$, where d_m is the typical dimension within a control volume (e.g. the diameter of the pore channel). We proceed to describe this procedure.

As we already mentioned, it is clear that the existence of the microstructure yields two different types of spatial changes of an arbitrary microscopical function of the variable $\mathbf{Y}$. One type is connected with the changes of the variable $\mathbf{Y}$ whose order of magnitude related to the characteristic macroscopical length L is much larger than the parameter ε. These are macroscopical changes. Another type is connected with the changes of the variable $\mathbf{Y}$ whose order of magnitude estimated in the same way is of the order of the parameter ε. These are microstructural changes. Consequently, we can consider two different *scales of spatial changes*. For this reason, we introduce a new variable $\mathbf{Z} = \mathbf{Y}/\varepsilon$, whose changes of the same order of magnitude as those of $\mathbf{X}$ mean much smaller, microstructural, changes of the variable $\mathbf{Y}$. In order to distinguish between microstructural changes and macroscopical changes, we consider the variables $\mathbf{X}$ and $\mathbf{Z}$ as two **independent** variables. The function of these two variables which we have to consider is the vector function $\mathbf{V}(\mathbf{X},\mathbf{Z},t)$. This function is called the **extension** of the microscopical velocity iff

$$\mathbf{V}\left(\mathbf{X} = \mathbf{Y}, \mathbf{Z} = \mathbf{Y}/\varepsilon, t\right) = \mathcal{H}(\mathbf{Y},t)\mathbf{V}^S_{micro}(\mathbf{Y},t). \tag{10.45}$$

This type of **multiscaling** is well known within the kinetic theory of gases where it is used for different time scales. Some details of the general procedure can be found in the paper by G. SANDRI (1965)*).

The extension $\mathbf{V}$ is assumed now to be approximated by a truncated regular perturbation series

$$\mathbf{V}(\mathbf{X},\mathbf{Z},t) = \mathbf{V}_0(\mathbf{X},\mathbf{Z},t) + \varepsilon\,\mathbf{V}_1(\mathbf{X},\mathbf{Z},t), \tag{10.46}$$

which is assumed to satisfy the extension of equation (10.44). Namely, we have for $\mathbf{X} = \mathbf{Y}$ and $\mathbf{Z} = \mathbf{Y}/\varepsilon$

$$\mathrm{div}_Y \mathbf{V} = \mathrm{Div}(\mathbf{V}_0 + \varepsilon \mathbf{V}_1) + \frac{1}{\varepsilon}\mathrm{div}_Z(\mathbf{V}_0 + \varepsilon \mathbf{V}_1), \tag{10.47}$$

where div_Y is the divergence with respect to the variable $\mathbf{Y}$, Div is the divergence with respect to the variable $\mathbf{X}$, and div_Z is the divergence with respect to the variable $\mathbf{Z}$. Equation (10.44) becomes

$$\frac{\partial n}{\partial t} \approx -\frac{1}{2}\frac{1}{V_c}\mathrm{Div}\int_{\mathcal{M}_X}(\mathbf{V}_0 + \varepsilon\,\mathbf{V}_1)\,d\,V_Z - \frac{1}{\varepsilon}\frac{1}{2}\frac{1}{V_c}\int_{\mathcal{M}_X}\mathrm{div}_Z(\mathbf{V}_0 + \varepsilon\,\mathbf{V}_1)\,d\,V_Z. \tag{10.48}$$

*) The method is commonly used in the theory of vibrations and it has an extensive mathematical literature connected with such problems. It is, however, usually limited to ordinary differential equations. For instance, see: J. KEVORKIAN (1966), ALI HASAN NAYFEH (1973).

The volume integration over the microstructure $\mathcal{M}_X$ is, certainly, the integration with respect to the „fast" (*fine grained*) variable **Z** which is indicated by dV_Z.

The application of the perturbation procedure to (10.48) yields for two subsequent powers of ε

$$\begin{aligned} &\varepsilon^{-1}: \quad \int_{\mathcal{M}_X} \operatorname{div}_Z \mathbf{V}_0 \, dV_Z = 0, \\ &\varepsilon^{0}: \quad \frac{\partial n}{\partial t} = -\frac{1}{2}\frac{1}{V_c} \operatorname{Div} \int_{\mathcal{M}_X} \mathbf{V}_0 \, dV_Z - \frac{1}{2}\frac{1}{V_c} \int_{\mathcal{M}_X} \operatorname{div}_Z \mathbf{V}_1 \, dV_Z. \end{aligned} \tag{10.49}$$

Higher-order terms are not reliable any more due to the form of the truncation (10.46).

The first relation (10.49) limits the dependence of $\mathbf{V}_0$ on the microscopical variable **Z**. For all practical purposes, we may most likely claim that the part $\mathbf{V}_0$ is solely dependent on the macroscopical variable **X**. It describes the bulk real velocity of a skeleton flowing through a domain of observation. Consequently, the part $\mathbf{V}_1$ is the velocity of real motions of a skeleton on the level of microstructure, it describes the small motions within a domain of observation.

Let us introduce the notation

$$\begin{aligned} &\mathbf{J}(\mathbf{X}, t) \equiv \frac{1}{2}\frac{1}{V_c} \int_{\mathcal{M}_X} \mathbf{V}_0 \, dV_Z, \\ &\hat{n}(\mathbf{X}, t) \equiv -\frac{1}{2}\frac{1}{V_c} \int_{\mathcal{M}_X} \operatorname{div}_Z \mathbf{V}_1 \, dV_Z. \end{aligned} \tag{10.50}$$

Then the equation (10.49) has the form

$$\frac{\partial n}{\partial t} + \operatorname{Div} \mathbf{J} = \hat{n}. \tag{10.51}$$

This is the **balance equation of porosity** which supplements our set of macroscopical field equations.

The relations (10.50) yield a simple semi-microscopical interpretation of the flux **J** and of the source $\hat{n}$. As indicated above, it is obvious that the flux **J** is primarily connected with the bulk micromotion of the microstructure through a control domain $\mathcal{M}_X$, i.e. it describes the in- and out-flow of material of a true skeleton to and from a domain of observation connected with the macroscopical point $\mathbf{X} \in \mathcal{B}_0$. This is connected with the above remark on the first relation (10.49). Simultaneously, the source $\hat{n}$ is produced by the micromotion within a control domain, i.e. it is primarily connected with the *relaxation processes* within the pores. It would appear, for instance, if a heterogeneous distribution of real stresses in a skeleton within a control domain caused a redistribution of the material of a skeleton from one part of $\mathcal{M}_X$ into some other part of $\mathcal{M}_X$ in order to relax these real stresses. Such motions are not observable on the macroscopical level of a continuum model of a porous material.

Both quantities **J** and $\hat{n}$ could be calculated from the solution of the microscopical problem if this was known. This is obviously only rarely the case.

Consequently, the flux **J** and the source $\hat{n}$ must be considered to be **macroscopical constitutive quantities**. We shall investigate this problem in the following sections.

Let us finally notice that the above motivation of the balance equation of porosity (10.51) is solely based on the **image analysis**. It bears no information on the constitutive character of true components such as the distribution of true mass densities. It is not even required that the pores carry any fluid components at all. The latter is connected with the assumption that it is a solid component, the true skeleton, which geometrically determines the changes of porosity.

The dependence on material properties is, however, hidden in the macroscopical constitutive form of **J** and $\hat{n}$ and it is solely the structure of the equation (10.51) which has a geometrical motivation.

10.4 Second Law of Thermodynamics for a Thermoelastic Skeleton and Ideal Fluid Components

The balance equations (10.11), (10.12) and (10.51) supplemented with proper constitutive relations form a set of governing field equations for the fields (10.19). In order to see the most fundamental properties of a porous material, which distinguish it from a single-component continuum, we consider an example of constitutive laws appropriate to describe the so-called **elastic skeleton** and the **ideal fluid** components. This means that if the porosity was zero all the time (the case of a single-component solid), the material would be elastic and, on the other hand, if the porosity was equal to one, and we had only a single fluid component (A=1), this fluid would be an ideal fluid. This, certainly, does not mean that there is no dissipation in the system. On the contrary, there is a dissipation connected with the relative motion of components (diffusion), there is a dissipation connected with the relaxation of the porosity and there is the thermal dissipation due to the thermal conductivity.

We assume that all components have the same temperature T. We already mentioned that the problem of a physical interpretation of this notion in models of multicomponent systems is far from clear. It is even worse in the case of systems with an individual temperature of components. We may, however, expect that the processes which are sufficiently slow do not require more than one temperature field.

Consequently, we want to consider a model which describes the following fields:

$$\left\{\rho^S, \rho^1, \dots \rho^A, \mathbf{f}^S, \mathbf{X}'^1, \dots, \mathbf{X}'^A, T, n\right\}. \qquad (10.52)$$

In this case we do not need the partial energy balance equations $(10.11)_3$ and $(10.12)_3$ but solely the total energy balance. This equation has the following form

$$\begin{aligned}
&\rho^S \frac{\partial \varepsilon^S}{\partial t} + \sum_{\alpha=1}^{A} \rho^\alpha \left(\frac{\partial \varepsilon^\alpha}{\partial t} + \mathbf{X}'^\alpha \cdot \operatorname{Grad} \varepsilon^\alpha \right) + \operatorname{Div} \mathbf{Q}_1 - \mathbf{P}^S \cdot \frac{\partial \mathbf{F}^S}{\partial t} - \\
&- \sum_{\alpha=1}^{A} \mathbf{P}^\alpha \cdot \operatorname{Grad} \mathbf{x}'^\alpha + \hat{\mathbf{p}}^S \cdot \mathbf{x}'^S + \sum_{\alpha=1}^{A} \hat{\mathbf{p}}^\alpha \cdot \mathbf{x}'^\alpha + \\
&+ \hat{\rho}^S \left(\varepsilon^S - \frac{1}{2} x'^{S2} \right) + \sum_{\alpha=1}^{A} \hat{\rho}^\alpha \left(\varepsilon^\alpha - \frac{1}{2} x'^{\alpha 2} \right) = 0,
\end{aligned} \qquad (10.53)$$

where the partial sources of the energy $\hat{\varepsilon}^S$, $\hat{\varepsilon}^\alpha$ and the energy radiation terms r^S, r^α have been ignored. The first simplification follows from the physical argument that the local exchange of energy between components, described by the energy sources, is connected with the local thermal relaxation due to the different partial temperatures of components. These processes do not appear in the case of a common temperature T. The second simplification is solely of a technical nature. The energy radiation terms do not influence the exploitation of the second law, and this is the main purpose of the remaining part of this chapter.

The inspection of the balance equations shows that in order to close the system we have to formulate constitutive relations for the following constitutive quantities:

$$\mathcal{Z}=\left\{\hat{\rho}^{S},\hat{\rho}^{\alpha},\mathbf{P}^{S},\mathbf{P}^{\alpha},\hat{\mathbf{p}}^{S}-\hat{\rho}^{S}\mathbf{x}'^{S},\hat{\mathbf{p}}^{\alpha}-\hat{\rho}^{\alpha}\mathbf{x}'^{\alpha},\mathbf{J},\hat{\mathrm{n}},\varepsilon^{S},\varepsilon^{\alpha},\mathbf{Q}_{\mathrm{I}}\right\} \tag{10.54}$$

where the momentum sources have been corrected for the reasons of Galilean invariance.

The above constitutive quantities are assumed to be functions of the following constitutive variables:

$$\mathcal{C}=\left\{\rho^{S},\rho^{\alpha},\mathrm{n},\mathbf{F}^{S},\mathbf{X}'^{\alpha},\mathrm{T},\mathbf{G}\right\},\qquad \mathbf{G}=\operatorname{Grad}\mathrm{T}. \tag{10.55}$$

In the above relations we used the short-hand notation replacing the sequences of quantities for all fluid components by a single symbol with the index α.

Let us notice that the presence of the deformation gradient $\mathbf{F}^{S}$ and the temperature gradient GradT indicates that we shall not use extended thermodynamics here. The work in this direction is in progress, but neither the structure of additional balance equations nor the consequences of the second law are available at present. Consequently, we investigate the thermodynamical admissibility of the classical non-equilibrium model which contains the derivatives of fields as constitutive variables.

The closure of the system can now be written in the following symbolic form:

$$\mathcal{Z}=\mathcal{Z}(\mathcal{C}), \tag{10.56}$$

and all these functions are assumed to be sufficiently smooth for further consideration. In most cases they must be twice continuously differentiable.

The set of constitutive variables (10.55) must be further restricted by the requirement of objectivity. We shall do this in the following.

We proceed to discuss the thermodynamical admissibility conditions of the constitutive relations (10.56). The construction of these conditions is based on the entropy balance equations for components which are assumed to have the following Lagrangian form:

$$\frac{\partial\rho^{S}\eta^{S}}{\partial t}+\operatorname{Div}\mathbf{H}^{S}=\hat{\eta}^{S},\qquad \frac{\partial\rho^{\alpha}\eta^{\alpha}}{\partial t}+\operatorname{Div}\left(\rho^{\alpha}\eta^{\alpha}\mathbf{X}'^{\alpha}+\mathbf{H}^{\alpha}\right)=\hat{\eta}^{\alpha}, \tag{10.57}$$

where $\rho^{S}\eta^{S}$ and $\rho^{\alpha}\eta^{\alpha}$ are the partial entropy densities per unit reference volume of the skeleton, $\mathbf{H}^{S}$ and $\mathbf{H}^{\alpha}$ are the partial entropy fluxes relative to the reference configuration $\mathcal{B}_0$ and $\hat{\eta}^{S}$, $\hat{\eta}^{\alpha}$ denote the intensities of the entropy sources per unit reference volume of the skeleton.

The second law of thermodynamics requires that any solution of the field equations yields a positive entropy production, i.e.

$$\forall(\text{solutions of field equations}):\quad \hat{\eta}^{S}+\sum_{\alpha=1}^{A}\hat{\eta}^{\alpha}\geq 0. \tag{10.58}$$

The source terms are determined by the balance equations (10.57) in which the entropy densities and the entropy fluxes are assumed to be constitutively determined

$$\eta^{S}=\eta^{S}(\mathcal{C}),\quad \eta^{\alpha}=\eta^{\alpha}(\mathcal{C}),\quad \mathbf{H}^{S}=\mathbf{H}^{S}(\mathcal{C}),\quad \mathbf{H}^{\alpha}=\mathbf{H}^{\alpha}(\mathcal{C}). \tag{10.59}$$

As usual, the above requirement can be transformed in such a way that we eliminate the limitation of the inequality (10.58) to hold only for the solutions of field equations. We use the method of the Lagrange multipliers. Then we obtain

$$
\begin{aligned}
&\rho^S \frac{\partial \eta^S}{\partial t} + \sum_{\alpha=1}^{A} \rho^\alpha \left[\frac{\partial \eta^\alpha}{\partial t} + \mathbf{X}'^\alpha \cdot \operatorname{Grad} \eta^\alpha \right] + \operatorname{Div} \mathbf{H}_I + \hat{\rho}^S \eta^S + \sum_{\alpha=1}^{A} \hat{\rho}^\alpha \eta^\alpha - \\
&- \lambda^S \left(\frac{\partial \rho^S}{\partial t} - \hat{\rho}^S \right) - \sum_{\alpha=1}^{A} \lambda^\alpha \left[\frac{\partial \rho^\alpha}{\partial t} + \operatorname{Div}\left(\rho^\alpha \mathbf{X}'^\alpha \right) - \hat{\rho}^\alpha \right] - \\
&- \lambda^n \left(\frac{\partial n}{\partial t} + \operatorname{Div} \mathbf{J} - \hat{n} \right) - \Lambda^S \cdot \left[\rho^S \frac{\partial \mathbf{x}'^S}{\partial t} - \operatorname{Div} \mathbf{P}^S - \hat{\mathbf{p}}^S + \hat{\rho}^S \mathbf{x}'^S \right] - \\
&- \sum_{\alpha=1}^{A} \Lambda^\alpha \cdot \left[\rho^\alpha \left(\frac{\partial \mathbf{x}'^\alpha}{\partial t} + \operatorname{Grad} \mathbf{x}'^\alpha \, \mathbf{X}'^\alpha \right) - \operatorname{Div} \mathbf{P}^\alpha - \hat{\mathbf{p}}^\alpha + \hat{\rho}^\alpha \mathbf{x}'^\alpha \right] - \\
&- \lambda^\varepsilon \left\{ \rho^S \frac{\partial \varepsilon^S}{\partial t} + \sum_{\alpha=1}^{A} \rho^\alpha \left(\frac{\partial \varepsilon^\alpha}{\partial t} + \mathbf{X}'^\alpha \cdot \operatorname{Grad} \varepsilon^\alpha \right) + \operatorname{Div} \mathbf{Q}_I - \mathbf{P}^S \cdot \frac{\partial \mathbf{F}^S}{\partial t} - \right. \\
&- \sum_{\alpha=1}^{A} \mathbf{P}^\alpha \cdot \operatorname{Grad} \mathbf{x}'^\alpha + \hat{\mathbf{p}}^S \cdot \mathbf{x}'^S + \sum_{\alpha=1}^{A} \hat{\mathbf{p}}^\alpha \cdot \mathbf{x}'^\alpha + \\
&\left. + \hat{\rho}^S \left(\varepsilon^S - \frac{1}{2} x'^{S2} \right) + \sum_{\alpha=1}^{A} \hat{\rho}^\alpha \left(\varepsilon^\alpha - \frac{1}{2} x'^{\alpha 2} \right) \right\} \geq 0,
\end{aligned}
\tag{10.60}
$$

where

$$
\mathbf{H}_I = \mathbf{H}^S + \sum_{\alpha=1}^{A} \mathbf{H}^\alpha, \tag{10.61}
$$

denotes the intrinsic entropy flux and the Lagrange multipliers are constitutive functions

$$
\begin{aligned}
&\lambda^S = \lambda^S(\mathcal{C}),\ \lambda^\alpha = \lambda^\alpha(\mathcal{C}),\ \Lambda^S = \Lambda^S(\mathcal{C}),\ \Lambda^\alpha = \Lambda^\alpha(\mathcal{C}), \\
&\lambda^n = \lambda^n(\mathcal{C}),\ \lambda^\varepsilon = \lambda^\varepsilon(\mathcal{C}).
\end{aligned}
\tag{10.62}
$$

Inequality (10.60) must hold for **arbitrary fields** and not only for the solutions of field equations. If we use the chain rule of differentiation, then it can be seen easily that the inequality is linear with respect to the following quantities:

- time derivatives

$$
\left\{ \frac{\partial \rho^S}{\partial t}, \frac{\partial \rho^\alpha}{\partial t}, \frac{\partial n}{\partial t}, \frac{\partial T}{\partial t}, \frac{\partial \mathbf{G}}{\partial t}, \frac{\partial \mathbf{x}'^S}{\partial t}, \frac{\partial \mathbf{X}'^\alpha}{\partial t}, \frac{\partial \mathbf{F}^S}{\partial t} \right\}, \tag{10.63}
$$

- spatial derivatives

$$
\left\{ \operatorname{Grad} \rho^S, \operatorname{Grad} \rho^\alpha, \operatorname{Grad} n, \operatorname{Grad} \mathbf{G}, \operatorname{Grad} \mathbf{X}'^\alpha, \operatorname{Grad} \mathbf{F}^S \right\}. \tag{10.64}
$$

This means, according to Liu's Theorem, that the coefficients of these derivatives must vanish. We obtain the thermodynamical identities defining the Lagrange multipliers in terms of other constitutive functions, and some other relations restricting the constitutive relations, i.e. the **thermodynamical admissibility relations**.

For the Lagrange multipliers we obtain

$$\lambda^S = \rho^S\left(\frac{\partial\eta^S}{\partial\rho^S} - \lambda^\varepsilon\frac{\partial\varepsilon^S}{\partial\rho^S}\right) + \sum_{\beta=1}^{A}\rho^\beta\left(\frac{\partial\eta^\beta}{\partial\rho^S} - \lambda^\varepsilon\frac{\partial\varepsilon^\beta}{\partial\rho^S}\right),$$
$$\lambda^\alpha = \rho^S\left(\frac{\partial\eta^S}{\partial\rho^\alpha} - \lambda^\varepsilon\frac{\partial\varepsilon^S}{\partial\rho^\alpha}\right) + \sum_{\beta=1}^{A}\rho^\beta\left(\frac{\partial\eta^\beta}{\partial\rho^\alpha} - \lambda^\varepsilon\frac{\partial\varepsilon^\beta}{\partial\rho^\alpha}\right), \tag{10.65}$$
$$\lambda^n = \rho^S\left(\frac{\partial\eta^S}{\partial n} - \lambda^\varepsilon\frac{\partial\varepsilon^S}{\partial n}\right) + \sum_{\beta=1}^{A}\rho^\beta\left(\frac{\partial\eta^\beta}{\partial n} - \lambda^\varepsilon\frac{\partial\varepsilon^\beta}{\partial n}\right),$$

$$\rho^S\Lambda^S = -\sum_{\alpha=1}^{A}\rho^\alpha\Lambda^\alpha,$$
$$\rho^\alpha\Lambda^\alpha = \mathbf{F}^{S-T}\left\{\rho^S\left(\frac{\partial\eta^S}{\partial\mathbf{X}'^\alpha} - \lambda^\varepsilon\frac{\partial\varepsilon^S}{\partial\mathbf{X}'^\alpha}\right) + \sum_{\beta=1}^{A}\rho^\beta\left(\frac{\partial\eta^\beta}{\partial\mathbf{X}'^\alpha} - \lambda^\varepsilon\frac{\partial\varepsilon^\beta}{\partial\mathbf{X}'^\alpha}\right)\right\}. \tag{10.66}$$

These relations determine constitutive relations for the multipliers provided the constitutive relations for partial internal energies and partial entropies are known. In addition we need a relation for the multiplier λ^ε. This is also determined and we discuss the problem in the following.

Apart from the above relations, we obtain from the inequality (10.60) the following relations for the partial Piola-Kirchhoff stresses:

$$\lambda^\varepsilon\mathbf{P}^S = -\rho^S\left(\frac{\partial\eta^S}{\partial\mathbf{F}^S} - \lambda^\varepsilon\frac{\partial\varepsilon^S}{\partial\mathbf{F}^S}\right) - \sum_{\beta=1}^{A}\rho^\beta\left(\frac{\partial\eta^\beta}{\partial\mathbf{F}^S} - \lambda^\varepsilon\frac{\partial\varepsilon^\beta}{\partial\mathbf{F}^S}\right) +$$
$$+2\sum_{\beta=1}^{A}\rho^\beta\Lambda^\beta\otimes\mathbf{X}'^\beta - \lambda^\varepsilon\sum_{\beta=1}^{A}\mathbf{P}^\beta,$$
$$\lambda^\varepsilon\mathbf{F}^{ST}\mathbf{P}^\alpha = \rho^\alpha\lambda^\alpha\mathbf{1} + \rho^\alpha\Lambda^\alpha\otimes\mathbf{X}'^\alpha - \sum_{\beta=1}^{A}\rho^\beta\left(\frac{\partial\eta^\beta}{\partial\mathbf{X}'^\alpha} - \lambda^\varepsilon\frac{\partial\varepsilon^\beta}{\partial\mathbf{X}'^\alpha}\right)\otimes\mathbf{X}'^\alpha - \tag{10.67}$$
$$-\left(\frac{\partial\mathbf{H}_I}{\partial\mathbf{X}'^\alpha} - \lambda^\varepsilon\frac{\partial\mathbf{Q}_I}{\partial\mathbf{X}'^\alpha} - \lambda^n\frac{\partial\mathbf{J}}{\partial\mathbf{X}'^\alpha}\right)^T -$$
$$-\left[\left(\frac{\partial\mathbf{P}^S}{\partial\mathbf{X}'^\alpha}\right)^{T^{13}}\Lambda^S + \sum_{\beta=1}^{A}\left(\frac{\partial\mathbf{P}^\beta}{\partial\mathbf{X}'^\alpha}\right)^{T^{13}}\Lambda^\beta\right]^T.$$

Inspection of these relations shows that these are not usual constitutive relations because they contain the derivatives of stresses with respect to the Lagrangian velocities. This is connected with the non-linearity of the model relative to the diffusion. This problem is immaterial in the special case considered in the next section.

The dependence on the temperature and its gradient is restricted by the following identities:

$$\rho^S\left(\frac{\partial\eta^S}{\partial T} - \lambda^\varepsilon\frac{\partial\varepsilon^S}{\partial T}\right) + \sum_{\beta=1}^{A}\rho^\beta\left(\frac{\partial\eta^\beta}{\partial T} - \lambda^\varepsilon\frac{\partial\varepsilon^\beta}{\partial T}\right) = 0,$$
$$\rho^S\left(\frac{\partial\eta^S}{\partial\mathbf{G}} - \lambda^\varepsilon\frac{\partial\varepsilon^S}{\partial\mathbf{G}}\right) + \sum_{\beta=1}^{A}\rho^\beta\left(\frac{\partial\eta^\beta}{\partial\mathbf{G}} - \lambda^\varepsilon\frac{\partial\varepsilon^\beta}{\partial\mathbf{G}}\right) = 0, \tag{10.68}$$

$$\operatorname{sym}\left\{\sum_{\beta=1}^{A}\rho^{\beta}\left(\frac{\partial\eta^{\beta}}{\partial\mathbf{G}}-\lambda^{\varepsilon}\frac{\partial\varepsilon^{\beta}}{\partial\mathbf{G}}\right)\otimes\mathbf{X}'^{\beta}+\frac{\partial\mathbf{H}_{\mathrm{I}}}{\partial\mathbf{G}}-\lambda^{\varepsilon}\frac{\partial\mathbf{Q}_{\mathrm{I}}}{\partial\mathbf{G}}-\lambda^{n}\frac{\partial\mathbf{J}}{\partial\mathbf{G}}+\right.$$
$$\left.+\left(\frac{\partial\mathbf{P}^{S}}{\partial\mathbf{G}}\right)^{T^{13}}\Lambda^{S}+\sum_{\beta=1}^{A}\left(\frac{\partial\mathbf{P}^{\beta}}{\partial\mathbf{G}}\right)^{T^{13}}\Lambda^{\beta}\right\}=0. \tag{10.69}$$

In the classical thermodynamics of a one-component material these relations yield the multiplier Λ^{ε} to be the coldness (the inverse of the absolute temperature T), and the entropy flux to be proportional to the heat flux with the coldness as the coefficient of proportionality. This is not the case for the model considered in this chapter. We investigate these identities from this point of view again under the simplifying assumptions of the next section.

Apart from the above relations, we also obtain the following identities restricting the constitutive relations:

$$\sum_{\beta=1}^{A}\rho^{\beta}\left(\frac{\partial\eta^{\beta}}{\partial\rho^{S}}-\lambda^{\varepsilon}\frac{\partial\varepsilon^{\beta}}{\partial\rho^{S}}\right)\mathbf{X}'^{\beta}+\frac{\partial\mathbf{H}_{\mathrm{I}}}{\partial\rho^{S}}-\lambda^{\varepsilon}\frac{\partial\mathbf{Q}_{\mathrm{I}}}{\partial\rho^{S}}-\lambda^{n}\frac{\partial\mathbf{J}}{\partial\rho^{S}}+$$
$$+\left(\frac{\partial\mathbf{P}^{S}}{\partial\rho^{S}}\right)^{T}\Lambda^{S}+\sum_{\beta=1}^{A}\left(\frac{\partial\mathbf{P}^{\beta}}{\partial\rho^{S}}\right)^{T}\Lambda^{\beta}=0,$$
$$\sum_{\beta=1}^{A}\rho^{\beta}\left(\frac{\partial\eta^{\beta}}{\partial\rho^{\alpha}}-\lambda^{\varepsilon}\frac{\partial\varepsilon^{\beta}}{\partial\rho^{\alpha}}\right)\mathbf{X}'^{\beta}+\frac{\partial\mathbf{H}_{\mathrm{I}}}{\partial\rho^{\alpha}}-\lambda^{\varepsilon}\frac{\partial\mathbf{Q}_{\mathrm{I}}}{\partial\rho^{\alpha}}-\lambda^{n}\frac{\partial\mathbf{J}}{\partial\rho^{\alpha}}+$$
$$+\left(\frac{\partial\mathbf{P}^{S}}{\partial\rho^{\alpha}}\right)^{T}\Lambda^{S}+\sum_{\beta=1}^{A}\left(\frac{\partial\mathbf{P}^{\beta}}{\partial\rho^{\alpha}}\right)^{T}\Lambda^{\beta}-\lambda^{\alpha}\mathbf{X}'^{\alpha}=0, \tag{10.70}$$

$$\sum_{\beta=1}^{A}\rho^{\beta}\left(\frac{\partial\eta^{\beta}}{\partial n}-\lambda^{\varepsilon}\frac{\partial\varepsilon^{\beta}}{\partial n}\right)\mathbf{X}'^{\beta}+\frac{\partial\mathbf{H}_{\mathrm{I}}}{\partial n}-\lambda^{\varepsilon}\frac{\partial\mathbf{Q}_{\mathrm{I}}}{\partial n}-\lambda^{n}\frac{\partial\mathbf{J}}{\partial n}+$$
$$+\left(\frac{\partial\mathbf{P}^{S}}{\partial n}\right)^{T}\Lambda^{S}+\sum_{\beta=1}^{A}\left(\frac{\partial\mathbf{P}^{\beta}}{\partial n}\right)^{T}\Lambda^{\beta}=0,$$

and

$$\operatorname{sym}^{23}\left\{\sum_{\beta=1}^{A}\rho^{\beta}\left(\frac{\partial\eta^{\beta}}{\partial\mathbf{F}^{S}}-\lambda^{\varepsilon}\frac{\partial\varepsilon^{\beta}}{\partial\mathbf{F}^{S}}\right)\otimes\mathbf{X}'^{\beta}+\left(\frac{\partial\mathbf{H}_{\mathrm{I}}}{\partial\mathbf{F}^{S}}-\lambda^{\varepsilon}\frac{\partial\mathbf{Q}_{\mathrm{I}}}{\partial\mathbf{F}^{S}}-\lambda^{n}\frac{\partial\mathbf{J}}{\partial\mathbf{F}^{S}}\right)^{T^{12}}+\right.$$
$$+\lambda^{\varepsilon}\sum_{\beta=1}^{A}\mathbf{P}^{\beta}\otimes\mathbf{X}'^{\beta}+\left[\left(\frac{\partial\mathbf{P}^{S}}{\partial\mathbf{F}^{S}}\right)^{T^{14}}\Lambda^{S}+\sum_{\beta=1}^{A}\left(\frac{\partial\mathbf{P}^{\beta}}{\partial\mathbf{F}^{S}}\right)^{T^{14}}\Lambda^{\beta}\right]^{T^{12}}- \tag{10.71}$$
$$\left.-\sum_{\beta=1}^{A}\rho^{\beta}\Lambda^{\beta}\otimes\mathbf{X}'^{\beta}\otimes\mathbf{X}'^{\beta}\right\}=0.$$

We shall discuss these relations in the following.

There remains the residual inequality which describes the **dissipation density $\mathcal{D}$**

$$\mathcal{D} \equiv \Lambda^\theta \cdot \mathbf{G} + \lambda^n \hat{n} + \sum_{\beta=1}^{A} \Lambda^{p\beta} \cdot \left(\hat{\mathbf{p}}^\beta - \hat{\rho}^\beta \mathbf{x}'^\beta\right) + \lambda^\varepsilon \sum_{\beta=1}^{A} \mu^\beta \hat{\rho}^\beta \geq 0, \tag{10.72}$$

where

$$\begin{aligned}
\Lambda^\theta &= \frac{\partial \mathbf{H}_I}{\partial T} - \lambda^\varepsilon \frac{\partial \mathbf{Q}_I}{\partial T} - \lambda^n \frac{\partial \mathbf{J}}{\partial T} + \sum_{\beta=1}^{A} \rho^\beta \left(\frac{\partial \eta^\beta}{\partial T} - \lambda^\varepsilon \frac{\partial \varepsilon^\beta}{\partial T}\right) \mathbf{X}'^\beta + \\
&+ \left(\frac{\partial \mathbf{P}^S}{\partial T}\right) \Lambda^S + \sum_{\beta=1}^{A} \left(\frac{\partial \mathbf{P}^\beta}{\partial T}\right) \Lambda^\beta, \\
\Lambda^{p\beta} &= -\lambda^\varepsilon \mathbf{F}^S \mathbf{X}'^\beta + \Lambda^\beta - \Lambda^S, \qquad (10.73) \\
\lambda^\varepsilon \mu^\beta &= \left[\left(\eta^\beta - \lambda^\varepsilon \varepsilon^\beta + \lambda^\beta\right) - \left(\eta^S - \lambda^\varepsilon \varepsilon^S + \lambda^S\right)\right] - \Lambda^S \cdot \mathbf{F}^S \mathbf{X}'^\beta - \\
&- \frac{1}{2} \lambda^\varepsilon \mathbf{C}^S \cdot \mathbf{X}'^\beta \otimes \mathbf{X}'^\beta, \qquad \mathbf{C}^S \equiv \mathbf{F}^{ST} \mathbf{F}^S.
\end{aligned}$$

As expected from the structure of the balance equations, we have four mechanisms of dissipation:

- thermal conduction connected with the temperature gradient **G**,
- change of porosity due to the sources $\hat{n}$,
- exchange of momentum between components described by $\hat{\mathbf{p}}^\alpha - \hat{\rho}^\alpha \mathbf{x}'^\alpha, \alpha = 1, \ldots, A$,
- exchange of mass between components described by $\hat{\rho}^\alpha, \alpha = 1, \ldots, A$.

It should be mentioned that the mass sources $\hat{\rho}^\alpha, \alpha = 1, \ldots, A$ do not have to yield a dissipation. To see this, let us consider the simple case of phase transformation in which the component α=1 transforms into the component α=2 and all other sources are identically zero. Then the contribution to the dissipation would consist of two terms corresponding to these components (phases). The relation

$$\hat{\rho}\left[\left(\eta^1 - \lambda^\varepsilon \varepsilon^1 + \lambda^1\right) - \left(\eta^2 - \lambda^\varepsilon \varepsilon^2 + \lambda^2\right)\right] = 0, \qquad \hat{\rho} \equiv \hat{\rho}^1 = -\hat{\rho}^2, \tag{10.74}$$

whose left-hand side corresponds to the last term in the dissipation (10.72) holds not only in the case of the vanishing mass source but also in the case of equal values of the functions in the parentheses. According to relation $(10.65)_2$, the multipliers λ^α are the prototypes of the partial pressure divided by the mass density and by the temperature. Hence, the condition of equal values of the above function would mean the continuity of the chemical potential (Gibbs free energy) on the interface between the two phases. This is the classical Maxwell condition for a phase equilibrium line. Consequently, the mass sources may not yield a dissipation. However, there are also physical situations in which the dissipation of mass sources is not equal to zero and the phase transformations in which the hysteresis effects appear to constitute an example of such a dissipative mass source.

Further on in this chapter, we consider solely such processes in which the mass sources do not contribute to the dissipation $\mathcal{D}$

$$\sum_{\beta=1}^{A} \mu^\beta \hat{\rho}^\beta = 0. \tag{10.75}$$

The case of the vanishing dissipation defines the so-called **state of the thermodynamical equilibrium**. According to our assumption concerning mass sources and the structure of the dissipation given by the relation (10.72), such a state appears if

$$\mathbf{G}\big|_E = 0, \quad \mathbf{X}'^{\alpha}\big|_E = 0, \quad n - n\big|_E = 0, \tag{10.76}$$

where $n|_E$ denotes the constant equilibrium value of the porosity. The last relation yields a vanishing source of porosity $\hat{n}$ as it follows from the analysis of the balance equation of porosity and from the assumption on the **isotropy** of the system. The equilibrium value of porosity $n|_E$ is a part of the constitutive problem for the source of porosity. It corresponds to equilibrium processes in which there is no relative motion of components and no temperature gradient. Simultaneously, the assumption of isotropy means that the vector fluxes $\mathbf{Q}_I$, $\mathbf{H}_I$, $\mathbf{J}$ as well as Λ^S and Λ^α must be homogeneous functions of vector variables $\mathbf{G}$ and $\mathbf{X}'^{\alpha}$, i.e.

$$\begin{gathered}\mathbf{Q}_I\big|_E = 0, \quad \mathbf{H}_I\big|_E = 0, \quad \mathbf{J}\big|_E = 0, \quad \Lambda^S\big|_E = 0, \quad \Lambda^\alpha\big|_E = 0,\\ \hat{n}\left(\rho^S, \rho^\alpha, n, \mathbf{F}^S, \mathbf{X}'^{\alpha}, T, \mathbf{G}\right) = 0 \quad \Rightarrow \quad n\big|_E = n\big|_E\left(\rho^S, \rho^\alpha, \mathbf{F}^S, T\right).\end{gathered} \tag{10.77}$$

The last relation can be established easily in static experiments. For instance, in the simplest case we could expect the universal relation $n|_E = n_0 J^S$ for contraction ($J^S<1$) and $n|_E = 1-(1-n_0)J^{S-1}$ for swelling ($J^S>1$) where n_0 is the constant value of porosity in the natural undeformed state ($J^S=1$). In such a case the **equilibrium changes** of the porosity would be controlled solely by the volume changes of a skeleton.

Here we rest the matter of a fully non-linear model and proceed to investigate the processes which do not deviate the system too much from the equilibrium state.

10.5 Small Deviations from the Thermodynamical Equilibrium

The complexity of the fully non-linear model presented in the previous sections hides many physical properties connected primarily with the interactions of components which in turn play an important role in the experimental verifications of the model. In order to see the structure of these interactions better, we now consider a class of processes which yield **small deviations** of the system from the thermodynamical equilibrium (10.76). Appropriate approximations of constitutive functions can be derived only in combination with stability requirements for the thermodynamical equilibrium. This follows from the fact that the dissipation $\mathcal{D}$ reaches its minimum value (equal to zero) in the state of the thermodynamical equilibrium. A straightforward analysis shows that the model must be linear with respect to the temperature gradient $\mathbf{G}$ and the relative velocities $\mathbf{X}'^{\alpha}$, $\alpha=1,\dots,A$, the partial internal energies and entropies must be quadratic, even functions with respect to deviations of porosity $\Delta=(n - n|_E)$ from the uniform equilibrium value $n|_E$, and that the source of porosity $\hat{n}$ must be linear with respect to Δ.

In addition we assume the **isotropy** of the components.

These assumptions yield, among other simplifications of the constitutive relations, the independence of all scalar constitutive functions from the vector constitutive variables.

We make also a simplifying assumption on the interactions described by the vector fluxes of the model. Namely, we assume the intrinsic heat flux and the intrinsic entropy flux to be independent of the relative velocities and the flux of porosity to be independent of the temperature gradient. Consequently,

$$\mathbf{Q}_I = \mathbf{Q}\mathbf{G}, \quad \mathbf{H}_I = \mathbf{H}\mathbf{G}, \quad \mathbf{J} = \sum_{\alpha=1}^{A} \Phi^\alpha \mathbf{X}'^\alpha,$$

$$\mathrm{Q} = \mathrm{Q}\left(\rho^S, \rho^\beta, \mathbf{F}^S, T; n|_E\right), \quad \mathrm{H} = \mathrm{H}\left(\rho^S, \rho^\beta, \mathbf{F}^S, T; n|_E\right), \tag{10.78}$$

$$\Phi^\alpha = \Phi^\alpha\left(\rho^S, \rho^\beta, \Delta, \mathbf{F}^S, T; n|_E\right), \qquad \Delta \equiv n - n|_E.$$

The independence of Q and H from Δ follows from $(10.70)_3$. We have not accounted for the objectivity relative to the deformation gradient for technical reasons. This will be corrected in the later stages of the analysis. The dependence on the uniform equilibrium porosity $n|_E$ is parametric.

It should be stressed that the lack of the direct thermodiffusional coupling between partial fluxes does not eliminate entirely such effects from the model. For instance, the total heat flux **Q** still contains the explicit dependence on the relative velocities. In the linear case considered in this section we have

$$\mathbf{Q} = \mathbf{Q}_I - \mathbf{P}^{ST}\left(\mathbf{x}'^S - \dot{\mathbf{x}}\right) - \sum_{\alpha=1}^{A} \mathbf{P}^{\alpha T}\left(\mathbf{x}'^\alpha - \dot{\mathbf{x}}\right) - \rho^S \varepsilon^S \dot{\mathbf{X}} + \sum_{\alpha=1}^{A} \rho^\alpha \varepsilon^\alpha \left(\mathbf{X}'^\alpha - \dot{\mathbf{X}}\right),$$
$$\dot{\mathbf{X}} \equiv \sum_{\alpha=1}^{A} \frac{\rho^\alpha}{\rho_0} \mathbf{X}'^\alpha, \quad \rho_0 \equiv \rho^S + \sum_{\alpha=1}^{A} \rho^\alpha. \tag{10.79}$$

For the linear two-component model considered in the PHILIP-DE VRIES *moisture transport theory* [see: D. A. DE VRIES (1987)], the above relation yields in the spatial representation of the heat flux vector

$$\mathbf{q} = \mathbf{q}_I + \frac{1}{\rho}\left(\rho_t^F \mathbf{T}^S - \rho_t^S \mathbf{T}^F\right)\mathbf{v}^F - \frac{\rho_t^S \rho_t^F}{\rho}\left(\varepsilon^S - \varepsilon^F\right)\mathbf{v}^F \approx$$
$$\approx -K \operatorname{grad} T + \rho_t^F c_V^F T \mathbf{v}^F, \tag{10.80}$$

where the influence of stresses has been totally ignored and it has been assumed that the specific heat of the fluid component c_V^F is much larger than the specific heat c_V^S of the solid. The absence of stresses has not been justified in these models at all and the last assumption is also not substantiated through measurements. For instance, we have for sandstones $c_V^S \approx 0.71 \frac{\mathrm{kJ}}{\mathrm{kg\,K}}$ and for the water (20°) $c_V^F \approx 4.182 \frac{\mathrm{kJ}}{\mathrm{kg\,K}}$. In spite of these flaws, the PHILIP-DE VRIES model works quantitatively very well in many cases of practical importance [e.g. see: D. A. DE VRIES (1987), Q. JIANG, R. K. N. D. RAJAPAKSE (1994)].

Bearing the above assumptions in mind, we obtain immediately from (10.66)

$$\Lambda^S = 0, \quad \Lambda^\alpha = 0. \tag{10.81}$$

On the other hand, the relations (10.69) and (10.78) yield

$$\mathbf{H}_I = \lambda^\varepsilon \mathbf{Q}_I. \tag{10.82}$$

Due to the linearity of the thermodynamic identities (10.70) and (10.71) with respect to the variables **G** and $\mathbf{X}'^\alpha$, the above result now leads easily to the conclusion that the multiplier λ^ε is solely a function of the absolute temperature. A standard argument based on the comparison with the classical Gibbs relation of thermostatics then yields

$$\lambda^{\varepsilon} = \frac{1}{T}, \tag{10.83}$$

i.e. this multiplier becomes identical with the coldness.

This important conclusion enables us to construct the thermodynamical potentials which, for the choice of variables in this chapter, are identical with the **partial Helmholtz free energy functions**

$$\begin{aligned} &\psi^S = \varepsilon^S - T\eta^S, \qquad \psi^\alpha = \varepsilon^\alpha - T\eta^\alpha, \\ &\psi^S = \psi_0^S\left(\rho^S, \rho^\beta, \mathbf{F}^S, T; n|_E\right) + \frac{1}{2}\psi_1^S\left(\rho^S, \rho^\beta, \mathbf{F}^S, T; n|_E\right)\Delta^2, \\ &\psi^\alpha = \psi_0^S\left(\rho^S, \rho^\beta, \mathbf{F}^S, T; n|_E\right) + \frac{1}{2}\psi_1^\alpha\left(\rho^S, \rho^\beta, \mathbf{F}^S, T; n|_E\right)\Delta^2, \end{aligned} \tag{10.84}$$

where we incorporated the assumption of a small deviation from the thermodynamical equilibrium related to the change of porosity Δ.

Bearing these definitions in mind, we can write the multipliers (10.65) in the following form:

$$\begin{aligned} &\lambda^S = -\frac{1}{T}\left\{\rho^S \frac{\partial \psi^S}{\partial \rho^S} + \sum_{\beta=1}^{A} \rho^\beta \frac{\partial \psi^\beta}{\partial \rho^S}\right\}, \quad \lambda^\alpha = -\frac{1}{T}\left\{\rho^S \frac{\partial \psi^S}{\partial \rho^\alpha} + \sum_{\beta=1}^{A} \rho^\beta \frac{\partial \psi^\beta}{\partial \rho^\alpha}\right\}, \\ &\lambda^n = -\frac{1}{T}\left\{\rho^S \frac{\partial \psi^S}{\partial \Delta} + \sum_{\beta=1}^{A} \rho^\beta \frac{\partial \psi^\beta}{\partial \Delta}\right\} = -\frac{1}{T}\left\{\rho^S \psi_1^S + \sum_{\beta=1}^{A} \rho^\beta \psi_1^\beta\right\}\Delta. \end{aligned} \tag{10.85}$$

The identities (10.68) now reduce to the following conditions:

$$\begin{aligned} &\rho\eta \equiv \rho^S\eta^S + \sum_{\beta=1}^{A} \rho^\beta \eta^\beta = -\rho^S \frac{\partial \psi^S}{\partial T} - \sum_{\beta=1}^{A} \rho^\beta \frac{\partial \psi^\beta}{\partial T} = -\rho \frac{\partial \psi}{\partial T}, \\ &\rho\psi \equiv \rho^S \psi^S + \sum_{\beta=1}^{A} \rho^\beta \psi^\beta. \end{aligned} \tag{10.86}$$

This is the classical result for the bulk entropy density η and the bulk Helmholtz free energy ψ.

The relations for stresses (10.67) have the form

$$\begin{aligned} &\mathbf{P}^S = \rho^S \frac{\partial \psi^S}{\partial \mathbf{F}^S} + \left\{\rho^S \psi_1^S + \sum_{\beta=1}^{A} \rho^\beta \psi_1^\beta\right\} \sum_{\beta=1}^{A} \frac{\partial \Phi^\beta}{\partial \mathbf{F}^S}\Delta, \\ &\mathbf{F}^{ST}\mathbf{P}^\alpha = -p^\alpha \mathbf{1}, \end{aligned} \tag{10.87}$$

where

$$p^\alpha \equiv \rho^\alpha \left\{\rho^S \frac{\partial \psi^S}{\partial \rho^\alpha} + \sum_{\beta=1}^{A} \rho^\beta \frac{\partial \psi^\beta}{\partial \rho^\alpha}\right\} + \left\{\rho^S \psi_1^S + \sum_{\beta=1}^{A} \rho^\beta \psi_1^\beta\right\} \Phi^\alpha \Delta. \tag{10.88}$$

Hence, the partial Cauchy stress tensors in fluid components are spherical in this approximation.

Simultaneously, the scalar functions

$$\left\{\psi^S, \psi^\alpha, \Phi^\alpha\right\}, \tag{10.89}$$

form a set of thermodynamical potentials for stresses. This set can be further reduced by the remaining identities. Namely,

$$\begin{aligned}
&\frac{\partial \Phi^\alpha}{\partial \rho^S} = -\frac{\rho^\alpha}{T\lambda^n}\frac{\partial \psi^\alpha}{\partial \rho^S}, \qquad \frac{\partial \Phi^\alpha}{\partial \rho^\beta} = -\frac{\rho^\alpha}{T\lambda^n}\frac{\partial \psi^\alpha}{\partial \rho^\beta}, \quad \alpha \neq \beta, \\
&\frac{\partial \Phi^\alpha}{\partial \rho^\alpha} = \frac{1}{T\lambda^n}\left\{-\rho^\alpha \frac{\partial \psi^\alpha}{\partial \rho^\alpha} + \rho^S \frac{\partial \psi^S}{\partial \rho^\alpha} + \sum_{\beta=1}^{A} \rho^\beta \frac{\partial \psi^\beta}{\partial \rho^\alpha}\right\}, \\
&\frac{\partial \Phi^\alpha}{\partial \Delta} = -\frac{\rho^\alpha}{T\lambda^n}\psi_I^\alpha \Delta, \qquad \frac{\partial \Phi^\alpha}{\partial \mathbf{F}^S} = \frac{1}{T\lambda^n}\left\{\mathbf{P}^\alpha - \rho^\alpha \frac{\partial \psi^\alpha}{\partial \mathbf{F}^S}\right\}.
\end{aligned} \tag{10.90}$$

Consequently, the functions Φ^α are determined up to functions of the temperature by the Helmholtz free energy functions.

Let us notice that the absence of the flux of porosity $\mathbf{J}$ in the model would yield severe limitations of interactions between components:

(i) the stress tensor in the skeleton would reduce to the form known from the nonlinear elasticity of single component media [see: $(6.85)_5$]

$$\mathbf{P}^S = \rho^S \frac{\partial \psi^S}{\partial \mathbf{F}^S}, \tag{10.91}$$

(ii) the partial pressures in the fluid components would have the form

$$p^\alpha = \left(\rho^\alpha\right)^2 \frac{\partial \psi^\alpha}{\partial \rho^\alpha}, \tag{10.92}$$

because the relations (10.90) yield in this case

$$\frac{\partial \psi^\alpha}{\partial \rho^\beta} = 0, \quad \alpha \neq \beta; \qquad \frac{\partial \psi^S}{\partial \rho^\alpha} = 0 \Rightarrow \psi^S = \psi^S\left(\mathbf{F}^S, \Delta, T; n|_E\right). \tag{10.93}$$

These are the properties of the so-called *simple mixtures* [see: I. MÜLLER (1985)]. Finally, the relation $(10.90)_4$ could solely be satisfied if the free energies ψ^α were dependent on $\mathbf{F}^S$ only through the third invariant J^S. An integration of this relation would yield

$$\psi^\alpha = \psi^\alpha\left(\rho_t^\alpha, T; n|_E\right). \tag{10.94}$$

The energetic interactions among components would reduce in this case to the dependence of ψ^S on the changes of porosity Δ. This is, certainly, too restrictive.

Consequently, we have obtained a macroscopical justification for the balance equation of porosity as an alternative to the evolution equation proposed in some papers on the subject of porous materials.

We proceed to analyse the dissipation $\mathcal{D}$ defined in the general case by (10.72). The relations (10.73) now have the form

$$\Lambda^{\theta} = -\frac{1}{T^2}QG - \frac{1}{T}\sum_{\beta=1}^{A}\left(\rho^{\beta}\frac{\partial\psi^{\beta}}{\partial T} + \rho^{\beta}\eta^{\beta} - T\lambda^{n}\frac{\partial\Phi^{\beta}}{\partial T}\right)\mathbf{X}'^{\beta},$$
$$\Lambda^{p\beta} = -\frac{1}{T}\mathbf{F}^{S}\mathbf{X}'^{\beta}, \qquad \mu^{\beta} = \left[\left(\psi^{\beta} - T\lambda^{\beta}\right) - \left(\psi^{S} - T\lambda^{S}\right)\right]. \tag{10.95}$$

Under the assumption (10.75), we obtain the following dissipation inequality:

$$-\frac{1}{T^2}QG\cdot G - \frac{1}{T}\sum_{\beta=1}^{A}\left(\rho^{\beta}\frac{\partial\psi^{\beta}}{\partial T} + \rho^{\beta}\eta^{\beta} - T\lambda^{n}\frac{\partial\Phi^{\beta}}{\partial T}\right)\mathbf{X}'^{\beta}\cdot \mathbf{G} +$$
$$+\lambda^{n}\hat{n} - \frac{1}{T}\sum_{\beta=1}^{A}\mathbf{X}'^{\beta}\cdot\mathbf{F}^{ST}\left(\hat{\mathbf{p}}^{\beta} - \hat{\rho}^{\beta}\mathbf{x}'^{\beta}\right) \geq 0, \tag{10.96}$$

which should hold for an arbitrary temperature gradient **G** and arbitrary relative velocities $\mathbf{X}'^{\alpha}$. Hence,

$$\frac{\partial\Phi^{\beta}}{\partial T} = -\frac{1}{T\lambda^{n}}\left\{\rho^{\beta}\frac{\partial\psi^{\beta}}{\partial T} + \rho^{\beta}\eta^{\beta}\right\}, \tag{10.97}$$

which supplements the relations (10.90) and determines Φ^{β} up to constants, i.e.

$$d\Phi^{\alpha} = -\frac{\rho^{\alpha}}{T\lambda^{n}}\left\{d\psi^{\alpha} + \eta^{\alpha}dT - \left(\rho^{S}\frac{\partial\psi^{S}}{\partial\rho^{\alpha}} + \sum_{\beta=1}^{A}\rho^{\beta}\frac{\partial\psi^{\beta}}{\partial\rho^{\alpha}}\right)d\rho^{\alpha} - \right.$$
$$\left. - \frac{1}{\rho^{\alpha}}\mathbf{P}^{\alpha}\cdot d\mathbf{F}^{\alpha}\right\}. \tag{10.98}$$

Simultaneously, under the assumption that the objective momentum sources are independent of the temperature gradient, we have

$$\mathbf{F}^{ST}\left(\hat{\mathbf{p}}^{\beta} - \hat{\rho}^{\beta}\mathbf{x}'^{\beta}\right) = -\pi^{\beta}\mathbf{X}'^{\beta}, \qquad \pi^{\beta} = \pi^{\beta}\left(\rho^{S}, \rho^{\alpha}, T, \mathbf{F}^{S}; n|_{E}\right). \tag{10.99}$$

The remaining term of the dissipation inequality describing the source of porosity is independent of **G** and $\mathbf{X}'^{\alpha}$. Consequently, bearing the quadratic dependence of the Helmholtz free energies on Δ in mind, we obtain

$$\lambda^{n}\hat{n} \geq 0, \qquad \lambda^{n} = -\frac{1}{T}\left\{\rho^{S}\psi_1^{S} + \sum_{\beta=1}^{A}\rho^{\beta}\psi_1^{\beta}\right\}\Delta. \tag{10.100}$$

Then the natural assumption

$$\hat{n} = -\mathcal{M}\left\{\rho^{S}\psi_1^{S} + \sum_{\beta=1}^{A}\rho^{\beta}\psi_1^{\beta}\right\}\Delta, \quad \mathcal{M} = \mathcal{M}\left(\rho^{S}, \rho^{\alpha}, T, \mathbf{F}^{S}; n|_{E}\right) > 0, \tag{10.101}$$

yields identically the inequality (10.100). We shall use the following notation:

$$\tau^{-1} \equiv \mathcal{M}\left\{\rho^{S}\psi_1^{S} + \sum_{\beta=1}^{A}\rho^{\beta}\psi_1^{\beta}\right\} \Rightarrow \hat{n} = -\frac{\Delta}{\tau}. \tag{10.102}$$

The parameter τ has the interpretation of the **relaxation time** for spontaneous changes in the porosity. It is positive if the thermodynamical equilibrium is stable.

The set of conditions following from this stability condition is as follows:

$$Q \leq 0, \quad \tau \geq 0, \quad \pi^{\beta} \geq 0. \tag{10.103}$$

These relations complete the set of thermodynamical restrictions for the model considered in this section.

As an illustration of the above thermodynamical results, we present here constitutive relations in the case of the absence of mass sources in the skeleton

$$\hat{\rho}^S = 0. \tag{10.104}$$

Then the mass density of the skeleton ρ^S is constant. The identities of this section yield after cumbersome but elementary calculations

$$\begin{aligned}
&\psi^S = \psi_0^S\left(\mathrm{I}, \mathrm{II}, \mathrm{III}, T; n|_E\right) + \frac{\Delta^2}{2\rho^S \tau \mathcal{M}}, \\
&\qquad\qquad \tau\mathcal{M} = \left(\rho^S \psi_1^S\right)^{-1} = \not{b}\left(\mathrm{I}, \mathrm{II}, \mathrm{III}, T; n|_E\right), \\
&\psi^\alpha = \psi^\alpha\left(\rho_t^\alpha, T; n|_E\right), \qquad \rho_t^\alpha \equiv \rho^\alpha J^{S-1}, \quad \rho_t^S \equiv \rho^S J^{S-1}, \\
&\Phi^\alpha = J^S \varphi^\alpha\left(n|_E\right), \\
&\mathbf{P}^\alpha = -p^\alpha \mathbf{F}^{S\text{-}T}, \qquad p^\alpha = J^S\left\{\rho_t^{\alpha 2} \frac{\partial \psi^\alpha}{\partial \rho_t^\alpha} + \frac{\Delta}{\tau\mathcal{M}} \varphi^\alpha\right\}, \\
&\mathbf{P}^S = J^S\left\{\mathfrak{I}_1 \mathbf{B}^S + \mathfrak{I}_0 \mathbf{1} + \mathfrak{I}_{-1} \mathbf{B}^{S\text{-}1} + \frac{\Delta}{\tau\mathcal{M}} \sum_{\alpha=1}^{A} \varphi^\alpha \mathbf{1}\right\} \mathbf{F}^{S\text{-}T},
\end{aligned} \tag{10.105}$$

where the following standard notation of non-linear elasticity has been used [see: (6.88])

$$\begin{aligned}
&\mathbf{B}^S = \mathbf{F}^S \mathbf{F}^{ST}, \quad \mathrm{I} = \mathbf{B}^S \cdot \mathbf{1}, \quad \mathrm{II} = \frac{1}{2}\left(\mathrm{I}^2 - \mathbf{B}^{S2} \cdot \mathbf{1}\right), \quad \mathrm{III} = \det \mathbf{B}^S = J^{S2}, \\
&\mathfrak{I}_1 = 2\rho_t^S \frac{\partial \psi_0^S}{\partial \mathrm{I}}, \quad \mathfrak{I}_0 = 2\rho_t^S\left(\mathrm{II}\frac{\partial \psi_0^S}{\partial \mathrm{II}} + \mathrm{III}\frac{\partial \psi_0^S}{\partial \mathrm{III}}\right), \quad \mathfrak{I}_{-1} = -2\rho_t^S \mathrm{III} \frac{\partial \psi_0^S}{\partial \mathrm{II}}.
\end{aligned} \tag{10.106}$$

We have used the assumption of isotropy in the derivation of these relations.

In addition, we have the following thermal conditions:

$$\eta^S = -\frac{\partial \psi^S}{\partial T}, \quad \eta^\alpha = -\frac{\partial \psi^\alpha}{\partial T}, \tag{10.107}$$

which specify the partial entropies, and consequently, the partial internal energies in terms of the partial Helmholtz free energies.

The most restrictive part of these results is connected with the free energies of fluid components which do not contain any coupling with other components. The free energy of the skeleton contains a non-equilibrium term connected with changes in the porosity but it is, otherwise, also independent of the presence of other components. These restrictions are similar to the restrictions of the theory of simple mixtures and can be eliminated, if needed, either by the extension of the set of constitutive variables or by the non-linearities with respect to the deviation from the thermodynamical equilibrium or both. It is also rather surprising that the constitutive relation for the flux of porosity is reduced to a constitutive constant and it is, for instance, independent of the temperature.

The set of field equations which we have constructed in the last two sections is **hyperbolic** for isothermal processes under the usual convexity assumptions which we shall not discuss in this work. As the analysis of the propagation conditions for the two-component porous materials clearly shows [e.g.: K. WILMANSKI (1995,1,2)] it describes the typical P1- and P2-waves (fast and slow longitudinal waves), S-waves (shear waves) as well as the additional wave of diffusivity. However, the latter wave has a very small amplitude in usual circumstances, it is of the order of the ratio $\left|\mathbf{X}'^F\right|/U^F$, where the index F denotes the fluid component and U^F is the speed of propagation of the P2-wave. For this reason, such waves have not been identified in experiments on soils, rocks or granular materials in spite of the considerable progress of dynamical measurements in this field [e.g.: T. BOURBIE, O. COUSSY, B. ZINSZNER (1987)]. In the next section we present a simple example of the analysis of the propagation condition for the present model.

Let us finally mention that the temperature changes have, in the present model, a classical parabolic character.

10.6 Propagation of Plane Waves of Small Amplitude

In this section we illustrate the above considerations for a very simple example of weak discontinuity waves under isothermal conditions. The purpose of this analysis is twofold. Firstly, we show that the above presented model yields this sort of waves which are indeed observed in porous materials. Secondly, we indicate the possibilities of dynamical measurements of macroscopical (effective) material properties necessary for practical applicability of the model.

We consider a two-component porous material with a **linearly elastic skeleton**. In the isothermal case of plane waves which we proceed to analyse, the set of unknown fields is as follows

$$\left\{\rho_t^F, \Delta, v^S, v^F, \varepsilon^S\right\}, \tag{10.108}$$

where the index F denotes the fluid component, v^S, v^F are the components of partial velocities in the direction of the x-axis, and ε^S denotes a small extension of the skeleton in the x-direction.

The field equations for these fields have the following form:

$$\begin{gathered}
\frac{\partial \rho_t^F}{\partial t} + \frac{\partial \rho_t^F v^F}{\partial x} = 0, \\
\frac{\partial \Delta}{\partial t} + v^S \frac{\partial \Delta}{\partial x} + n_0 \gamma \frac{\partial \left(v^F - v^S\right)}{\partial x} = -\frac{\Delta}{\tau}, \\
\rho^S \frac{\partial v^S}{\partial t} = \frac{\partial \sigma^S}{\partial x} + \pi^F\left(v^F - v^S\right), \quad \frac{\partial \varepsilon^S}{\partial t} = \frac{\partial v^S}{\partial x}, \\
\rho_t^F \left(\frac{\partial v^F}{\partial t} + \frac{\partial v^F}{\partial x} v^F\right) = -\frac{\partial}{\partial x}\left(p^F + \frac{n_0}{\tau \mathcal{M}} \Delta\right) - \pi^F\left(v^F - v^S\right),
\end{gathered} \tag{10.109}$$

$$\sigma^S = \left(\lambda^S + 2\mu^S\right)\varepsilon^S + \frac{n_0}{\tau \mathcal{M}}\Delta, \quad p^F = \rho_t^{F2}\frac{\partial \psi^F}{\partial \rho_t^F} = p^F\left(\rho_t^F; n_0\right),$$

where σ^S is the normal component of the partial Cauchy stress tensor in the skeleton in the x-direction, λ^S, and μ^S are the effective Lamé constants of the skeleton solely dependent on the equilibrium porosity $n|_E$ and the constant φ^α, α=F, which defines the flux of porosity, has been assumed to be identical with the equilibrium porosity itself. The remaining material parameters τ, $\mathcal{M}$, and π^F are also assumed to be constant.

Obviously, the constitutive relation for the skeleton is the linear counterpart of the relation $(10.105)_5$.

We shall seek the solution of the above set of equations in the form of a small dynamical disturbance of the uniform static initial state, i.e.

$$\begin{aligned} &\rho_t^F = \rho_0^F + \varepsilon R^F \exp i\left(\omega t - k^* x\right), \quad \Delta = \varepsilon D \exp i\left(\omega t - k^* x\right), \\ &v^S = \varepsilon V^S \exp i\left(\omega t - k^* x\right), \quad v^F = \varepsilon V^F \exp i\left(\omega t - k^* x\right), \\ &\varepsilon^S = \varepsilon E^S \exp i\left(\omega t - k^* x\right), \quad \varepsilon << 1, \end{aligned} \tag{10.110}$$

where ω is a (given) frequency of the disturbance, R^F, D, V^S, V^F, E^S are constant amplitudes of the disturbance, ε is a small parameter and

$$k^* = k + i\alpha, \tag{10.111}$$

with the real part k being the wave number and the imaginary part α being the attenuation coefficient.

Substitution of these relations in the set of equations (10.109) yields a set of homogeneous algebraic equations for the amplitudes. The condition of existence of non-trivial solutions gives rise to the following **dispersion relation**:

$$\begin{aligned} &\left(i\frac{\pi\omega}{\rho_0^F}\frac{\rho_0^F}{\rho^S} - \omega^2 + U^{S2}k^{*2} + \frac{n_E^2\gamma^2 i\omega\tau}{(1+i\omega\tau)\tau\mathcal{M}\rho_0^F}\frac{\rho_0^F}{\rho^S}k^{*2}\right)\times\left(i\frac{\pi\omega}{\rho_0^F} - \omega^2 + U^{F2}k^{*2} + \right. \\ &\left. + \frac{n_E^2\gamma^2 i\omega\tau}{(1+i\omega\tau)\tau\mathcal{M}\rho_0^F}k^{*2}\right) + \left(\frac{\pi\omega}{\rho_0^F} + \frac{n_E^2\gamma^2\omega\tau}{(1+i\omega\tau)\tau\mathcal{M}\rho_0^F}k^{*2}\right)^2\frac{\rho_0^F}{\rho^S} = 0. \end{aligned} \tag{10.112}$$

where

$$U^{S2} \equiv \frac{\lambda^S + 2\mu^S}{\rho^S}, \qquad U^{F2} \equiv \left.\frac{\partial p^F}{\partial \rho_t^F}\right|_0, \tag{10.113}$$

correspond to the speeds of propagation of weak discontinuity surfaces in such a two-component system in two extreme cases of n=0 and n=1, respectively [see: K. WILMANSKI (1995,1,2)].

The dispersion relation (10.112) is a biquadratic equation for k*. Consequently, we obtain two essential solutions for the wave numbers k and for the attenuation coefficients α. Each pair corresponds to different longitudinal waves which are called P1- and P2-waves in the theories of porous materials [compare: T. BOURBIE, O. COUSSY, B. ZINSZNER (1987)]. The second longitudinal wave, the P2-wave, is also called the **Biot's**

wave and it is characteristic for multicomponent systems. Its appearance may be considered to be a test for the applicability of a model of porous materials in the description of dynamical phenomena. An example of a model which does not describe such a wave is one of Bowen's models of porous materials [see: R. M. BOWEN (1980)] - the model with incompressible true components.

The shear waves do not appear in this example due to the one-dimensional character of the example.

In order to see the most important features of the above-mentioned waves we plotted in Fig.10.2 and in Fig.10.3 the attenuation coefficients, the wave numbers and the phase velocities as functions of the frequency ω (in MHz) for the material constant γ=1.

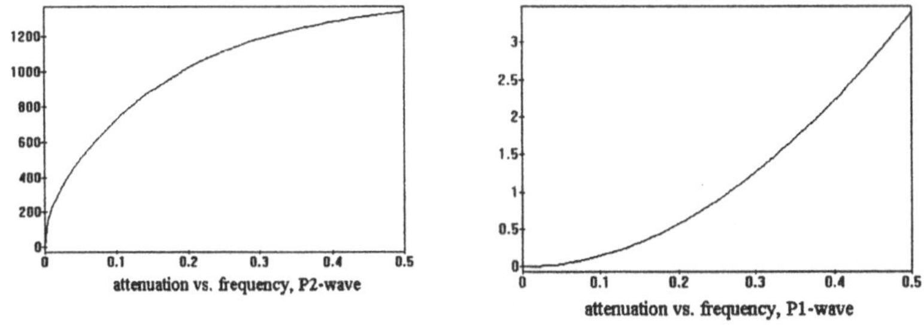

Fig. 10.2. *Attenuation of the plane* P2- *and* P1-*waves*

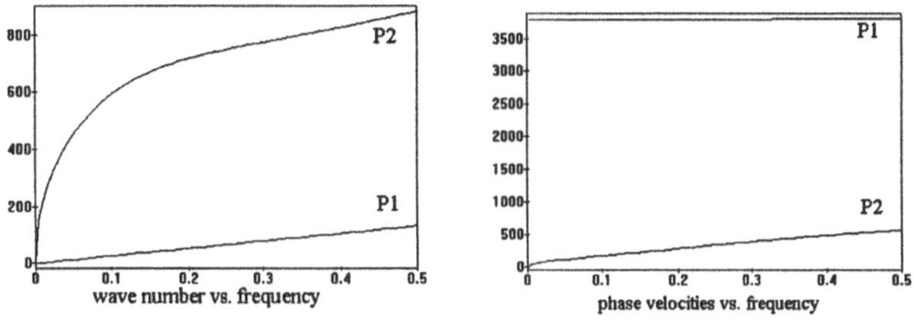

Fig. 10.3. *Wave number and phase velocities for the plane* P1- *and* P2-*waves*

The above curves have been obtained for the following values of material parameters

$$n_0 = 0.23, \quad \rho_0^F = 0.23\times10^3\frac{kg}{m^3}, \quad \rho^S = 2.4\times10^3\frac{kg}{m^3},$$

$$U^F = 1.0\times10^3\frac{m}{s}, \quad U^S = 3.8\times10^3\frac{m}{s}, \tag{10.114}$$

$$\pi^F = 2.602\times10^9\frac{kg}{m^3 s}, \quad \tau = 3.7\times10^{-6}s, \quad \tau\mathcal{M}\rho_0^F = 7.347\times10^{-8}\frac{s^2}{m^2}.$$

These values were estimated on the basis of measurements reported by T. BOURBIE, O. COUSSY, B. ZINSZNER (1987) for Massillon sandstone saturated with water. The estimates are rather rough and should solely be treated as reflecting the qualitative behaviour of this material.

The fast monochromatic P1-waves are almost not attenuated within the frame of the present model. They all propagate with almost the same phase velocity (~3800 m/s). The slow monochromatic P2-waves (Biot's waves) have a growing phase velocity up to ~1000 m/s with growing frequency. This property is confirmed by experiments. The same observation concerns the attenuation coefficient whose inverse is proportional to the so-called *quality factor*. The values of this factor calculated for the present data are higher than observed, which is most likely connected with the rough estimate for the relaxation time τ.

The above example shows that the present model can be verified experimentally by means of dynamical measurements. However, the available data are still very much dispersed and far from being complete.

10.7 Final Remarks

The continuous thermodynamical model of porous materials discussed in this chapter demonstrates clearly that many important properties of such materials can be successfully described within this theoretical framework. However, it also shows that numerous fundamental problems are still open. This concerns particularly the question of measurability of the quantities appearing in the model. We have seen that the conditions on the contact surfaces do not admit the usual interpretation of such notions as temperature or partial pressures. Consequently, here we have an open problem of identifying the macroscopical quantities. The same conditions on the surface yield the problem of boundary conditions on free surfaces, such as the boundary of the skeleton with the outflow of fluid components or the problem of contact conditions on the boundary between two different layers of porous materials. Some of these questions are presently being investigated but it is still too early to report on these new results.

Appendix A: Thermostatics

Thermodynamic theories of macroscopic processes, which were developed some 100 years ago by Carnot, Helmholtz, Mayer, Calusius, Duhem, Gibbs and others, are concerned with *uniform* systems. For this reason the fields appearing in the non-equilibrium thermodynamics may only coincide with the corresponding quantities of these classical models in the particular case of the thermodynamical equilibrium state. In this state the spatial changes of fields, yielding the dissipation can be neglected. Hence, if an equilibrium process is considered as a sequence of such uniform states, it must proceed slowly enough to let the system relax to each subsequent uniform state after a small change of control parameters. In other words, the relaxation times of the non-equilibrium thermodynamics must be much shorter than the characteristic times of macroscopic observations in order to yield not only the local in space equilibrium (in small vicinities of the points of a body) but also the global equilibrium in macroscopic domains. Such processes are called **quasistatic**, and classical thermostatics deals solely with such processes.

We do not intend to go into any details about equilibrium thermodynamics (thermostatics) since many extensive textbooks on this subject are easily available*). We would rather demonstrate the most important notions of the theory using the example of an *ideal fluid.*

Thermostatics is based on four axioms concerning the interactions of uniform systems with the surroundings as well as with each other. To formulate these axioms, we introduce a few notions appropriate for uniform systems. The basic variables describing a chosen system are its volume V and the energy E. Dealing with very slow processes, we can neglect the kinetic part of the energy and, consequently, E describes the potential energy of the system. Let us assume that the volume V contains n moles of a substance with the molecular mass M. Then,

$$\begin{aligned} &m=nM=\text{const.} && \textit{- total mass of the system,} \\ &\rho=m/V && \textit{- (uniform!) mass density,} \\ &v=1/\rho && \textit{- specific (uniform!) volume,} \\ &\varepsilon=E/m && \textit{- specific (uniform!) energy per unit mass.} \end{aligned} \tag{A.1}$$

An arbitrary (equilibrium!) state of the ideal fluid is described by the pair (ε,ρ) or (E,V). This state can be changed, for instance, by the work done from the external world (e.g. a piston changing the volume of the cylinder containing the fluid). If the interaction of the system is limited to such mechanical actions an arbitrary quasistatic process must satisfy the purely mechanical energy conservation law. In terms of continuum mechanics it would have the following form:

*) e.g.: J. W. GIBBS (1872) [*Dover* (1961)], M. PLANCK (1922) [*Dover* (1945)], D. ELWELL, A. J. POINTON (1972), A. SOMMERFELD (1977), K. HUTTER (1991), I. MÜLLER (1994).

$$\rho\dot{\varepsilon} + p \operatorname{div} \mathbf{v} = 0, \qquad \operatorname{div} \mathbf{v} \equiv -\frac{\dot{\rho}}{\rho}, \tag{A.2}$$

where $p=p(\varepsilon,\rho)$ denotes the pressure. The divergence of the velocity describes for the uniform systems the changes of the volume, or as indicated in the above formula, the changes of the uniform mass density due to the infinitesimal change of state in the quasistatic process. Consequently, in terms of the global quantities introduced above, this law can be written in the form

$$dE + p\,dV = 0, \qquad p = p(E, V), \tag{A.3}$$

where the increments dE, dV have been introduced instead of time increments. The time is only a parameter along the paths of the considered quasistatic processes identifying the states appearing in such processes.

A system satisfying (A.3) is said to be isolated by **adiabatic walls**. If we allow an exchange of energy with the surroundings without, however, any changes of the volume V and the mass m, the wall is called **diathermal**. In such a case the conservation law (A.3) does not hold any longer because $dE \neq 0$ and $dV=0$.

Bearing this in mind, we postulate the **first law of thermodynamics** (the **law of energy conservation**) in a general case of arbitrary interactions with the surroundings,

$$dE = W + Q, \qquad W \equiv -p\,dV, \tag{A.4}$$

where W is the **mechanical working** (i.e. the work done in an infinitesimal change of state), and Q denotes the **heat exchange** corresponding to a change of state $E \to E+dE$, $V \to V+dV$.

Obviously, the simple formula $(A.4)_2$ for the mechanical working is solely connected with our choice of the ideal fluids whose uniform mechanical reactions are described by the spherical stress tensor $\mathbf{T}=-p\mathbf{1}$. The generalizations are straightforward and we shall not discuss them here, instead we will focus on the more important parts of thermostatics without running into technical difficulties. Both W and Q appearing in (A.4) are the so-called **differential 1-forms**, which are usually not integrable (compare: the exercise 2.1.). To stress their difference from the quantities appearing in cases of finite processes (e.g. $E_1 \to E_2$, $V_1 \to V_2$), they are sometimes denoted as $\dot{W}$ and $\dot{Q}$, or $đW$ and $đQ$., but we will not use this notation in this Appendix.

Now let us consider two systems, A and B, in adiabatic walls whose states are given by the pairs (E_A, V_A) and (E_B, V_B), respectively. The corresponding pressures are $p_A=p(E_A, V_A)$, $p_B=p(E_B, V_B)$. Without changing their volumes we bring these two systems into contact replacing the adiabatic wall separating them by a diathermal wall. The systems will exchange energy according to the first law of thermodynamics,

$$dE_A = Q_A, \qquad dE_B = Q_B. \tag{A.5}$$

We assume the energy to be **additive**, i.e. the energy E of both systems is the sum of E_A and E_B, and the first law implies ($W_A = W_B = 0$) that

$$dE = dE_A + dE_B = 0 \;\Rightarrow\; Q_A + Q_B = 0, \tag{A.6}$$

as the combination of both systems is isolated from the surroundings by the adiabatic walls. The additivity assumption also yields that the diathermal wall does not store any energy. Such walls are called **ideal**.

We expect the systems A and B to reach a new state of equilibrium (E'_A, V_A), (E'_B, V_B) for which Q_A and Q_B vanish. Then we say that the systems are in **thermal equilibrium**. This definition yields the result that only one of the variables E'_A, E'_B can be independent. Hence, we have established the existence of a function θ_{AB} determining the condition of a thermal equilibrium of two systems A and B,

$$\theta_{AB}(E_A, V_A) = \theta_{AB}(E_B, V_B). \tag{A.7}$$

The systems A and B do not exchange energy under this condition if the mechanical working is zero.

Now we introduce the **zeroth law of thermodynamics** (thermostatics!) which requires that the thermal equilibrium is **transitive**. Let us choose three arbitrary systems denoted by A, B, C. If A, B and B, C are mutually in the thermal equilibrium we conclude that A, C also are in a thermal equilibrium

$$\begin{aligned} &\theta_{AB}(E_A, V_A) = \theta_{AB}(E_B, V_B) \wedge \theta_{BC}(E_B, V_B) = \theta_{BC}(E_C, V_C) \quad \Rightarrow \\ &\qquad\qquad \Rightarrow \quad \theta_{AC}(E_A, V_A) = \theta_{AC}(E_C, V_C). \end{aligned} \tag{A.8}$$

Hence, there exists a function for an arbitrary system

$$\theta : (E, V) \rightarrow \Re, \tag{A.9}$$

called the **empirical temperature** which determines the existence or the lack of existence of thermal interactions of this system with another system through an ideal diathermal wall of contact.

According to this law, the state of a system with constant mass can be described by a pair (θ,V), instead of (E,V), provided (A.9) is invertible for any V. We shall justify this invertibility condition in the sequel.

We proceed to formulate the second law of thermodynamics (thermostatics!). Among many available formulations we choose this of C. CARATHÉODORY (1909) [compare: K. WILMANSKI (1992)]. Again, we consider a system whose changes of states (quasistatic processes) satisfy the first law of thermodynamics (A.4). If Q=0 for the entire process, then it is called **adiabatic**.

The **second law of thermodynamics** requires that in any neighbourhood of a chosen state there exist states which cannot be reached in an adiabatic process.

In the case of our simple system described by the two variables (E,V) [or (θ,V)] this statement seems to be rather obvious. It is much less obvious for systems whose states are described by more than two variables.

In the general case, the above assumption yields the existence of a multiplier Λ for the Pfaffian form dE+W which turns this form into an integrable one [for a proof see, for instance: K. WILMANSKI (1992)]

$$dS = \Lambda(dE - W). \tag{A.10}$$

Generally the multiplier Λ depends on all variables defining the state of a system. Let us notice in passing, that the case of two variables is mathematically trivial because the Pfaffian forms of two independent variables are always integrable, i.e. the existence of the multiplier Λ does not require any additional assumptions.

We proceed to investigate the properties of this multiplier. Again, let us consider two systems A and B in a thermal equilibrium whose states are described by (θ, V_A) and (θ, V_B), respectively. Then we have

$$dS_A = \Lambda_A(dE_A + p_A dV_A), \quad dS_B = \Lambda_B(dE_B + p_B dV_B), \tag{A.11}$$

for the infinitesimal changes of these states. Simultaneously, the state of the system A∪B, which is defined as the sum of both systems A and B in mutual diathermal contact, is described by the parameters (θ, V_A, V_B), and it satisfies

$$\begin{aligned} &dE = dE_A + dE_B = -(p_A dV_A + p_B dV_B) \Rightarrow \\ &\Rightarrow dS_{A\cup B} = \Lambda_{A\cup B}(dE_A + p_A dV_A + dE_B + p_B dV_B). \end{aligned} \tag{A.12}$$

Now we change the state variables for the systems A and B

$$(\theta, V_A) \to (\theta, S_A), \quad (\theta, V_B) \to (\theta, S_B). \tag{A.13}$$

Then, bearing (A.11) in mind, the relation (A.12) assumes the form

$$dS_{A\cup B} = \Lambda_{A\cup B}\left(\frac{1}{\Lambda_A} dS_A + \frac{1}{\Lambda_B} dS_B\right). \tag{A.14}$$

The left-hand side of this relation is the full differential of $S_{A\cup B}(\theta, S_A, S_B)$. Hence,

$$\frac{\partial S_{A\cup B}}{\partial \theta} = 0, \quad \frac{\partial S_{A\cup B}}{\partial S_A} = \frac{\Lambda_{A\cup B}}{\Lambda_A}, \quad \frac{\partial S_{A\cup B}}{\partial S_B} = \frac{\Lambda_{A\cup B}}{\Lambda_B}. \tag{A.15}$$

The integrability conditions for these relations immediately yield

$$\frac{\partial}{\partial \theta}\left(\frac{\Lambda_{A\cup B}}{\Lambda_A}\right) = 0, \quad \frac{\partial}{\partial \theta}\left(\frac{\Lambda_{A\cup B}}{\Lambda_B}\right) = 0. \tag{A.16}$$

Hence, we obtain

$$\frac{1}{\Lambda_{A\cup B}} \frac{\partial \Lambda_{A\cup B}}{\partial \theta} = \frac{1}{\Lambda_A} \frac{\partial \Lambda_A}{\partial \theta} = \frac{1}{\Lambda_B} \frac{\partial \Lambda_B}{\partial \theta} \equiv \mu(\theta). \tag{A.17}$$

These relations mean that each multiplier splits into a product of a function of θ and a function of the remaining variables:

$$\begin{aligned} &\Lambda_A(\theta, S_A) = \nu_A(S_A) \exp\left(\int \mu(\theta)\, d\theta\right), \\ &\Lambda_B(\theta, S_B) = \nu_B(S_B) \exp\left(\int \mu(\theta)\, d\theta\right), \\ &\Lambda_{A\cup B}(\theta, S_A, S_B) = \nu_{A\cup B}(S_A, S_B) \exp\left(\int \mu(\theta)\, d\theta\right). \end{aligned} \tag{A.18}$$

Let us introduce the notation

$$T(\theta) = \left\{C \exp \int \mu(\theta)\, d\theta\right\}^{-1}, \tag{A.19}$$

where C is an arbitrary positive constant. This quantity is called the **absolute temperature**. According to its definition (A.19), it is positive and **universal**, i.e. the same for all systems described by the state variables (θ, V), regardless of the particular **material** (**constitutive**) relations for p, E, and S.

The above construction of the absolute temperature can be carried out for a higher number of state variables. In this sense, the absolute temperature is not only universal for ideal fluids but also for arbitrary systems which can be described within the above presented frame of thermostatics.

Substituting of (A.18) and (A.19) into $(A.11)_1$ yields

$$dS'_A = \frac{1}{T(\theta)}\left(dE_A + p_A dV_A\right), \tag{A.20}$$

where

$$S'_A \equiv \frac{1}{C}\int \frac{dS_A}{\nu_A} + \text{const.} \tag{A.21}$$

is called the **entropy** of the system A.

Hence, in thermostatics, the most important consequence of the second law of thermodynamics can be written as the following **Gibbs equation**:

$$dS = \frac{1}{T}\left(dE - W\right), \tag{A.22}$$

where we dropped the prime in the entropy for the sake of a simple notation.

Let us notice that the equation (A.22) has been derived for systems in the adiabatic isolation, i.e. for such systems which may interact with each other but the states of all these systems are principally known. If we account for interactions with the external world, we have to assume that the system and its exterior form together an adiabatically isolated system. In such a case one has to split the changes of the entropy dS into these which follow from processes inside of the system, and these which are caused by the external flux of the entropy

$$dS = dS^{in} + dS^{ext}, \qquad dS^{ext} = \frac{Q^{ext}}{T^{ext}}, \tag{A.23}$$

where the second formula follows from the first law of thermodynamics for the whole system, Q^{ext} denotes the heat exchange of the system with the external world, which is principally, measurable (or even controllable), and T^{ext} denotes the temperature at the external world on the contact wall with the system.

The Gibbs equation (A.22) has important consequences for the construction of constitutive relations in the equilibrium states of a given material. As an example, we consider an ideal fluid again. In this case, the equation (A.22) has the following form:

$$dS = \frac{\partial S}{\partial T}dT + \frac{\partial S}{\partial V}dV = \frac{1}{T}\left(\frac{\partial E}{\partial T}dT + \left(\frac{\partial E}{\partial V} + p\right)dV\right), \tag{A.24}$$

where we replaced the empirical temperature θ by the absolute temperature T as a state variable. The latter is also one of the admissible empirical tempratures. It follows that

$$\frac{\partial E}{\partial T} = T\frac{\partial S}{\partial T}, \qquad p = -\left(\frac{\partial E}{\partial V} - T\frac{\partial S}{\partial V}\right). \tag{A.25}$$

If we introduce the **Helmholtz free energy** Ψ,

$$\Psi \equiv E - TS = \Psi(T,V), \tag{A.26}$$

we obtain

$$S = -\frac{\partial \Psi}{\partial T}, \quad p = -\frac{\partial \Psi}{\partial V}, \quad E = \Psi - T\frac{\partial \Psi}{\partial T}. \tag{A.27}$$

Hence, for a given function (A.26), we can find the remaining constitutive relations by differentiating Ψ. Such functions are called **thermodynamical potentials** and their existence within thermostatics follows from the Gibbs equation.

Let us mention that in the classical literature on thermostatics, the Helmholtz free energy is frequently denoted by F.

The other examples of such thermodynamical potentials follow from (A.24) after a transformation of state variables. For the transformation of state variables given by

$$(T,V) \to (T,p), \tag{A.28}$$

we have

$$d(E - TS + pV) = -S\,dT + V\,dp, \tag{A.29}$$

i.e. the function

$$G \equiv E - TS + pV = G(T,p), \tag{A.30}$$

is the potential for S, V, and consequently, for E. G is called the **Gibbs free energy (free enthalpy)**. We obtain

$$S = -\frac{\partial G}{\partial T}, \quad V = \frac{\partial G}{\partial p}, \quad E = G - T\frac{\partial G}{\partial T} - p\frac{\partial G}{\partial p}. \tag{A.31}$$

In the case of the following transformation

$$(T,V) \to (S,p), \tag{A.32}$$

we obtain

$$d(E + pV) = T\,dS + V\,dp, \tag{A.33}$$

i.e. it is reasonable to define the function

$$H \equiv E + pV = H(S,p), \tag{A.34}$$

called the **enthalpy**. This function is the potential for T, V, and E

$$T = \frac{\partial H}{\partial S}, \quad V = \frac{\partial H}{\partial p}, \quad E = H - p\frac{\partial H}{\partial p}. \tag{A.35}$$

The above examples illustrate the role of the **Legendre transformations** in thermostatics which determine – by means of the Gibbs equation – the thermodynamical potentials for a chosen set of state variables.

Let us return to the equation (A.24). It is obvious that for the left-hand side to be a full differential, the following integrability condition is required

$$\frac{\partial}{\partial V}\left(\frac{1}{T}\frac{\partial E}{\partial T}\right)=\frac{\partial}{\partial V}\left[\frac{1}{T}\left(\frac{\partial E}{\partial V}+p\right)\right], \tag{A.36}$$

i.e.

$$\frac{\partial E}{\partial V}=T\frac{\partial p}{\partial T}-p. \tag{A.37}$$

This is an example of the so-called **Maxwell equations** which follow as integrability conditions from the Gibbs equation for each particular choice of state variables. The equation (A.37) shows that the derivative $\frac{\partial E}{\partial V}$ does not have to be sought by caloric measurements if the **thermal state equation**

$$p=p(T,V), \tag{A.38}$$

is known. The measurements of the latter are much simpler than the caloric measurements of $\frac{\partial E}{\partial V}$.

However, in order to find experimentally the energy E, one has to perform **some calorimetric measurements,** as the following argument shows. Bearing the first law of thermodynamics in mind, we have

$$Q=mc_V\,dT+\left(\frac{\partial E}{\partial V}+p\right)dV, \qquad c_V\equiv\frac{1}{m}\frac{\partial E}{\partial T}, \tag{A.39}$$

where c_V is called the **specific heat under a constant volume**. According to (A.37), we obtain

$$m\frac{\partial c_V}{\partial V}=T\frac{\partial^2 p}{\partial T^2} \quad\Rightarrow\quad mc_V=mc_V^0(T)+T\frac{\partial^2}{\partial T^2}\int_{V_0}^{V}p\,dV. \tag{A.40}$$

This relation shows that c_V follows from the thermal state equation up to a function of the temperature $c_V{}^0(T)$. This function must be determined in a **single** calorimetric measurement, fo instance for $V=V_0$.

The relations (A.37) and $(A.40)_2$ give both derivatives of E, and consequently, the **caloric state equation**

$$E=E(T,V), \tag{A.41}$$

can be determined up to an arbitrary constant.

Let us mention in passing, that instead of direct measurements in order to find the thermal state equation (A.38), it is sometimes convenient to measure other quantities as demonstrated in the case of pressure control. The first law of thermodynamics has the form

$$Q = d(E + pV) - V\,d p = \frac{\partial(E + pV)}{\partial T}\,d T + \left[\frac{\partial(E + pV)}{\partial p} - V\right] d p. \tag{A.42}$$

The quantity H=E+pV is reminiscent of the enthalpy (A.34) with the important difference that it is, in the case (A.42), the function of (T,p) rather than (S,p). The quantity

$$mc_p \equiv \frac{\partial(E + pV)}{\partial T}, \quad E = E(T,p), \quad V = V(T,p), \tag{A.43}$$

is called the **specific heat under a constant pressure** and can be measured in the calorimetric experiments under constant pressure, which can easily be seen from (A.42).

Bearing the change of state variables (T,V)→(T,p) in mind, we easily deduce from (A.43) that

$$c_p - c_V = \frac{1}{m}\left(\frac{\partial E}{\partial V} + p\right)\frac{\partial V}{\partial T}, \tag{A.44}$$

i.e.

$$\gamma \equiv \frac{c_p}{c_V} = 1 + \frac{\left[\left(\frac{\partial E}{\partial V}\right)\Big|_{T=\text{const}} + p\right]\frac{\partial V}{\partial T}}{\left(\frac{\partial E}{\partial T}\right)\Big|_{V=\text{const}}}, \tag{A.45}$$

where γ is called the **adiabatic exponent** and appears frequently in gas dynamics.

Apart from the above fundamental axioms of thermostatics, it is frequently assumed that the state of the thermodynamical equilibrium should be insensitive to small variations of state variables. We demonstrate this **thermodynamical stability condition** using a simple example. It follows from the assumption that the equilibrium state of a system in adiabatic isolation (i.e. Q=0 ⇒ dS=0) appears in the state of a **maximum value** of the entropy. This is, certainly, an additional assumption of thermostatics which does not follow from the second law in the form of the Gibbs equation (A.22). It is also known that certain **metastable** equilibrium states, appearing in real physical systems, do not satisfy this condition.

In order to realize the consequences of this assumption, we consider two identical systems each having the energy E and the volume V. We confine both systems to the common adiabatic isolation and consider an initial state with small variations of E and V: (E+δE, V+δV) for the first system, and (E-δE, V-δV) for the second system. We let the system establish the equilibrium state of equal energies and volumes. According to the above assumption, the final state must have an entropy greater than that of the initial state

$$S(2E,2V) > S(E + \delta E, V + \delta V) + S(E - \delta E, V - \delta V), \tag{A.46}$$

where the additivity of the entropy is used.

The Taylor expansion of this relation yields

$$\frac{\partial^2 S}{\partial E^2}(\delta E)^2 + 2\frac{\partial^2 S}{\partial E \partial V}\delta E \delta V + \frac{\partial^2 S}{\partial V^2}(\delta V)^2 < 0, \tag{A.47}$$

i.e. the negative definiteness of the Hessian matrix of the entropy S.

Bearing in mind the relations following from the Gibbs equation (A.22) for the above choice of variables

$$\frac{\partial S}{\partial E} = \frac{1}{T}, \quad \frac{\partial S}{\partial V} = \frac{p}{T}, \tag{A.48}$$

and changing the variables (E,V)→(T,V), we easily arrive at the following form of the inequality (A.47)

$$\frac{mc_V}{T^2}(\delta T)^2 - \frac{1}{T}\frac{\partial p}{\partial V}(\delta V)^2 > 0, \tag{A.49}$$

where the relation (A.37) was used.

Exercise A.1. Prove the relation (A.49)•

Hence, we obtain the following **thermodynamical stability conditions** of the equilibrium state

$$c_V > 0, \quad \kappa_T > 0, \qquad \kappa_T \equiv -V\frac{\partial p}{\partial V}, \tag{A.50}$$

where κ_T is called the **isothermal volume compressibility**.

Let us notice that the first condition justifies the transformation of variables (E,T)→(T,V) ($\left(\frac{\partial E}{\partial T}\right) > 0$) which has been used already to invert the variables in (A.9).

We complete our considerations with a brief remark concerning the **third law of thermodynamics**. According to the Gibbs equation in isochoric processes ($dV \equiv 0$), the changes of the entropy of an adiabatically isolated system satisfy the relation

$$dS = \frac{mc_V}{T}dT. \tag{A.51}$$

In the case of low temperatures, these changes will become very large (singular for T=0) if the changes of c_V are not at least proportional to T. Simultaneously, experience with macroscopical systems shows that the lowering of the temperature increases the microscopical order of a system (condensation of vapours, crystallization of melts, etc.), which according to the macroscopical interpretations of the entropy as a measure of disorder leads to a function S decreasing with the temperature. According to the definition (A.19), the value T=0 is the minimum value of T. The third law of thermodynamics requires that this state is simultaneously the state of the minimum value of the entropy S.

Consequently, the specific heat c_V must vanish at T=0. Some thermodynamical properties connected with this requirement are discussed in Chap. 9. The third law imposes, in this way, an infimum condition on the extent of cooling which may be realised in nature.

Some applications of thermostatics have led to the conclusion that negative absolute temperatures may appear in nature. If we disregard all other requirements presented in this Appendix and select (A.48) as valid for an arbitrary system in the equilibrium, then the entropy S, decreasing with the increasing energy E, would certainly give negative values of T. This seems to be the case for a system of magnetic spins. Apart from the formal doubts concerning the applicability of this formula by itself, in such a case the range of negative temperatures cannot be reached by crossing the value T=0. It is rather attainable via T=∞ corresponding in this system to the maximum value of the entropy S ($\frac{\partial S}{\partial E} = 0$). We do not consider such „pathological" situations in this book.

Appendix B: Curvilinear Coordinates

Cylindrical coordinates	Spherical coordinates
Position vector:	
$\mathbf{r} = y^1 \cos y^2 \mathbf{e}_1 +$ $+ y^1 \sin y^2 \mathbf{e}_2 + y^3 \mathbf{e}_3$	$\mathbf{r} = y^1 \cos y^2 \sin y^3 \mathbf{e}_1 +$ $+ y^1 \sin y^2 \sin y^3 \mathbf{e}_2 + y^1 \cos y^3 \mathbf{e}_3$
Covariant base vectors:	
$\mathbf{g}_1 = \cos y^2 \mathbf{e}_1 + \sin y^2 \mathbf{e}_2.$ $\mathbf{g}_2 = y^1\left(-\sin y^2 \mathbf{e}_1 + \cos y^2 \mathbf{e}_2\right),$ $\mathbf{g}_3 = \mathbf{e}_3$	$\mathbf{g}_1 = \cos y^2 \sin y^3 \mathbf{e}_1 + \sin y^2 \sin y^3 \mathbf{e}_2 +$ $+ \cos y^3 \mathbf{e}_3,$ $\mathbf{g}_2 = -y^1 \sin y^2 \sin y^3 \mathbf{e}_1 + y^1 \cos y^2 \sin y^3 \mathbf{e}_2,$ $\mathbf{g}_3 = y^1 \cos y^2 \cos y^3 \mathbf{e}_1 + y^1 \sin y^2 \cos y^3 \mathbf{e}_2 -$ $- y^1 \sin y^3 \mathbf{e}_3$
Contravariant base vectors	
$\mathbf{g}^1 = \cos y^2 \mathbf{e}_1 + \sin y^2 \mathbf{e}_2.$ $\mathbf{g}^2 = \frac{1}{y^1}\left(-\sin y^2 \mathbf{e}_1 + \cos y^2 \mathbf{e}_2\right),$ $\mathbf{g}^3 = \mathbf{e}_3$	$\mathbf{g}^1 = \cos y^2 \sin y^3 \mathbf{e}_1 + \sin y^2 \sin y^3 \mathbf{e}_2 +$ $+ \cos y^3 \mathbf{e}_3,$ $\mathbf{g}^2 = \frac{1}{y^1 \sin^2 y^3}\left(-\sin y^2 \sin y^3 \mathbf{e}_1 + \cos y^2 \sin y^3 \mathbf{e}_2\right)$ $\mathbf{g}^3 = \frac{1}{y^1}\left(\cos y^2 \cos y^3 \mathbf{e}_1 + \sin y^2 \cos y^3 \mathbf{e}_2 -\right.$ $\left. - \sin y^3 \mathbf{e}_3\right)$
Metric tensors	
$\mathbf{g} = \mathbf{g}_1 \otimes \mathbf{g}_1 + \left(y^1\right)^2 \mathbf{g}_2 \otimes \mathbf{g}_2 +$ $+ \mathbf{g}_3 \otimes \mathbf{g}_3,$ $\mathbf{g} = \mathbf{g}^1 \otimes \mathbf{g}^1 + \left(\frac{1}{y^1}\right)^2 \mathbf{g}^2 \otimes \mathbf{g}^2 +$ $+ \mathbf{g}^3 \otimes \mathbf{g}^3$	$\mathbf{g} = \mathbf{g}_1 \otimes \mathbf{g}_1 + \left(y^1 \sin y^3\right)^2 \mathbf{g}_2 \otimes \mathbf{g}_2 +$ $+ \left(y^1\right)^2 \mathbf{g}_3 \otimes \mathbf{g}_3,$ $\mathbf{g} = \mathbf{g}^1 \otimes \mathbf{g}^1 + \left(\frac{1}{y^1 \sin y^3}\right)^2 \mathbf{g}^2 \otimes \mathbf{g}^2 +$ $+ \left(\frac{1}{y^1}\right)^2 \mathbf{g}^3 \otimes \mathbf{g}^3$
Christoffel symbols	
$\Gamma^2_{12} = \frac{1}{y^1}, \quad \Gamma^1_{22} = -y^1.$	$\Gamma^1_{22} = -y^1 \sin^2 y^3, \quad \Gamma^1_{33} = -y^1, \quad \Gamma^2_{12} = \Gamma^3_{13} = \frac{1}{y^1},$ $\Gamma^3_{22} = -\cos y^3 \sin y^3, \quad \Gamma^2_{32} = \cot y^3.$

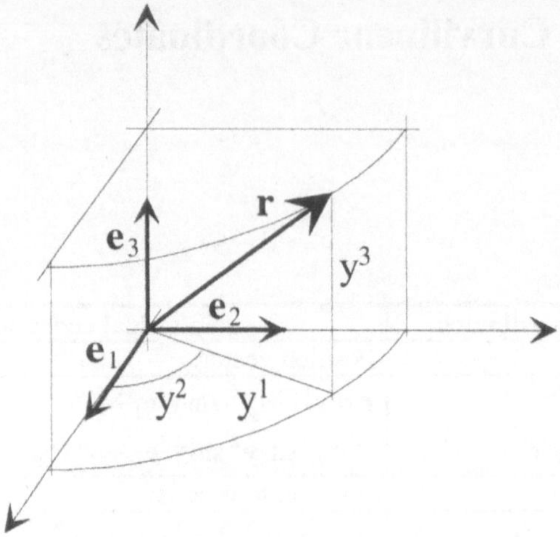

Fig. B.1. *Cylindrical coordinates*
another notations: (r,ϑ,z), (R,Θ,Z).

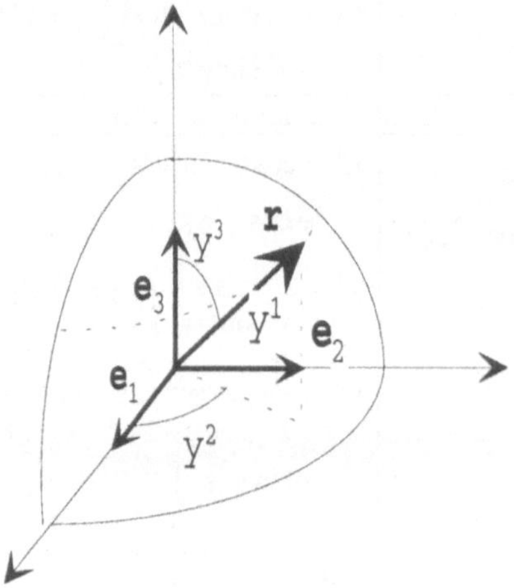

Fig. B.2. *Spherical coordinates*
another notations: (r,ϑ,φ), (R,Θ,Φ)

Appendix C: Hyperbolic Systems of PDE

C.1. Preliminaries

In this book, we only consider physical continuous models whose fields are described by partial differential equations. Therefore, it seems natural to expect that at least some boundary value problems of these equations possess a unique local classical solution which depends continuously on the boundary data. The proofs of this **well-posedness** of the boundary value problems lead usually to complex mathematical considerations. Rather exceptional and physically very important situations arise in the case of the well-posedness of the Cauchy initial value problem for hyperbolic systems of PDS. The theory of such problems has been investigated substantially and the local classical solutions have wave character which matches nicely with physical expectations.

We do not intend to present any self consistent and full theory of hyperbolic systems of PDE in this Appendix. The reader interested in a rigorous mathematical treatment should consult one of many books on this subject*).

The purpose of this Appendix is to present basic notions connected with hyperbolic systems and to illustrate those using simple examples. It was written with the intention of explaining these basic concepts to readers without a professional mathematical background.

C.2. Single Equation of Two Independent Variables

We begin our presentation with the simplest possible case of one unknown scalar field, u, described by the following partial differential equation:

$$a(u,t,x)\frac{\partial u}{\partial t} + b(u,t,x)\frac{\partial u}{\partial x} = c(u,t,x), \tag{C.1}$$

where a, b, c are continuously differentiable and a, b are different from zero.

We seek a solution of the equation (C.1) which is at least once differentiable with respect to t and x, and such that it satisfies the condition

$$u(t,x)\big|_{\mathcal{C}} = U(s), \qquad \mathcal{C}:\ t = \chi(s),\ x = \xi(s), \tag{C.2}$$

*) e.g.: R. COURANT, D. HILBERT; *Methods of Mathematical Physics*, vol.II: R. COURANT; *Partial Differential Equations*, N.Y.-London (1962),
F. JOHN; *Partial Differential Equations*, Springer, N.Y. (1971),
A. JEFFREY, T. TANIUTI; *Non-linear Wave Propagation with Applications to Physics and Magnetohydrodynamics*, Academic Press, N.Y.-London (1964),
W. F. AMES; *Nonlinear Partial Differential Equations in Engineering*, Academic Press (1965).

where U is a given function, $\mathcal{C}$ and s denote an „initial" curve and its parametrization parameter, respectively.

To construct this solution we replace the above problem with an equivalent set of ordinary differential equations. Namely, let us choose a family of curves in the (t,x,u)-space such that a, b, c become components of their tangent vector

$$\frac{dt}{d\tau}=a(u,t,x),\quad \frac{dx}{d\tau}=b(u,t,x),\quad \frac{du}{d\tau}=c(u,t,x), \tag{C.3}$$

where τ is the parameter along each curve and t, x, u are evaluated along these curves.

If we denote by s an arbitrary parameter, then a solution of (C.3) can be written in the form

$$t=t(\tau;s),\quad x=x(\tau;s),\quad u=u(\tau;s). \tag{C.4}$$

A curve defined by (C.4), for an arbitrary but fixed value of s, is called a **characteristic** of the equation (C.1) (see: Fig. C.1.).

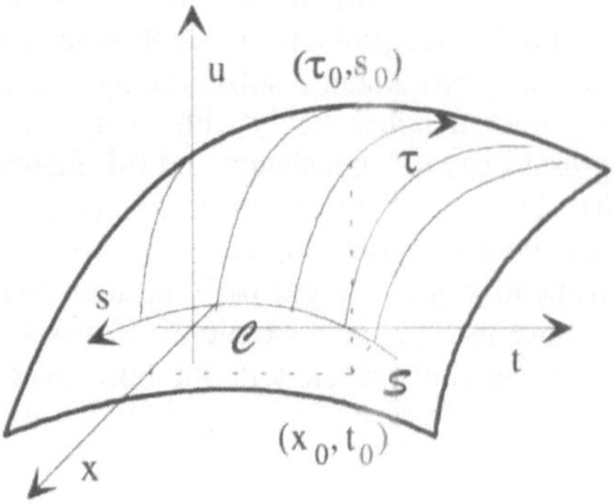

Fig. C.1. *Characteristic curves on the integral surface of equation* (C.1)

It is easy to check that every solution (C.4) of (C.3) satisfies also (C.1). Hence, (C.3) is **equivalent** to (C.1) according to our construction.

Let us assume that the first two relations in (C.4) can be solved **locally** with respect to τ and s, i.e.

$$\tau=\tau(t,x;\tau_0,s_0),\quad s=s(t,x;\tau_0,s_0),$$
$$I\equiv\begin{vmatrix}\frac{\partial t}{\partial\tau} & \frac{\partial t}{\partial s}\\ \frac{\partial x}{\partial\tau} & \frac{\partial x}{\partial s}\end{vmatrix}_{\tau=\tau_0,s=s_0}\neq 0, \tag{C.5}$$

where (τ_0,s_0) are chosen values of parameters.

Then the last relation in (C.4) can be written in the form

$$u-g(t,x;t_0,x_0)=0, \tag{C.6}$$

where

$$t_0 \equiv t(\tau_0, s_0), \quad x_0 \equiv x(\tau_0, s_0), \tag{C.7}$$

$$g(t, x; t_0, x_0) \equiv u\left[\tau(t, x; t_0, x_0), s(t, x; t_0, x_0)\right], \tag{C.8}$$

in the vicinity of the point (t_0, x_0) given by (C.7).

The surface $\mathcal{S}$ satisfying (C.6) and shown also in Fig. C.1. is called the **surface of solutions** or the **integral surface** of the equation (C.1).

The problem of finding such solutions of (C.1), which satisfy the condition (C.2), is called a **Cauchy initial value problem** for the equation (C.1). According to our considerations the curve $\mathcal{C}$ must lie on the surface $\mathcal{S}$.

Now we are in a position to formulate the following

Theorem. Consider the first order quasi-linear partial differential equation

$$a\frac{\partial u}{\partial t} + b\frac{\partial u}{\partial x} = c, \tag{C.9}$$

where a, b, c are continuously differentiable with respect to t, x, u. Suppose that the initial values $u=U(s)$ are prescribed along an initial curve $t=\chi(s)$, $x=\xi(s)$ with χ, ξ, U being continuously differentiable functions of s. Furthermore, let

$$I = \frac{d\xi}{ds} a(\chi(s), \xi(s), U(s)) - \frac{d\chi}{ds} b(\chi(s), \xi(s), U(s)) \neq 0. \tag{C.10}$$

Then there exists a unique solution u(t,x) defined in some neighbourhood of the initial curve which satisfies (C.9) and the initial condition

$$u(\xi(s), \chi(s)) = U(s). \tag{C.11}$$

In such a case we say that the Cauchy problem is **well-posed**. Otherwise, we say that the Cauchy problem is **ill-posed** ■

The transformation of variables (C.5) indicates an interesting property of the derivatives of u appearing in the equation (C.1) and it turns out that it is of particular importance for systems of equations in higher-space dimensions. We will briefly present this property in the following. Namely, if we perform the transformation (C.5) in the equation (C.1), we obtain

$$\frac{\partial u}{\partial \tau}\left(\tilde{a}\frac{\partial \tau}{\partial t} + \tilde{b}\frac{\partial \tau}{\partial x}\right) + \frac{\partial u}{\partial s}\left(\tilde{a}\frac{\partial s}{\partial t} + \tilde{b}\frac{\partial s}{\partial x}\right) = \tilde{c}, \tag{C.12}$$

where

$$\begin{gathered} \tilde{a} \equiv a\left[t(\tau,s), x(\tau,s), u(\tau,s)\right], \quad \tilde{b} \equiv b\left[t(\tau,s), x(\tau,s), u(\tau,s)\right], \\ \tilde{c} \equiv c\left[t(\tau,s), x(\tau,s), u(\tau,s)\right]. \end{gathered} \tag{C.13}$$

Under the condition $(C.5)_3$, we have

$$\left. \begin{aligned} dt &= \frac{\partial t}{\partial \tau} d\tau + \frac{\partial t}{\partial s} ds \\ dx &= \frac{\partial x}{\partial \tau} d\tau + \frac{\partial x}{\partial s} ds \end{aligned} \right\} \Rightarrow \left\{ \begin{aligned} d\tau &= \frac{1}{I}\left(\frac{\partial x}{\partial s} dt - \frac{\partial t}{\partial s} dx \right), \\ ds &= -\frac{1}{I}\left(\frac{\partial x}{\partial \tau} dt - \frac{\partial t}{\partial \tau} dx \right). \end{aligned} \right. \tag{C.14}$$

Hence,

$$\begin{aligned} \frac{\partial \tau}{\partial t} &= \frac{1}{I}\frac{\partial x}{\partial s}, \qquad & \frac{\partial \tau}{\partial s} &= -\frac{1}{I}\frac{\partial t}{\partial s}, \\ \frac{\partial s}{\partial t} &= -\frac{1}{I}\frac{\partial x}{\partial \tau}, \qquad & \frac{\partial s}{\partial x} &= \frac{1}{I}\frac{\partial t}{\partial \tau}. \end{aligned} \tag{C.15}$$

It follows that

$$\tilde{a}\frac{\partial \tau}{\partial t} + \tilde{b}\frac{\partial \tau}{\partial x} = 1, \qquad \tilde{a}\frac{\partial s}{\partial t} + \tilde{b}\frac{\partial s}{\partial x} = 0. \tag{C.16}$$

Therefore, the derivative $\frac{\partial u}{\partial s}$ in the equation (C.12) can be **arbitrary** and is consequently not limited by the equation. This means that for a given u and $\frac{\partial u}{\partial t}$ at an arbitrary point (t,x), $\frac{\partial u}{\partial x}$ is **indeterminate** in the characteristic directions. This property allows us to use an arbitrary initial function U(s) provided that the initial curve is not a characteristic itself. Otherwise, either the solution is not unique or does not exist at all.

We illustrate the above considerations by a few simple examples.

Example C.1. *The Equation with Constant Coefficients*
Consider the following equation

$$\frac{\partial u}{\partial t} + \frac{\partial u}{\partial x} = u^2, \tag{C.17}$$

with

$$u(t = 0, x) = U(x), \tag{C.18}$$

where U is given.

The characteristics of this equation are given by the following equations:

$$\frac{dt}{d\tau} = 1, \quad \frac{dx}{d\tau} = 1, \quad \frac{du}{d\tau} = u^2. \tag{C.19}$$

Hence,

$$t = \tau, \quad x = \tau + s, \quad u = -\frac{1}{\tau + C(s)}. \tag{C.20}$$

It can be seen that the initial curve t=0 is not a characteristic. Bearing the initial condition (C.18) in mind, we obtain the following solution to our problem,

$$u(t,x)=\frac{U(x-t)}{1-tU(x-t)}. \tag{C.21}$$

This solution exists globally except of the points of singularity

$$1-tU(x-t)=0\blacksquare \tag{C.22}$$

Example C.2. *Ill- and Well-posed Cauchy Problems*
Consider the equation

$$t\frac{\partial u}{\partial t}+x\frac{\partial u}{\partial x}=u, \tag{C.23}$$

with

$$u(t=t_0,x)=U(x). \tag{C.24}$$

The characteristics are now given by the equations

$$\frac{dt}{d\tau}=t,\quad \frac{dx}{d\tau}=x,\quad \frac{du}{d\tau}=u. \tag{C.25}$$

Hence,

$$t=A(s)e^{\tau},\quad x=B(s)e^{\tau},\quad u=C(s)e^{\tau}, \tag{C.26}$$

i.e.

$$\frac{t}{x}=\frac{A(s)}{B(s)}\ \text{- constant for each s.} \tag{C.27}$$

This means that the characteristics form a fan of straight lines in the (t,x)-plane which include the origin (0,0). Hence, the line t=0 is a characteristic which implies that no unique solution exists if $t_0=0$. Otherwise, B(s) can be chosen as the parameter defining the characteristic and we have

$$u(t,x)=\frac{t}{t_0}U\left(t_0\frac{x}{t}\right). \tag{C.28}$$

Thus, the Cauchy problem is ill-posed for $t_0=0$ and well-posed for $t_0\neq 0$■

Example C.3. *Non-uniqueness*
We consider the Cauchy problem

$$x\frac{\partial u}{\partial t}+t\frac{\partial u}{\partial x}=u^2,\quad u(t=0,x)=U(x). \tag{C.29}$$

Then the characteristics are defined by

$$\frac{dt}{d\tau}=x,\quad \frac{dx}{d\tau}=t,\quad \frac{du}{d\tau}=u^2. \tag{C.30}$$

The first two equations have two families of solutions,

$$\left.\begin{aligned} t&=s\sinh\tau\\ x&=s\cosh\tau\end{aligned}\right\}\quad \text{for}\quad s^2=x^2-t^2>0, \tag{C.31}$$

and

$$\left.\begin{aligned} t&=s\cosh\tau\\ x&=s\sinh\tau\end{aligned}\right\}\quad \text{for}\quad s^2=t^2-x^2>0. \tag{C.32}$$

They are shown in Fig. C.2.

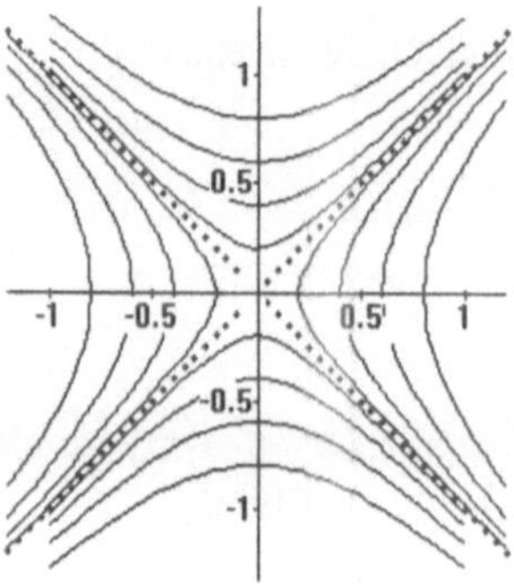

Fig. C.2. *The characteristics for the example* C.3.

The initial line t=0 does not cross the characteristics of the family (C.32). Consequently, this family, which is not lying on the integral surface, cannot be used for the construction of the solution of the problem (C.29). Bearing the relations (C.31) in mind, we easily solve the equations (C.30) satisfying the condition $(C.29)_2$. It follows that

$$u(t,x)=\frac{U\left(\sqrt{x^2-t^2}\right)}{1-\ln\sqrt{\frac{x+t}{x-t}}\,U\left(\sqrt{x^2-t^2}\right)},\quad x^2-t^2>0. \tag{C.33}$$

Hence, the function u is undetermined for all points (x,t) satisfying the condition $x^2-t^2<0$. The Cauchy problem (C.29) has only a local solution in a neighbourhood of the initial line. The extension of this solution to points beyond the domain $x^2-t^2>0$ is not unique. For instance, we can assume that

$$u(t,x)=0 \quad \text{for} \quad x^2-t^2<0, \tag{C.34}$$

or

$$u(t,x)=\frac{U\left(\sqrt{x^2-t^2}\right)}{1-\ln\sqrt{\dfrac{x+t}{x-t}}\,U\left(\sqrt{x^2-t^2}\right)} \quad \text{for} \quad x^2-t^2<0. \tag{C.35}$$

Obviously, both extensions of (C.33) satisfy the equation $(C.29)_1$ and do not violate the initial condition because the regions of (C.33) and either of (C.34) or (C.35) are disjoint■

Example C.4. *Quasilinear Equation; Weak Solutions*
We proceed to a problem in which the coefficients of the derivatives depend on the unknown function u. Let us consider the simplest Cauchy problem of this type

$$\frac{\partial u}{\partial t}+u\frac{\partial u}{\partial x}=0, \quad u(t=0,x)=U(x), \quad t>0. \tag{C.36}$$

The characteristics are now described by the equations

$$\frac{dt}{d\tau}=1, \quad \frac{dx}{d\tau}=u, \quad \frac{du}{d\tau}=0, \tag{C.37}$$

i.e. the function u remains constant along each characteristic carrying its initial value without any changes. The solution of the set (C.37) has the following form:

$$t=\tau, \quad x=s+u\tau, \quad u(\tau,s=\text{const.})=\text{const.} \tag{C.38}$$

The initial line t=0 is not a characteristic and the implicit form of the solution of (C.36) is as follows:

$$u(t,x)=U(x-ut). \tag{C.39}$$

Let us illustrate this result by the following three examples:

$$1)\ U(x)=\begin{cases}0 & \text{for} \quad x<0,\\ x & \text{for} \quad 0\le x\le 1,\\ 1 & \text{for} \quad x>1,\end{cases} \qquad 2)\ U(x)=\begin{cases}1 & \text{for} \quad x<0,\\ 1-x & \text{for} \quad 0\le x\le 1,\\ 0 & \text{for} \quad x>1,\end{cases} \tag{C.40}$$

$$3)\ U(x)=\begin{cases}0 & \text{for} \quad x<0,\\ 1 & \text{for} \quad x>1.\end{cases}$$

The corresponding characteristics on the (t,x) - plane are shown in Fig. C.3.

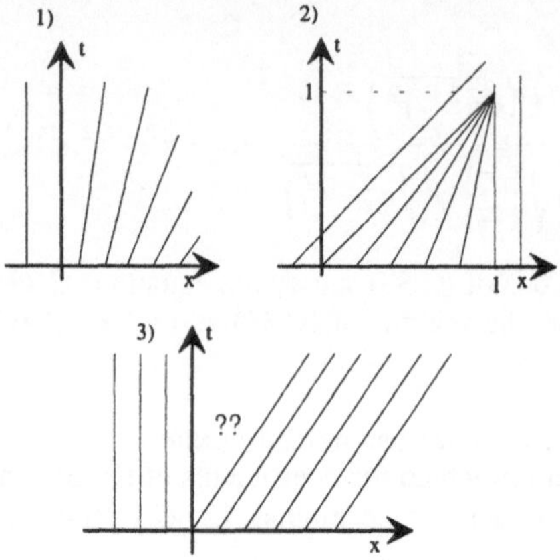

Fig. C.3. *The characteristics for the Example* C.4.

In case 1) there are no difficulties constructing the solution and we easily obtain

$$u(t,x)=\begin{cases}0 \quad \text{for} \quad x<0, \\ x(1-t) \quad \text{for} \quad 0\le x\le 1, \\ 1-t \quad \text{for} \quad x>1,\end{cases} \qquad t>0. \tag{C.41}$$

In case 2) the difficulties start occurring from t=1. It can be seen that at the point (1,1) all characteristics starting in the interval <0,1> at the initial line have a common point. Simultaneously, each characteristic brings its own value of u to this point [compare $(C.38)_3$]. This means that we are not able to define u at this point in any unique manner. The classical solution of (C.36) ceases to exist. In Fig. C.4. we show the plots of u for three different instants of time. It can be observed that for t=1 the derivative $\frac{\partial u}{\partial x}$ becomes infinite. This property allows us to find the maximum time of existence of the classical solution of the form (C.39). This is not only possible for our simple example but also in a quite general case discussed further on in this Appendix.

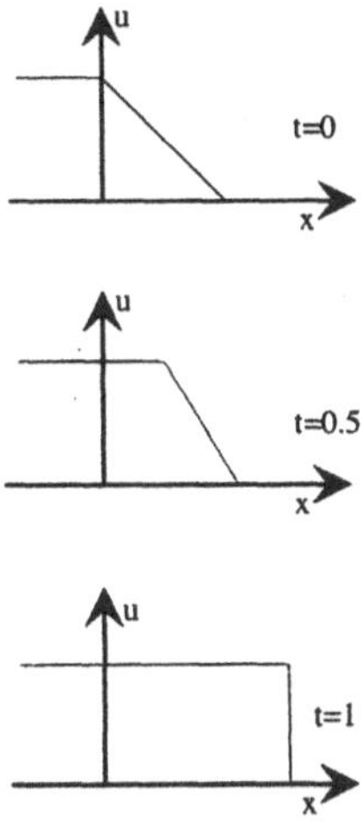

Fig. C.4. *Plots of* u(t,x) *of the example* C.4. *for different times* t

Differentiation of (C.36) with respect to x yields for each s

$$\frac{d a}{d\tau}+a^2=0, \qquad a\equiv\frac{\partial u}{\partial x}, \quad \frac{d a}{d\tau}=\frac{\partial a}{\partial t}+u\frac{\partial a}{\partial x}, \tag{C.42}$$

with

$$a(\tau=0;s)=U'(s)\equiv\frac{dU}{ds}(s). \tag{C.43}$$

Hence,

$$a(\tau;s)=\frac{U'(s)}{1+\tau U'(s)}. \tag{C.44}$$

This solution for the **amplitude** a has a singularity at the point

$$\tau=-\frac{1}{U'(s)}.$$

Hence, we expect the above described **critical time** t_c to be

$$t_c=\min\left\{-\frac{1}{U'(s)}\,\middle|\,U'(s)<0\right\}. \tag{C.46}$$

In our three examples we have

$$\begin{array}{ccc} 1) & 2) & 3) \\ U'(s)\geq 0, & U'(s)=\begin{cases}0 \text{ for } s<0,\\ -1 \text{ for } 0\leq s\leq 1,\\ 0 \text{ for } s>0,\end{cases} & U'(s)=0.\end{array} \tag{C.47}$$

Consequently, only case 2) leads to a critical time which is equal to 1 ■

Let us mention in passing that for $t>t_c$ we can construct the so-called **weak solutions** which satisfy a global balance equation following from the differential equation by integrating over the domain of space which is of physical interest. The construction is based on Kotchine's jump conditions which we discuss in Chap. 4. This procedure is possible if the differential equation is of the form of a balance law. For instance, in our example C.4. we have

$$\frac{\partial u}{\partial t}+\frac{\partial}{\partial x}\left(\tfrac{1}{2}u^2\right)=0, \tag{C.48}$$

and the above condition is satisfied.

Such weak solutions are known in the gas dynamics as **shock waves**. In the case of shear waves in the non-Newtonian fluids they appear as **vortex sheets**. Similar structures also appear in structural plasticity.

Due to the lack of space in this book we do not go into any details concerning weak solutions despite their strong connection with thermodynamics. The latter should indicate a criterion for the choice of solutions.

Let us return to case 3) of our example C.4. It is apparent that the critical time does not appear for the initial condition $(C.40)_3$ in the range of interest. However, here we face a problem similar to that discussed in the example C.3. The solution u cannot be specified in the triangular region indicated by the question marks in Fig. C.3. We can extend the solution which is of the form

$$u(t,x)=\begin{cases}0 & \text{for} \quad x<0,\\ 1 & \text{for} \quad x-t>0,\end{cases} \tag{C.49}$$

to this region in an infinite number of ways, i.e. the solution for the whole (t,x)-space is not unique. For instance, we can assume the following continuous extension:

$$u(t,x)=\frac{x}{t} \quad \text{for} \quad x>0 \quad \text{and} \quad x-t<0, \tag{C.50}$$

which, certainly, satisfies the equation $(C.36)_1$ and does not violate the initial condition. The choice of such an extension is not connected with the mathematical formulation of the problem and must be done, if needed, by means of some additional arguments.

C.3. Set of Equations with Two Independent Variables

Let us turn our attention to the **sets of quasilinear differential equations** for a vector-valued function $\mathbf{u}=\mathbf{u}(t,x)\in\mathcal{V}^N$. This function is assumed to satisfy the equation

$$\mathbf{A}^0\frac{\partial \mathbf{u}}{\partial t}+\mathbf{A}\frac{\partial \mathbf{u}}{\partial x}=\mathbf{B}, \tag{C.51}$$

where the square matrices $\mathbf{A}^0$, $\mathbf{A}\in\mathcal{V}^N\times\mathcal{V}^N$ and the vector $\mathbf{B}\in\mathcal{V}^N$ are assumed to be continuously differentiable functions of **u** alone. For the sake of simplicity, we shall not discuss the general case in which these coefficients also depend on t and x in an explicit manner.

The **method of characteristics** in this case is constructed in a manner which is a direct generalization of the method for a scalar function u presented in the previous section. Namely, we shall try to combine the equations (C.51) in such a way that the combination has the form of a **directional derivative** along a certain curve.

To this aim, let us form a scalar product of (C.51) with a vector $\mathbf{l} \in \mathcal{V}^N$. We get

$$\left(\mathbf{A}^{0T}\mathbf{l}\right) \cdot \frac{\partial \mathbf{u}}{\partial t} + \left(\mathbf{A}^T\mathbf{l}\right) \cdot \frac{\partial \mathbf{u}}{\partial x} = \mathbf{l} \cdot \mathbf{B}. \tag{C.52}$$

The left-hand side of this equation is a directional derivative if

$$\mathbf{A}^{0T}\mathbf{l} = \mathbf{m}\alpha, \qquad \mathbf{A}^T\mathbf{l} = \mathbf{m}\beta, \qquad \mathbf{m} \in \mathcal{V}^N, \tag{C.53}$$

where α, β are the components of the tangent vector to a curve in the (t,x)-plane along which we want to differentiate

$$\alpha = \frac{dt}{d\tau}, \qquad \beta = \frac{dx}{d\tau}, \qquad \mathbf{m} \cdot \frac{d\mathbf{u}}{d\tau} = \mathbf{l} \cdot \mathbf{B}, \tag{C.54}$$

and **m** is for the time being an unspecified vector.

Multiplying the relation $(C.53)_1$ by β, $(C.53)_2$ - by α and subtracting the results we arrive at

$$\left(\mathbf{A}^{0T}\beta - \mathbf{A}^T\alpha\right)\mathbf{l} = 0. \tag{C.55}$$

This certainly is the generalized eigenvalue problem with **l** as the **left eigenvector**. For α and β we have the following relation:

$$\det\left(\mathbf{A}^0\beta - \mathbf{A}\alpha\right) = 0, \tag{C.56}$$

which is obviously a condition for the existence of non-trivial solutions of (C.55).

After solving the above eigenvalue problem, we can find the vector **m** from one of the equations (C.53).

Simultaneously, each solution of the algebraic problem (C.55), (C.56) yields the corresponding set of equations (C.54). The family of curves given by each solution of these equations is called **characteristic** for the set (C.51).

In order to reproduce the set of equations (C.51) by the set of ordinary differential equations (C.54), we need N linearly independent vectors **l** and N pairs of the real components α, β of the tangent vector to the characteristic curves in the (t,x)-plane. This is, of course, not the case for arbitrary matrices $\mathbf{A}^0$ and **A**. We shall specify the appropriate conditions in the definition of hyperbolicity.

Before we do so, let us investigate another aspect of the characteristics. Namely, we perform the transformation of variables

$$\xi = \xi(t,x), \quad \eta = \eta(t,x), \qquad \begin{vmatrix} \xi_t & \xi_x \\ \eta_t & \eta_x \end{vmatrix} \neq 0, \tag{C.57}$$

with

$$\xi_t = \frac{\partial \xi}{\partial t}, \quad \xi_x = \frac{\partial \xi}{\partial x}, \quad \eta_t = \frac{\partial \eta}{\partial t}, \quad \eta_x = \frac{\partial \eta}{\partial x}. \tag{C.58}$$

Then we obtain

$$\frac{\partial \mathbf{u}}{\partial t} = \xi_t \frac{\partial \mathbf{u}}{\partial \xi} + \eta_t \frac{\partial \mathbf{u}}{\partial \eta}, \quad \frac{\partial \mathbf{u}}{\partial x} = \xi_x \frac{\partial \mathbf{u}}{\partial \xi} + \eta_x \frac{\partial \mathbf{u}}{\partial \eta}. \tag{C.59}$$

Substitution in (C.51) yields

$$\left(\xi_t \mathbf{A}^0 + \xi_x \mathbf{A}\right)\frac{\partial \mathbf{u}}{\partial \xi} + \left(\eta_t \mathbf{A}^0 + \eta_x \mathbf{A}\right)\frac{\partial \mathbf{u}}{\partial \eta} = \mathbf{B}. \tag{C.60}$$

If any of the matrices was degenerate, i.e

$$\det \mathbf{A}^0 = 0 \quad \text{and / or} \quad \det \mathbf{A} = 0, \tag{C.61}$$

then the above transformation would eliminate such a degeneracy by an appropriate choice of the transformation (C.57). However this could destroy the invariance of the time-like and space-like directions which is certainly, undesirable for physical reasons. We can prevent this by assuming that the relation

$$\det\left(\lambda \mathbf{A}^0 + \mu \mathbf{A}\right) \neq 0 \tag{C.62}$$

is satisfied for some λ and μ, which are not simultaneously equal to zero. These non-zero λ and μ determine the transformation (C.53).

Bearing the above remarks in mind, we can define the notion of hyperbolicity in the following manner.

The set of equations (C.51) subject to the condition (C.62) for some non-zero λ, μ is called **hyperbolic** if there exist N linearly independent vectors $\mathbf{l}^{(A)}$, A=1,...,N, such that

$$\left(\beta^{(A)} \mathbf{A}^{0T} - \alpha^{(A)} \mathbf{A}^{T}\right)\mathbf{l}^{(A)} = 0, \tag{C.63}$$

for any $A \in \{1,...,N\}$ and the directions $(\alpha^{(A)}, \beta^{(A)})$ are real and such that

$$\left(\alpha^{(A)}\right)^2 + \left(\beta^{(A)}\right)^2 \neq 0. \tag{C.64}$$

If not only all $\mathbf{l}^{(A)}$ are linearly independent (i.e. they span the space of solutions) but also all directions $(\alpha^{(A)}, \beta^{(A)})$ are distinct (N different families of characteristics!) then the system (C.51) is called **strongly hyperbolic**.

In the particular case

$$\det \mathbf{A}^0 \neq 0, \tag{C.65}$$

we say that the set (C.51) is **hyperbolic in the t-direction**. Then the set can be written in a **normal form**

$$\frac{\partial \mathbf{u}}{\partial t} + A\frac{\partial \mathbf{u}}{\partial x} = B, \qquad A \equiv \mathbf{A}^{0-1}\mathbf{A}, \quad B \equiv \mathbf{A}^{0-1}\mathbf{B}. \tag{C.66}$$

Further considerations in this section are restricted to such sets.

Now we reformulate the method of characteristics for this particular case. The linear combination of the equations (C.66) by means of the scalar multiplication with a vector $\mathbf{l} \in \mathcal{V}^N$ yields

$$\mathbf{l}\cdot\left(\frac{\partial \mathbf{u}}{\partial t}+A\frac{\partial \mathbf{u}}{\partial x}\right)=\mathbf{l}\cdot B. \tag{C.67}$$

Let us choose $\mathbf{l}$ in such a way that

$$A^T\mathbf{l}=\mu\mathbf{l}. \tag{C.68}$$

Then

$$\frac{dt}{d\tau}=1, \qquad \frac{dx}{d\tau}=\mu, \qquad \mathbf{l}\cdot\frac{d\mathbf{u}}{d\tau}=\mathbf{l}\cdot B, \tag{C.69}$$

define the **characteristics** of (C.66). It is obvious that $\mathbf{l}$ and μ are the **left eigenvector** of A and the **eigenvalue** of A, respectively. If μ is real then it is called the **characteristic speed**.

Now we can reformulate the definition of hyperbolicity for the set (C.66). It is called **hyperbolic** if the matrix A has N **real** (not necessarily distinct) eigenvalues $\mu^{(A)}$

$$\det\left(A-\mu^{(A)}\mathbf{1}\right)=0, \qquad A=1,\cdots,N, \tag{C.70}$$

corresponding to N **linearly independent** left eigenvectors $\mathbf{l}^{(A)}$. We have immediately:

Lemma: If A is symmetric and real then the set (C.66) is hyperbolic ■

Such systems are called **symmetric hyperbolic.**

Let us consider two examples of Cauchy initial value problems.

Example C.5. *A Cauchy Problem for Two Unknown Functions*

We investigate the following set of equations:

$$\frac{\partial u}{\partial t}+v\frac{\partial u}{\partial x}=0, \qquad \frac{\partial v}{\partial t}+u\frac{\partial v}{\partial x}=0, \tag{C.71}$$

with the initial conditions

$$u(t=0,x)=U(x), \qquad v(t=0,x)=V(x). \tag{C.72}$$

The characteristics of this system are determined by the following eigenvalue problem:

$$\det(A-\mu\mathbf{1})=\begin{vmatrix} v-\mu & 0 \\ 0 & u-\mu \end{vmatrix}=0 \;\Rightarrow\; \mu^{(1)}=v, \quad \mu^{(2)}=u, \tag{C.73}$$

i.e. u and v are characteristic speeds.

The left eigenvectors defined by the equations

$$\left(A^T-\mu^{(A)}\mathbf{1}\right)\mathbf{l}^{(A)}=0, \quad A=1,2, \tag{C.74}$$

follow in the form

$$\mathbf{l}^{(1)}=(1,0)^T, \quad \mathbf{l}^{(2)}=(0,1)^T. \tag{C.75}$$

Hence, they are linearly independent and the system (C.71) is hyperbolic.

The equations for the characteristics (C.69) are now

$$\begin{aligned} &1)\ \frac{dt}{d\alpha}=1, \quad \frac{dx}{d\alpha}=v, \qquad \text{i.e.}\ \ dx=v\,dt, \quad \frac{du}{d\alpha}=0, \\ &2)\ \frac{dt}{d\beta}=1, \quad \frac{dx}{d\beta}=u, \qquad \text{i.e.}\ \ dx=u\,dt, \quad \frac{dv}{d\beta}=0, \end{aligned} \tag{C.76}$$

α, β being the parameters along the characteristic curves in the (t,x)-plane.

It can be seen that u is constant along the first family, and v is constant along the second family of characteristics. If these were known then we could construct the solution at an arbitrary point (t,x). However this is not the case.

Let us consider three different possibilities for (C.73) and (C.76).

1. $U(x)=U_0=\text{const.}$

Then we have

$$u(t,x)=U_0=\text{const.} \quad \text{for all} \quad t \quad \text{and} \quad x. \tag{C.77}$$

Hence, the equation $(C.71)_2$ has the solution

$$v(t,x)=\tilde{v}(x-U_0t), \tag{C.78}$$

with

$$v(t=0,x)=V(x). \tag{C.79}$$

Finally,

$$u(t,x)=U_0, \quad v(t,x)=V(x-U_0t), \tag{C.80}$$

is the solution of the problem.

2. $V(x)=V_0=\text{const.}$ Similar considerations yield the solution

$$u(t,x)=U(x-V_0t), \quad v(t,x)=V_0, \qquad u(t=0,x)\equiv U(x). \tag{C.81}$$

3. Neither U nor V is constant. Then locally

$$x=U^{-1}(u) \ \Rightarrow v=V(x)=V\circ U^{-1}(u), \tag{C.82}$$

on the initial line t=0.

It is convenient to identify α with v and β with u. By doing so we obtain the equations of characteristics

$$\begin{aligned} dx=v\,dt \quad &\text{on} \quad u=\text{const.}, \\ dx=u\,dt \quad &\text{on} \quad v=\text{const.}, \end{aligned} \tag{C.83}$$

i.e.

$$\frac{\partial x}{\partial v} = v\frac{\partial t}{\partial v}, \quad \frac{\partial x}{\partial u} = u\frac{\partial t}{\partial u}, \tag{C.84}$$

with the following initial conditions:

$$x = U^{-1}(u), \qquad t = 0 \quad \text{on} \quad v = V \circ U^{-1}(u). \tag{C.85}$$

This system is **linear**. We construct the solution in the following manner. Elimination of x from (C.84) yields

$$\frac{\partial^2 t}{\partial u \partial v} = 0. \tag{C.86}$$

Hence, we deduce

$$t = t^1(u) + t^2(v), \tag{C.87}$$

with the initial conditions following from (C.85)

$$\begin{aligned} &t^1(u) + t^2\left[V \circ U^{-1}(u)\right] = 0, \\ &u\frac{dt^1}{du} = \frac{d}{du}U^{-1}(u). \end{aligned} \tag{C.88}$$

Let us consider a simple example for the functions U and V

$$\begin{aligned} U(x) &= \begin{cases} 1 - x & \text{for } 0 \le x \le 1, \\ 0 & \text{otherwise}, \end{cases} \\ V(x) &= \begin{cases} x & \text{for } 0 \le x \le 1, \\ 0 & \text{otherwise}. \end{cases} \end{aligned} \tag{C.89}$$

This implies

$$U^{-1}(u) = 1 - u, \tag{C.90}$$

and it means that the initial conditions for t=t(u,v) should be formulated on the curve $\mathcal{C}^0$ given by

$$u + v = 1. \tag{C.91}$$

On this curve we have

$$t|_{\mathcal{C}^0} = 0, \quad x|_{\mathcal{C}^0} = 1 - u. \tag{C.92}$$

Hence, the directional derivative of t and x on $\mathcal{C}^0$ has the form

$$\begin{aligned} \operatorname{grad} t \cdot (1,-1)^T &= \left.\frac{\partial t}{\partial u}\right|_{\mathcal{C}^0} - \left.\frac{\partial t}{\partial v}\right|_{\mathcal{C}^0} = 0, \\ \operatorname{grad} x \cdot (1,-1)^T &= \left.\frac{\partial x}{\partial u}\right|_{\mathcal{C}^0} - \left.\frac{\partial x}{\partial v}\right|_{\mathcal{C}^0} = -1. \end{aligned} \tag{C.93}$$

Bearing the equations (C.84) in mind, we easily get

$$\left.\frac{\partial t}{\partial u}\right|_{e^0}=\frac{1}{1-2u},\quad \left.\frac{\partial t}{\partial v}\right|_{e^0}=\frac{1}{2v-1}\equiv\frac{1}{1-2u},$$
$$\left.\frac{\partial x}{\partial u}\right|_{e^0}=\frac{1}{1-2u},\quad \left.\frac{\partial x}{\partial v}\right|_{e^0}=\frac{v}{2v-1}\equiv\frac{1-u}{1-2u}. \tag{C.94}$$

On the other hand, the relation (C.87) combined with $(C.94)_{1,2}$ gives rise to the following relations:

$$\frac{dt^1}{du}=\frac{1}{1-2u}\quad\Rightarrow\quad t^1=-\tfrac{1}{2}\ln\left(u-\tfrac{1}{2}\right)+c_1,$$
$$\frac{dt^2}{du}=\frac{1}{2v-1}\quad\Rightarrow\quad t^2=\tfrac{1}{2}\ln\left(\tfrac{1}{2}-v\right)+c_2. \tag{C.95}$$

Hence, using $(C.92)_2$, we obtain

$$t=\tfrac{1}{2}\ln\left(\frac{1-2v}{2u-1}\right). \tag{C.96}$$

In the similar way we get the solution for x

$$x=\tfrac{1}{2}(v-u+1)+\tfrac{1}{4}\ln\left(\frac{1-2v}{2u-1}\right). \tag{C.97}$$

Inversion of (C.96) and (C.97) yields the solution for u and v

$$u=\tfrac{1}{2}-\frac{2x-t-1}{1+e^{2t}},$$
$$v=\tfrac{1}{2}+e^{2t}\frac{2x-t-1}{1+e^{2t}}. \tag{C.98}$$

The equations for the characteristics (C.83) can now be written in the form

$$\frac{dx}{dt}=\tfrac{1}{2}+e^{2t}\frac{2x-t-1}{1+e^{2t}}\quad\Rightarrow\quad x=s\left(1+e^{2t}\right)+\tfrac{1}{2}(t+1)\quad\text{with } u=\text{const.},$$
$$\frac{dx}{dt}=\tfrac{1}{2}-\frac{2x-t-1}{1+e^{2t}}\quad\Rightarrow\quad x=s\frac{1+e^{2t}}{e^{2t}}+\tfrac{1}{2}(t+1)\quad\text{with } v=\text{const.}, \tag{C.99}$$
$$s\in\left\langle-\tfrac{1}{4},\tfrac{1}{4}\right\rangle.$$

The first family of these characteristics is shown in Fig. C.5. as solid lines. The values of the parameter s have been chosen to be -0.25, -0.125, 0, 0.125, 0.25. The horizontal axis corresponds to the variable t and due to the form of the initial conditions (C.89), all characteristics start from t=0 in the x-interval <0,1>. The dotted lines in this figure correspond to the second family of characteristics with the same values of the parameter s as for the first family ■

Example C.6. *A Non-hyperbolic System for Two Unknown Functions*
Let us consider the two sets of equations

$$\begin{aligned} &1)\ \frac{\partial u}{\partial t}+u^2\frac{\partial v}{\partial x}=0, \qquad \frac{\partial v}{\partial t}-v^2\frac{\partial u}{\partial x}=0,\\ &2)\ \frac{\partial u}{\partial t}+u^2\frac{\partial v}{\partial x}=0, \qquad \frac{\partial v}{\partial t}+v^2\frac{\partial u}{\partial x}=0. \end{aligned} \tag{C.100}$$

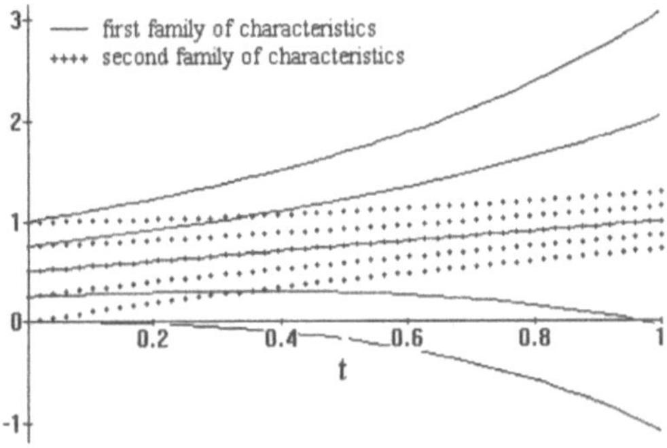

Fig. C.5. *Characteristics to the example* C.5.

The corresponding eigenvalue problems are:

$$\begin{aligned} &1)\ \begin{vmatrix} -\mu & u^2 \\ -v^2 & -\mu \end{vmatrix} = \mu^2 + u^2v^2 = 0,\\ &2)\ \begin{vmatrix} -\mu & u^2 \\ v^2 & -\mu \end{vmatrix} = \mu^2 - u^2v^2 = 0. \end{aligned} \tag{C.101}$$

Hence, the first system cannot be hyperbolic due to the imaginary eigenvalues μ. It is easy to check that the second system is hyperbolic. The left eigenvectors are

$$\mathbf{l}^{(1)} = (u, v)^T, \qquad \mathbf{l}^{(2)} = (u, -v)^T, \tag{C.102}$$

and they are linearly independent ■

C.4. Time of Existence of Classical Solutions

We have indicated in the previous sections that classical solutions of the Cauchy problem may cease to exist after a finite (critical) time. By means of the geometrical properties of characteristics, one can find the range of independent variables in which the classical solutions do exist. Beyond that range the system of equations may have only weak solutions, if any.

In this section we present only one aspect of this problem. Namely, we show how the time t_c of existence of classical solutions can be found in the case of one spatial variable.

The set of equations for the unknown vector function $\mathbf{u} \in \mathcal{V}^N$ is assumed to have the normal form

$$\frac{\partial \mathbf{u}}{\partial t}+A\frac{\partial \mathbf{u}}{\partial x}=B, \tag{C.103}$$

where the matrix A and the production B are differentiable functions of **u** alone. It is assumed that only the derivatives of **u** may have finite discontinuities when crossing a curve in the (t,x)-plane given by the relation

$$\mathcal{C}(t,x)=0. \tag{C.104}$$

The function **u** is assumed to be twice continuously differentiable except at the points on $\mathcal{C}$ where it possesses one-sided derivatives. In particular we have

$$[[\mathbf{u}]]\equiv \mathbf{u}^{+}-\mathbf{u}^{-}=0 \quad \text{on } \mathcal{C}, \tag{C.105}$$

where $\mathbf{u}^+$ and $\mathbf{u}^-$ are the limits of **u** on both sides of $\mathcal{C}$.

If c denotes the speed of propagation along that curve then the derivatives of **u** satisfy the following compatibility conditions:

$$\left[\left[\frac{\partial \mathbf{u}}{\partial t}\right]\right]=-c\left[\left[\frac{\partial \mathbf{u}}{\partial x}\right]\right], \tag{C.106}$$

$$\left[\left[\frac{\partial^2 \mathbf{u}}{\partial x\partial t}\right]\right]=\frac{d}{d\tau}\left[\left[\frac{\partial \mathbf{u}}{\partial x}\right]\right]-c\left[\left[\frac{\partial^2 \mathbf{u}}{\partial x^2}\right]\right], \tag{C.107}$$

τ ($\equiv$t) being the parameter along the curve and the time derivative of the jump in (C.107) is calculated in the reference system moving with the point of the curve.

Exercise C.1. Prove the compatibility conditions (C.106), (C.107) (the so-called Maxwell relations)•

The discontinuity

$$\mathbf{a}\equiv\left[\left[\frac{\partial \mathbf{u}}{\partial x}\right]\right], \tag{C.108}$$

is called the **amplitude of the weak discontinuity**.

Taking the limits of (C.103) on both sides of the curve $\mathcal{C}$ and bearing (C.105) in mind, we obtain

$$(A-c\mathbf{1})\mathbf{a}=0, \tag{C.109}$$

which follows from the assumption that A and B are functions of **u** alone.

The equation (C.109) shows that c coincides with the eigenvalues of the matrix A (the **characteristic speeds of propagation**) and the amplitude **a** is the right eigenvector of A. This means that the curve $\mathcal{C}$ must be a characteristic.

We expect that the classical solution of (C.103) ceases to exist when the magnitude of the amplitude **a** becomes infinite. To find the instant of time t_c in which this happens, we differentiate the equation (C.103) with respect to x and then calculate the limits on the curve. The result written in Cartesian coordinates has the form

$$\frac{d a_A}{d\tau} - c\left[\left[\frac{\partial^2 u_A}{\partial x^2}\right]\right] + \frac{\partial A_{AB}}{\partial u_C}\left[\left[\frac{\partial u_C}{\partial x}\frac{\partial u_B}{\partial x}\right]\right] + \\ + A_{AB}\left[\left[\frac{\partial^2 u_B}{\partial x^2}\right]\right] = \frac{\partial B_A}{\partial u_C} a_C, \quad A, B, C = 1, \cdots, N. \tag{C.110}$$

Multiplying this equation by the left eigenvector **l** of A corresponding to the eigenvalue c, we obtain

$$l_A \frac{d a_A}{d\tau} + l_A \frac{\partial A_{AB}}{\partial u_C}\left\{ - a_C a_B + \left(\frac{\partial u_C}{\partial x}\right)^+ a_B + \left(\frac{\partial u_B}{\partial x}\right)^+ a_C \right\} = l_A \frac{\partial B_A}{\partial u_C} a_C. \tag{C.111}$$

We assume the solution **u** to be known on the positive side of $\mathcal{C}$. Then bearing the fact that **a** is the right eigenvector of A in mind, we have

$$\frac{d\pi}{d\tau} + \alpha_1 \pi + \alpha_2 \pi^2 = 0, \tag{C.112}$$

where, due to the t-hyperbolicity, the parameter τ is to be identified with t and

$$a_A = \pi(\tau) r_A, \tag{C.113}$$

i.e. π determines the amplitude of the discontinuity. Furthermore, we have

$$\alpha_1 = \left[l_A \frac{\partial A_{AB}}{\partial u_C}\left(r_B \left(\frac{\partial u_C}{\partial x}\right)^+ + r_C \left(\frac{\partial u_B}{\partial x}\right)^+ \right) + \right. \\ \left. + l_A \frac{\partial r_A}{\partial u_B}\left(\frac{d u_B}{d\tau}\right) - l_A \frac{\partial B_A}{\partial u_B} r_B \right] \frac{1}{\mathbf{r}\cdot\mathbf{l}}; \\ \alpha_2 = \left[- l_A \frac{\partial A_{AB}}{\partial u_C} r_B r_C \right] \frac{1}{\mathbf{r}\cdot\mathbf{l}}, \tag{C.114}$$

r being the right eigenvector of A calculated on, for example, the positive side of $\mathcal{C}$.

The Bernoulli equation (C.112) can be solved easily for a given initial value of the amplitude **a**. Let us denote the corresponding initial value of the magnitude by π_0. Then we obtain

$$\frac{1}{\pi} = \left\{ \frac{1}{\pi_0} + \int_0^t \alpha_2 \exp\left(- \int_0^\tau \alpha_1 d\eta \right) d\tau \right\} \exp \int_0^t \alpha_1 d\tau, \tag{C.115}$$

and the critical time shall be reached for $1/\pi = 0$, i.e.

$$\frac{1}{\pi_0} + \int_0^{t_c} \alpha_2 \exp\left(- \int_0^\tau \alpha_1 d\eta \right) d\tau = 0. \tag{C.116}$$

We shall illustrate these considerations by a simple example.

Example C.7. *The Critical Time for an Isothermal Flow of an Ideal Gas*
Let us consider the isothermal one-dimensional flow of an ideal gas described by the following field equations:

$$\begin{aligned} &\frac{\partial \rho}{\partial t}+v\frac{\partial \rho}{\partial x}+\rho\frac{\partial v}{\partial x}=0, \\ &\frac{\partial v}{\partial t}+\frac{c^2}{\rho}\frac{\partial \rho}{\partial x}+v\frac{\partial v}{\partial x}=0, \qquad c^2=\frac{RT}{M}=\text{const.}, \end{aligned} \tag{C.117}$$

for the fields of the mass density ρ and the velocity v. R, T, and M denote the universal gas constant (≈8.3 kJ/kg·K), a constant (absolute) temperature, and the molecular weight, respectively.

The matrix A is then given by

$$(A)=\begin{pmatrix} v & \rho \\ \frac{c^2}{\rho} & v \end{pmatrix}. \tag{C.118}$$

Its eigenvalues (characteristic speeds) follow immediately

$$U^{(1),(2)}=v\pm c, \tag{C.119}$$

and the corresponding right and left eigenvectors are

$$\begin{aligned} &\mathbf{r}^{(1)}=(\rho,c)^T, \quad \mathbf{r}^{(2)}=(\rho,-c)^T, \\ &\mathbf{l}^{(1)}=(c,\rho)^T, \quad \mathbf{l}^{(2)}=(-c,\rho)^T. \end{aligned} \tag{C.120}$$

The initial state of the gas for $t<0^-$ is assumed to be ρ_0=const., $v_0=0$. The coefficients (C.114) follow as

$$\begin{aligned} &\alpha_1=0, \\ &\alpha_2^{(1)}=\alpha_2^{(2)}=-c. \end{aligned} \tag{C.121}$$

Hence, the critical time can be calculated using the relation $(C.116)_2$ and we obtain

$$t_c=\frac{1}{c\pi_0}, \tag{C.122}$$

where π_0 is the initial value of the magnitude π(t). Generally, we have on the characteristics

$$\begin{aligned} &a_1=\left[\left[\frac{\partial \rho}{\partial x}\right]\right]=r_1\pi=\rho\pi, \\ &a_2=\left[\left[\frac{\partial v}{\partial x}\right]\right]=r_2\pi=c\pi. \end{aligned} \tag{C.123}$$

Hence,

$$\pi_0 = \frac{1}{\rho_0}\left[\left[\frac{\partial \rho}{\partial x}\right]\right]_0 = \frac{1}{c}\left[\left[\frac{\partial v}{\partial x}\right]\right]_0 = -\frac{1}{c^2}\left[\left[\frac{\partial v}{\partial t}\right]\right]_0 = \frac{1}{c^2}\left(\frac{\partial v}{\partial t}\right)_0^-, \tag{C.124}$$

where $[[\cdot]]_0$ denotes the initial value of the jumps at $t=0^+$. Finally, we get

$$t_c = \frac{c}{\left(\frac{\partial v}{\partial t}\right)_0^-}. \tag{C.125}$$

For instance, if $c \approx 300$ m/s and $\left(\frac{\partial v}{\partial t}\right)^- \sim g = 10$ m/s^2, we see that the critical time is $t_c \approx 30$s. This corresponds to the creation of a shock wave in a long cylinder (~ 10 km!) in which the acceleration wave with an initial amplitude of 10 m/s^2 propagates ■

C.5. Systems with Many Independent Variables

The method of characteristics seeks the solutions by an appropriate choice of coordinates, when the original system of first-order partial differential equations is replaced by a system involving the characteristic coordinates in terms of which the differentiation is simplified considerably.

We saw that in the case of two independent variables the linear combinations of the original system transformed it into ordinary differential equations along characteristics.

Such combinations are not applicable in the case of more than two variables

$$\mathbf{A}^0 \frac{\partial \mathbf{u}}{\partial t} + \sum_{k=1}^{m} \mathbf{A}^k \frac{\partial \mathbf{u}}{\partial x^k} = \mathbf{B}, \qquad \mathbf{u}, \mathbf{B} \in \mathcal{V}^N. \tag{C.126}$$

We can, however, use another property of **hyperbolic** systems to simplify the integration. Namely, we established that the derivative of **u** orthogonal to a characteristic is not determined by the system of equations. This indeterminacy was the reason for ill-posedness of the Cauchy problem in the case of characteristic initial curves. This is also a reason for using predominantly hyperbolic systems to describe the waves, as such an indeterminacy admits jumps of the derivatives across characteristics, which is the main property expected from a **wave front**.

We proceed to discuss these properties within the framework defined by the set (C.126). Let $\mathcal{S}$ be an m-dimensional manifold embedded in the (t,**x**)-space across which the normal derivative of **u** is indeterminate and let $\mathcal{S}$ be defined by the equation

$$\varphi\left(t, x^1, \cdots, x^m\right) = 0, \qquad \left(x^1, \cdots, x^m\right) \in \Re^m. \tag{C.127}$$

Now let us introduce the new coordinates

$$\begin{aligned} \xi^0 &= \xi^0\left(t, x^1, \cdots, x^m\right), \\ \xi^r &= \xi^r\left(t, x^1, \cdots, x^m\right), \qquad r = 1, \cdots, m, \end{aligned} \tag{C.128}$$

such that $\mathcal{S}$ coincides with a parametric surface, e.g. $\xi^0=0$.

The system (C.126) has then the following form:

$$\left(\mathbf{A}^0\frac{\partial\xi^0}{\partial t}+\sum_{k=1}^{m}\mathbf{A}^k\frac{\partial\xi^0}{\partial x^k}\right)\frac{\partial\mathbf{u}}{\partial\xi^0}+\sum_{r=1}^{m}\left(\mathbf{A}^0\frac{\partial\xi^r}{\partial t}+\sum_{k=1}^{m}\mathbf{A}^k\frac{\partial\xi^r}{\partial x^k}\right)\frac{\partial\mathbf{u}}{\partial\xi^r}=\mathbf{B}, \tag{C.129}$$

where $\mathbf{A}^0$, $\mathbf{A}^k$ and $\mathbf{B}$ are considered to be functions of ξ and $\mathbf{u}$.

The indeterminacy of the normal derivative across $\mathcal{S}$, in our new coordinates $\frac{\partial\mathbf{u}}{\partial\xi^0}$, gives rise at once to the condition that the **characteristic determinant** must be zero,

$$\Delta=\det\left(\mathbf{A}^0\frac{\partial\xi^0}{\partial t}+\sum_{k=1}^{m}\mathbf{A}^k\frac{\partial\xi^0}{\partial x^k}\right)=0. \tag{C.130}$$

This is the equation determining the **characteristic manifold** $\mathcal{S}$. According to our definition of ξ-coordinates, the vector

$$\mathbf{p}\equiv(p_0,p_1,\cdots,p_m)\equiv\left(\frac{\partial\xi^0}{\partial t},\frac{\partial\xi^0}{\partial x^1},\cdots,\frac{\partial\xi^0}{\partial x^m}\right)_{\xi^0=0}\equiv\left(\frac{\partial\varphi}{\partial t},\frac{\partial\varphi}{\partial x^1},\cdots,\frac{\partial\varphi}{\partial x^m}\right) \tag{C.131}$$

is a **normal vector** to $\mathcal{S}$. Such vectors are called the **characteristic directions**. According to the definition, any infinitesimal element normal to **p** is an element of the characteristic manifold.

The characteristic determinant Δ is a homogeneous polynomial of degree N with respect to **p**. This means, according to the equation (C.130), that at every point P of the (t,**x**)-space we may have **at most** N real characteristic directions.

If we consider the equation (C.129) at the points P^+ and P^- on either side of $\mathcal{S}$ and arbitrarily close to a point P of $\mathcal{S}$ and take the difference of (C.126) between the points P^+ and P^-, we obtain in the limit as P^+ and P^- tend towards P

$$\left(1\frac{\partial\xi^0}{\partial t}+\sum_{k=1}^{m}\mathrm{A}^k\frac{\partial\xi^0}{\partial x^k}\right)\left[\left[\frac{\partial\mathbf{u}}{\partial\xi^0}\right]\right]=0,\qquad \mathrm{A}^k\equiv\mathrm{A}^{0-1}\mathrm{A}^k, \tag{C.132}$$

where it has been **assumed** that $\mathbf{A}^0$ is non-degenerated, i.e.

$$\det\mathbf{A}^0\neq0. \tag{C.133}$$

This result follows because the other derivatives are continuous across $\mathcal{S}$, the discontinuity exists only in the **normal** derivative.

Let us examine the nature of the quantity $\left[\left[\frac{\partial\mathbf{u}}{\partial\xi^0}\right]\right]$. We expand **u** in the following way

$$\mathbf{u}(\xi^0\pm h)=\mathbf{u}(\xi^0)\pm h\left.\frac{\partial\mathbf{u}}{\partial\xi^0}\right|^{\pm}+\text{remainder}, \tag{C.134}$$

and let

$$\delta\mathbf{u}=\mathbf{u}(\xi^0+h)-\mathbf{u}(\xi^0-h), \tag{C.135}$$

where $\left.\frac{\partial \mathbf{u}}{\partial \xi^0}\right|^{\pm}$ denotes the values of $\frac{\partial \mathbf{u}}{\partial \xi^0}$ at opposite sides of $\mathcal{S}$. Then for $\mathbf{u}$ continuous across $\mathcal{S}$ we get

$$\delta \mathbf{u} = h\left(\left.\frac{\partial \mathbf{u}}{\partial \xi^0}\right|^{+} + \left.\frac{\partial \mathbf{u}}{\partial \xi^0}\right|^{-} \right). \tag{C.136}$$

If we consider a wave front, i.e. such a characteristic surface $\mathcal{S}$ that the region ahead of $\mathcal{S}$ is undisturbed $\left(\left.\frac{\partial \mathbf{u}}{\partial \xi^0}\right|^{+} = 0 \right)$, then

$$\left[\left[\frac{\partial \mathbf{u}}{\partial \xi^0} \right]\right] = -\left.\frac{\partial \mathbf{u}}{\partial \xi^0}\right|^{-} = -\frac{1}{h}\delta \mathbf{u}, \tag{C.137}$$

and the set (C.132) implies

$$\left(\mathbf{1}\frac{\partial \xi^0}{\partial t} + \sum_{k=1}^{m} A^k \frac{\partial \xi^0}{\partial x^k} \right)\delta \mathbf{u} = 0. \tag{C.138}$$

It is convenient to introduce the normalized unit vector perpendicular to the characteristic surface $\mathcal{S}$:

$$\lambda \equiv -\frac{\frac{\partial \varphi}{\partial t}}{|\mathrm{grad}\,\varphi|}, \quad n_k \equiv \frac{\frac{\partial \varphi}{\partial x^k}}{|\mathrm{grad}\,\varphi|}, \qquad k = 1, \cdots, m, \tag{C.139}$$

where

$$\mathrm{grad}\,\varphi = \left(\frac{\partial \varphi}{\partial t}, \frac{\partial \varphi}{\partial x^1}, \cdots, \frac{\partial \varphi}{\partial x^m} \right). \tag{C.140}$$

Then (C.138) becomes

$$\left(\sum_{k=1}^{m} A^k n_k - \lambda \mathbf{1} \right)\delta \mathbf{u} = 0, \tag{C.141}$$

which is the form of the characteristic equations used to determine the **infinitesimal** disturbances of the field $\mathbf{u}$. λ of the equation (C.141) is interpreted as the **velocity of propagation** of the wave front and $\mathbf{n}$ is the **wave front normal**.

For the above considerations to hold true, we see that the following conditions must be fulfilled:

1. The matrix $\mathbf{A}^0$ must be non-degenerated; we can then transform the set (C.126) in such a way that (C.132) holds.

2. The eigenvalues λ of $\sum_{k=1}^{m} A^k n_k$ must be real for any choice of $\mathbf{n}$ if they are to be interpreted as the velocities and if the eigenvectors are to be defined.

3. If the set (C.129) is to be equivalent to (C.126) then the set of the right eigenvectors $\left(\frac{\partial \mathbf{u}}{\partial \xi^0}\right)$ of the matrix

$$\mathbf{A}^0 \frac{\partial \xi^0}{\partial t} + \sum_{k=1}^{m} \mathbf{A}^k \frac{\partial \xi^0}{\partial x^k} \tag{C.142}$$

relative to the matrix $\mathbf{A}^0$ must form the **basis** in the N-dimensional space, i.e. these eigenvectors must be linearly independent. According to (C.130) the matrix (C.142) is singular.

The above remarks indicate the following **definition of the hyperbolicity**:

The system of equations (C.126) with $\mathbf{A}^0$, $\mathbf{A}^k$ being differentiable functions of $\mathbf{u}$ is said to be t-**hyperbolic** if

1) $\det \mathbf{A}^0 \neq 0,$ (C.143)

2) for any unit vector $\nu=(\nu_1,\ldots,\nu_m)$ the matrix

$$\mathbf{A}_\nu \equiv \sum_{k=1}^{m} \mathbf{A}^k \nu_k \tag{C.144}$$

has N real eigenvalues $\lambda^{(A)}$ relative to $\mathbf{A}^0$:

$$\det\left(\mathbf{A}_\nu - \lambda^{(A)}\mathbf{A}^0\right) = 0, \tag{C.145}$$

and the corresponding left eigenvectors $\mathbf{l}^{(A)}$

$$\left(\mathbf{A}_\nu^T - \lambda^{(A)}\mathbf{A}^{0T}\right)\mathbf{l}^{(A)} = 0, \tag{C.146}$$

form a basis in $\Re^N$ (i.e. they are linearly independent).

It is worthwhile mentioning that a t-hyperbolic system does not have to be hyperbolic with respect to the other independent variables. The time direction has been chosen in the above definition to ensure well-posedness for physical **initial value** Cauchy problems. For this reason, it is convenient to have a definition of time-like directions at our disposal.

An infinitesimal displacement from a point P in the direction $\mathbf{n}$ is called **time-like** if the matrix

$$\mathbf{A}^0 n_0 + \sum_{k=1}^{m} \mathbf{A}^k n_k \tag{C.147}$$

is positive definite. The element of the hypersurface at P to which $\mathbf{n}$ is normal is said to be **space-like**. If, on the other hand, the matrix (C.147) is degenerated, an infinitesimal displacement in the direction $\mathbf{n}$ is called **space-like** and the element of the hypersurface at P to which $\mathbf{n}$ is normal is said to be **time-like**.

In the case of the characteristic direction $\mathbf{p}$ defined by (C.131), we have the matrix

$$\mathbf{A}^0 p_0 + \sum_{k=1}^{m} \mathbf{A}^k p_k, \tag{C.148}$$

which, according to (C.130), is degenerated. Hence, it follows that the characteristic directions are **space-like**. On the other hand, the direction

$$\mathbf{n} = (1,0,\cdots,0) \tag{C.149}$$

is time-like, provided the matrix

$$\mathbf{A}^0\mathbf{1} + \sum_{k=1}^{m} \mathbf{A}^k\mathbf{0} = \mathbf{A}^0 \tag{C.150}$$

is positive definite.

Let us return to the characteristic determinant (C.130). It is convenient to introduce the following function:

$$\mathrm{H} \equiv \det \sum_{k=0}^{m} \mathbf{A}^k p_k = \mathrm{H}\left(t, x^1,\cdots,x^m, p_0, p_1,\cdots,p_m,\mathbf{u}\right). \tag{C.151}$$

As we mentioned, the characteristic determinant and consequently also H are homogeneous polynomials of degree N with respect to **p**. This means that

$$\sum_{k=0}^{m} \frac{\partial \mathrm{H}}{\partial p_k} p_k = \mathrm{NH} = 0,^{*)} \tag{C.152}$$

the second part of this relation following from (C.130). This means that the vector

$$\nabla_p \mathrm{H} \equiv \left(\frac{\partial \mathrm{H}}{\partial p_0},\cdots,\frac{\partial \mathrm{H}}{\partial p_m}\right),$$

is tangential to the characteristic manifold $\mathcal{S}$. Let us find the equations of a curve such that

$$\frac{d x^k}{d\tau} = \frac{\partial \mathrm{H}}{\partial p_k}, \quad k = 0,\cdots,m, \quad x^0 \equiv t, \tag{C.153}$$

i.e. lying on the characteristic manifold. Such curves are called **rays**. The equation H=0 implies

---

*) Let us denote

$$\mathbf{A} \equiv \sum_{k=1}^{m} \mathbf{A}^k p_k.$$

Then we have

$$\sum_{k=0}^{m} \frac{\partial \mathrm{H}}{\partial p_k} p_k = \sum_{k=0}^{m} \frac{\partial \mathrm{H}}{\partial \mathrm{A}^{ij}} \frac{\partial \mathrm{A}^{ij}}{\partial p_k} = \sum_{k=0}^{m} \mathrm{A}^{kij} \frac{\partial \mathrm{H}}{\partial \mathrm{A}^{ij}} p_k = \sum_{i=1}^{m}\sum_{j=1}^{m} \mathrm{A}^{ij} \frac{\partial \mathrm{H}}{\partial \mathrm{A}^{ij}} = \mathrm{NH},$$

$\frac{\partial \mathrm{H}}{\partial \mathrm{A}^{ij}}$ being minors of H≡det **A**.

$$\frac{dH}{d\tau}d\tau = 0 = \sum_{k=0}^{m}\left(\frac{\partial H}{\partial x^k}\frac{dx^k}{d\tau} + \frac{\partial H}{\partial p_k}\frac{dp_k}{d\tau}\right)d\tau =$$
$$= \sum_{k=0}^{m}\left(\frac{\partial H}{\partial x^k}\frac{dx^k}{d\tau} + \frac{\partial H}{\partial p_k}\sum_{l=0}^{m}\frac{\partial p_k}{\partial x^l}\frac{dx^l}{d\tau}\right)d\tau = \tag{C.154}$$
$$= \sum_{k=0}^{m}\left(\frac{\partial H}{\partial x^k} + \sum_{l=0}^{m}\frac{\partial H}{\partial p_l}\frac{\partial p_l}{\partial x^k}\right)\frac{dx^k}{d\tau}d\tau,$$

where the definition (C.131) of **p** has been taken into account. It follows that along any ray

$$\frac{\partial H}{\partial x^k} = -\sum_{l=0}^{m}\frac{\partial p_l}{\partial x^k}\frac{\partial H}{\partial p_l}. \tag{C.155}$$

On the other hand,

$$\frac{dp_k}{d\tau} = \sum_{l=0}^{m}\frac{\partial p_k}{\partial x^l}\frac{dx^l}{d\tau} = \sum_{l=0}^{m}\frac{\partial p_k}{\partial x^l}\frac{\partial H}{\partial p_l} = -\frac{\partial H}{\partial x^k}, \tag{C.156}$$

where (C.153) has been used.

Bearing the above considerations in mind, we arrive at the following equations for rays:

$$\begin{aligned} &\frac{dx^k}{d\tau} = \frac{\partial H}{\partial p_k}, \qquad x^0 \equiv t, \\ &\frac{dp_k}{d\tau} = -\frac{\partial H}{\partial x^k}, \qquad k = 0, \cdots, m, \\ &\sum_{k=0}^{m}\frac{\partial H}{\partial p_k}p_k = 0. \end{aligned} \tag{C.157}$$

If these equations were solved, we could construct the characteristic manifold $\mathcal{S}$, i.e. the wave fronts. In contrast to the original equation (C.130) of $\mathcal{S}$, the equations (C.157) are ordinary differential equations. In general, the form of those equations depends on the solution **u**. Hence, the initial value problem must be solved simultaneously with (C.157). We have already seen it in the case of the characteristics for two independent variables.

The above-described method of characteristics for many independent variables forms the basis of the so-called **ray optics,** which we shall not investigate any further in this book.

References

(1824)
S. CARNOT; Réflexions sur la puissance, motrice du feu et sur les machines propres a développer cette puissance (Reflections on the motive power of fire, *Dover Publ.*, N.Y. (1960)), *Bachelier, Libraire*, Paris

(1842)
J. R. MAYER; Bemerkungen über die Kräfte der unbelebten Natur, *Annalen der Chemie und Pharmacie*, Vol. 42, 233-240

(1868)
L. BOLTZMANN; Studien über das Gleichgewicht der lebend. Kraft zwischen bewegten materiellen Punkten, *Wien Ber.* 58^2, 517; also: Weitere Studien über Wärmegleichgewicht unter Gasmolekülen (H-Theorem), (1872) *Wien Ber.* 66^2, 275

(1872)
J. WILLARD GIBBS; On the equilibrium of heterogeneous substance, *Transactions of the Connecticut Academy*, III, 108-248 (1875-1876), 343.524 (1877-1877); see also: The scientific papers of J.Willard Gibbs, vol.I (Thermodynamics), (1961) *Dover Publ.*, N.Y.

(1907)
P. EHRENFEST, T. EHRENFEST; Über zwei bekannte Einwände gegen das Boltzmannsche H-Theorem, *Physik. Z.*, **8**, 311-314

(1909)
C. CARATHÉODORY; Untersuchungen über die Grundlagen der Thermodynamik, *Mathematische Annalen*, **67**, 3, 355-386

(1912)
M. BORN, TH. VON KÁRMÁN; *Physik. Z.*, **13**, 297
P. DEBYE; *Ann. Physik*, **39**, 789
P. EHRENFEST, T. EHRENFEST; The conceptual foundations of the statistical approach in mechanics (In German), *Encyklopädie der mathematischen Wissenschaften*, No.6 of Volume IV2II; English Translation: (1959). *Cornell University Press*, Ithaca, N.Y.

(1913)
M. BORN, TH. VON KÁRMÁN; *Physik. Z.*, **14**, 15

(1922)
M. PLANCK; Treatise on Thermodynamics, *English Translation from 1926, Longmans, Green, & Co.*, N.Y., see also: (1945).*Dover Publ.*, N.Y.

(1948)
C. CATTANEO; Sulla conduzione del calore, *Atti Sem. Mat. Fis. Univ. Modena*, **3**, 83-101
R. S. RIVLIN; Large elastic deformations of isotropic materials,
II. Some uniqueness theorems for pure homogeneous deformations, *Philos. Trans. Roy. Soc., A.*, **240**, 491-508
IV. Further developments of the general theory, *Philos. Trans. Roy. Soc., A.*, **241**, 379-397
V. The problem of flexture, (1949) *Proc. Roy.Soc., A.*, **195**, 463-473
VI. Further results in the theory of torsion, shear and flexture, (1949) *Philos. Trans. Roy. Soc., A.*, **242**, 173-195

(1952)
H. GRAD; Statistical mechanics, thermodynamics, and fluid dynamics of systems with an arbitrary number of integrals, *Comm. Pure Appl. Math.*, **5**, 455-494

(1954)
J. L. ERICKSEN; Deformations possible in every isotropic, incompressible, perfectly elastic body, *ZAMP*, **5**, 466-489
J. L. ERICKSEN, R. S. RIVLIN; Large elastic deformations of homogeneous anisotropic materials, *J.Rational Mech. Anal.*,**3**, 281-301

(1956)
T. C. DOYLE, J. ERICKSEN; Nonlinear theory of elasticity, in: H. L. DRYDEN, TH. VON KÁRMÁN (eds.), Advances in Applied Mechanics, vol.IV, *Academic Press*, N.Y.
M. KAC; Some stochastic problems in physics and mathematics, *Magnolia Petroleum Company*, Dallas

(1957)
A. E. GREEN, R. S. RIVLIN; The mechanics of non-linear materials with memory, Part I, *Arch.Rat.Mech.Anal.*, **1**
C. TRUESDELL; Sulle basi della termomeccanica, *Accademia Nazionale dei Lincei, Dendiconti della Classe di Scienze Fisiche, Matematiche e Naturali*, (8) **22**, 33-38,158-166

(1959)
M. KAC; Probability and related topics in physical sciences, *Interscience Publ., Ltd.*, London
W. NOLL; The foundations of classical mechanics in the light of recent advances in continuum mechanics. The Axiomatic Method, with special reference to geometry and physics (1957). Amsterdam: *North- Holland Co.*, 266-281

(1960)
B. D. COLEMAN, W. NOLL; An approximation theorem for functionals, with applications in continuum mechanics, *Arch.Rat.Mech.Anal.*, 355-370, **6**
C. TRUESDELL, R. A. TOUPIN; The Classical Field Theories, in: S. FLÜGGE (ed.) *Handbuch der Physik*, III/1, *Springer-Verlag*, Berlin, Heidelberg, New York

(1962)
R. COURANT, D. HILBERT; Methods of mathematical physics, vo.II: R.COURANT; Partial differential equations, *Interscience*, N.Y.-London
C. TRUESDELL; Mechanical basis for diffusion, *J.Chem.Phys.*, **37**

(1964)
J. DE BOER; Phonons in liquids, in: T. A. BAK (ed.), Phonons and phonon interactions, Aarhus Summer School Lectures (1963), *W.A.Benjamin*, N.Y.
A. JEFFREY, T. TANIUTI; Non-linear wave propagation with applications to physics and magnetohydrodynamics, *Academic Press*, N.Y.London

(1965)
W. F. AMES; Nonlinear partial differential equations in engineering, *Academic Press*
G. SANDRI; A New Method of Expansion in Mathematical Physics, I, *Nuovo Cimento*, **36**, 1, 67-93
G. F. SMITH; On isotropic integrity bases, *Arch.Rat.Mech.Anal.*, **18**, 282-292
C. TRUESDELL, W. NOLL; The Non-Linear Field Theories of Mechanics, in: S. FLÜGGE (ed.) *Handbuch der Physik*, III/3, *Springer*, Berlin, Heidelberg, New York

(1966)
J. KEVORKIAN; The two variable expansion procedure for the approximate solution of certain nonlinear differential equations, J.B.ROSSER (ed.), Space Mathematics, Part 3, 206-275, *American Mathematical Society*, Providence, R.I.
I. MÜLLER; Zur Ausbreitungsgeschwindigkeit von Störungen in kontinuierlichen Medien, *Dissertation* TH Aachen (Germany)
C. TRUESDELL; Six lectures on modern natural philosophy, *Springer*, Berlin, Heidelberg, New York
C. TRUESDELL; The elements of continuum mechanics, *Springer*, Berlin, Heidelberg, New York

(1967)
I. MÜLLER; Zum Paradoxon der Wärmeleitungstheorie, *Z. Physik*, **198**, 329-344

(1968)
W. NOLL, R. A. TOUPIN, C.-C. WANG; Continuum Theory of Inhomogeneities in Simple Bodies, *Springer*, Berlin, Heidelberg, New York

(1969)
C. TRUESDELL; Rational Thermodynamics, *The first edition, McGraw-Hill*, N.Y.

(1970)
J. D. ESHELBY; Energy relations and the energy-momentum tensor in continuum mechanics, in: M. F. KANNINEN, W. F. ALDER, A. R. ROSENFIELD, R. I. JAFFE (eds.), *Inelastic Behavior of Solids*, *McGraw-Hill*, N.Y.
D. ROGULA; Moment stresses and the symmetry of the stress tensor in bodies with non-local structure, *Bull.Acad.Polon.Sci.*, *Ser.Sci.Techn.* **18**, 4, 159-164
W. SLEBODZINSKI; Exterior forms and their applications, *Polish Scientific Publishers*, Warsaw
C.-C. WANG; A new representation theorem for isotropic functions, Part I and II, *Arch.Rat.Mech.Anal.*,**36**, 166-197, 198-223. Corrigendum, *ibid.* (1971) **43**, 392-395

(1971)

H. E. JACKSON, C. T. WALKER; Thermal conductivity, second sound and Phonon-Phonon Interactions in NaF, *Phys.Rev. B*, **3**, 4, 1428

F. JOHN; Partial differential equations, *Springer-Verlag*, Berlin, Heidelberg, New York

(1972)

J. BEAR; Dynamics of fluids in porous media, *Dover Publ.*,N.Y.

D. ELWELL and A. J. POINTON; Classical Thermodynamics, *Penguin Books Ltd.*, Harmondsworth, Middlesex

M. A. GOODMAN, S. C. COWIN; A continuum theory of granular materials, *Arch.Rat. Mech.Anal.*, **44**, 249-266

M. E. GURTIN; The linear theory of elasticity, in: S. FLÜGGE (ed.) *Handbuch der Physik,* VIa/2, Mechanics of solids II, *Springer*, Berlin, Heidelberg, New York

I-SHIH LIU; Method of Lagrange multipliers for exploitation of the entropy principle, *Arch.Rat.Mech.Anal.*,**46**

I. MÜLLER; On the frame dependence of stress and heat flux, *Arch.Rat.Mech.Anal.*, **45**, 241-250

C. TRUESDELL; A first course in rational continuum mechanics, *The Johns Hopkins University*, Baltimore, Maryland

K. WILMANSKI;On thermodynamics and functions of states of nonisolated systems, *Arch.Rat.Mech.Anal.*, **45**, 4, 251-281

(1973)

M. E. GURTIN, P. PODIO-GUIDUGLI; The thermodynamics of constrained materials, *Arch.Rat.Mech.Anal.*, **55**, 192-208

A. H. NAYFEH; Perturbation methods, *J.Wiley*, N.Y

D. ROGULA; On nonlocal continuum theory of elasticity, *Arch.Mech.Stos.*, **25**, 2

(1974)

E. DUNN, R. L. FOSDICK; Thermodynamics, stability and boundedness of fluids of comp-lexity 2 and fluids of second grade, *Arch.Rat.Mech.Anal.*, **56**,191-252

A. MÜNSTER; Statistical thermodynamics, Vol.II, *Springer*, Berlin, Heidelberg, New York

K. WILMANSKI; Foundations of phenomenological thermodynamics (in Polish), *Polish Scientific Publishers*, Warsaw

(1975)

I. A. KUNIN; Theory of elastic continua with microstructure (in Russian), *Nauka*, Moscow

L. R. G. TRELOAR; The physics of rubber elasticity, Third Edition, *Clarendon Press*, Oxford

(1976)
I. MÜLLER; On the frame dependence of electric current and heat flux in a metal, *Acta Mechanica*, **24**, 117-128
A. J. M. SPENCER; Theory of invariants, in: *Continuum Physics*, Vol.I, Part III, A. C. ERINGEN (ed.), *Academic Press*
K. WILMANSKI; Foundations of neoclassical thermodynamics: metrization of direct thermodynamic processes, in: G. FICHERA (ed.), Trends in applications of pure mathematics to mechanics, 425-445, *Pitman*, Boston
N. W. ASHCROFT, N. D. MERMIN; Solid State Physics, *Holt, Reinhart and Winston*, New York

(1977)
A. SOMMERFELD; Thermodynamik und Statistik, 2.Auflage, *Deutsch-Verlag*, Thun-Frankfurt/M.

(1980)
R. M. BOWEN; Incompressible porous media models by use of the theory of mixtures, *Int.J.Engn.Sci.*, **18**, 1129-1148
K. WILMANSKI; Thermodynamic foundations of thermoelasticity, in: G. LEBON, P. PERZYNA (eds.), Recent developments in thermomechanics of solids, CISM 262,1-94, *Springer*, Berlin, Heidelberg, New York

(1981)
D. JOSEPH; Instability of the rest state of fluids of arbitrary grade greater than one, *Arch. Rat. Mech. Anal.*, **75**, 251-256

(1982)
R. M. BOWEN; Compressible porous media models by use of the theory of mixtures, *Int. J. Engn. Sci.*, **20**, 697-763
N. CRISTESCU, I. SULICIU; Viscoplasticity, *Martinus Nijhoff Publ.*, The Hague, Boston, London
I-SHIH LIU; On representations of anisotropic invariants, *Int. J. Engn. Sci.*, **20**, 1099-1109

(1983)
I-SHIH LIU, I. MÜLLER; Extended thermodynamics of classical and degenerate gases, *Arch.Rat.Mech.Anal.*, **83**, 4, 285-332
J. E. MARSDEN, T. J. R. HUGHES; Mathematical Foundations of Elasticity, *Prentice-Hall*, New Jersey

(1984)
J. BEAR, Y. BACHMAT; Transport phenomena in porous media - basic equations, in: A. BEAR, J. CORDPOIOGLU (eds.), Fundamentals of transport phenomena in porous media, NATO ASI Series E: Applied Sciences, #82, 3-62, *Martinus Nijhoff Publ.*, Amsterdam
C. TRUESDELL; Rational Thermodynamics, *2nd edn.*, *Springer*, Berlin, Heidelberg, New York.
H. GIESEKUS; On configuration-dependent generalized Oldroyd derivatives, *J. Non-Newtonian Fluid Mechanics*, **14**, 47-65

(1985)
W. HEIDUG, F. K. LEHNER; Thermodynamics of coherent phase transformations in non-hydrostatically stressed solids, *Pure Appl. Geophys.*, **123**, 91-98
I. MÜLLER; Thermodynamics; *Pitman Publ.*, Boston
G. F. SMITH; Lectures on constitutive expressions, in: W.FISZDON, K.WILMANSKI (eds.), *Mathematical models and methods in mechanics, Banach Center, Vol.15, Polish Scientific Publishers*, 645-678, Warsaw

(1986)
G. M. KREMER; Extended thermodynamics of ideal gases with 14 fields, *Annales de l'Institut Henri Poincaré*, **45**
I. MÜLLER, K. WILMANSKI; Extended thermodynamics of a non-Newtonian fluid, *Rheologica Acta*, **22**, 335-349
K. WILMANSKI; Non-Newtonian fluids of second grade - rheology, thermodynamics and extended thermodynamics, in: E. KRÖNER, K. KIRCHGÄSSNER (eds.), Trends in Applications of Pure Mathematics to Mechanics, 376-383, *Springer*, Berlin, Heidelberg, New York

(1987)
R. ABEYARATNE, J. K. KNOWLES; Non-elliptic elastic materials and the modeling of elastic-plastic behavior for finite deformation, *J. Mech. Phys. Solids*, **35**, 343-365
T. BOURBIE, O. COUSSY, B. ZINSZNER; Acoustics of porous media, *Editions Technip*, Paris
I. MÜLLER, T. RUGGERI; Symposium on Kinetic Theory and Extended Thermodynamics, Bologna, *Pitagora Editrice*
R. S. RIVLIN, K. WILMANSKI; The passage from memory functionals to Rivlin-Ericksen Constitutive Equations, *ZAMP*, **38**, 624-629
D. A. DE VRIES; The theory of heat and moisture transfer in porous media revisited, *Int.J. Heat Mass Transfer*, **30**, 7, 1343-1350

(1988)
G. CAPRIZ; Continui con microstruttura, *Ets Editrice Pisa*
I-SHIH LIU; Introduction to continuum mechanics, *University of Rio de Janeiro, Brasil.*
K. WILMANSKI; Thermodynamics of a heat conducting Maxwellian fluid, *Arch. Mech.*, **40**, 2-3, 217-232

(1989)
G. CAPRIZ; Continua with microstructure, *Springer Tracts in Natural Philosophy, Springer*, Berlin, Heidelberg, NewYork
I-SHIH LIU; Extended thermodynamics of viscoelastic materials, *Continuum Mech. Thermodyn.*, **1**, 143-164
T. RUGGERI; Galilean invariance and entropy principle for systems of balance laws, *Continuum Mech. Thermodyn.* **1**, 3-20

(1990)
M. E. GURTIN, A. STRUTHERS; Multiphase thermomechanics with interfacial structure, 3. Evolving phase boundaries in the presence of bulk deformation, *Arch. Rat. Mech. Anal.*, **112**, 97-160

(1991)
A. M. ANILE, S. PENNISI, M. SAMMARTINO; A thermodynamical approach to Eddington factors, *J. Math. Physics*, (2) **32**
F. DOBRAN; Theory of structured multiphase mixtures, *Springer*, Berlin, Heidelberg, New York
K. HUTTER; Einführung in die Fluid- und Thermodynamik, im *Selbstverlag*
G. M. KREMER, I. MÜLLER; Thermodynamics of light and sound, *Le Matematiche*, XLVI, 1, 213-228
W. LARECKI; Symmetric conservative form of low-temperature phonon gas hydrodynamics, II Equations of heat transport and thermal waves in the case of linear isotropic approximation of phonon frequency spectrum, *Il Nuovo Cimento*, **1868 D**

(1992)
G. M. KREMER, I. MÜLLER; Radiation thermodynamics, *J. Math. Physics*, **33**, 2265-2268
I-SHIH LIU; On interface equilibrium and inclusion problems, *Continuum Mech. Thermodyn.* , **4**, 177-186
K. WILMANSKI; Phenomenological Thermodynamics, in: Henryk Zorski (ed.), Foundations of mechanics, 485-589, *Polish Scientific Publishers*, Warsaw, *Elsevier*, Amsterdam

(1993)
D. R. AXELRAD; Stochastic mechanics of discrete media, *Springer*, Berlin, Heidelberg, New York
W. DREYER, H. STRUCHTRUP; Heat pulse experiments revisited, *Continuum Mech. Thermodyn.*, **5**, 3-50
I. MÜLLER, T. RUGGERI; Extended thermodynamics, *Springer Tracts in Natural Philosophy*, *Springer*, Berlin, Heidelberg, New York.

(1994)
Q. JIANG, R. K. N. D. RAJAPAKSE; On coupled heat-moisture transfer in deformable porous media, *Quat.Mech.Appl.Math.*, **47**, 1, 53-68
J. E. MARSDEN, T. S. RATIU; Introduction to mechanics and symmetry, *Springer*, Berlin, Heidelberg, New York
I. MÜLLER; Grundzüge der Thermodynamik mit historischen Anmerkungen, *Springer*, Berlin, Heidelberg, New York

(1995)
K. WILMANSKI; Lagrangean model of two-phase porous material, *J. Non-Equilib. Thermodyn.*, **20**, 50-77
K. WILMANSKI; On weak discontinuity waves in porous materials, in: M. MARQUES, J. RODRIGUES (eds.), Trends in applications of mathematics to mechanics, 71-83, *Longman*, Harlow, Essex

(1996)
K. WILMANSKI; Porous media at finite strains - the new model with the balance equation for porosity, *Arch. Mech.*, **48**, 4, 591-628
short version:
K. WILMANSKI; Porous media at finite strains, in: K.MARKOV (ed.), Continuum models and discrete systems, 317-324, *World Scientific*, Singapore

K. WILMANSKI; Dynamics of porous materials under large deformations and changing porosity, in: R.C.BATRA, M.F.BEATTY (eds.), Contemporary research in the mechanics and mathematics of materials, 343-358, *CIMNE*, Barcelona

(1997)

K. WILMANSKI; The thermodynamical model of compressible porous materials with the balance equation of porosity, *WIAS, Preprint No. 310* [to appear in: *Transport in Porous Media (1998)]*

Subject Index

Springer and the environment